PERGAMON INTERNATIONAL LIBRARY
of Science, Technology, Engineering and Social Studies
The 1000-volume original paperback library in aid of education,
industrial training and the enjoyment of leisure
Publisher: Robert Maxwell, M.C.

D1268384

Applied Geotechnology

A Text for Students and Engineers
on Rock Excavation and Related Topics

THE PERGAMON TEXTBOOK
INSPECTION COPY SERVICE

An inspection copy of any book published in the Pergamon International Library will
gladly be sent to academic staff without obligation for their consideration for course
adoption or recommendation. Copies may be retained for a period of 60 days from
receipt and returned if not suitable. When a particular title is adopted or recommended
for adoption for class use and the recommendation results in a sale of 12 or more copies,
the inspection copy may be retained with our compliments. The Publishers will be
pleased to receive suggestions for revised editions and new titles to be published in this
important International Library.

Related Pergamon Titles of Interest

Books

GRIFFITHS & KING
Applied Geophysics for Geologists and Engineers (The Elements of
Geophysical Prospecting)
[A new edition of Applied Geophysics for Engineers and Geologists]

HOEK*
KWIC Index of Rock Mechanics Literature Part One: 1870–1968

JENKINS & BROWN*
KWIC Index of Rock Mechanics Literature Part Two: 1969–1976

MAURER
Novel Drilling Techniques

ROBERTS
Geotechnology (An Introductory Text for Students and Engineers)

*Journals***

International Journal of Rock Mechanics and Mining Sciences (and
Geomechanics Abstracts)

Underground Space

  * *Not available under the Pergamon textbook inspection copy service.*
** *Free specimen copy of any Pergamon journal available on request.*

Applied Geotechnology

A Text for Students and Engineers
on Rock Excavation and Related Topics

by

A. Roberts

Mackay School of Mines
The University of Nevada, USA

PERGAMON PRESS

OXFORD · NEW YORK · TORONTO · SYDNEY · PARIS · FRANKFURT

Preface

We should encourage those ideas which look towards underground construction not merely to provide storage sites and shelters for civil and military defence needs, but also more generally to serve as an extension of the surface environment. Not hiding away like troglodytes, to emerge only when the "all clear" sounds, but looking and working forwards to add new dimensions to living. This is not a new idea, for there is hardly a major city in the world, finding its arteries choked by automobiles, that is not thinking about constructing an underground rapid transport system, if it has not already got one.

This is an age of innovation in underground technology. Some of the directions in which these ideas may take us are outlined in an epilogue to this text, and the author hopes that this may stimulate the interest of students into reading more about them and, who knows?, possibly adding some original ideas of their own.

The author acknowledges his indebtedness for much of the material in this book to friends and colleagues amongst the students and staff of the Mackay School of Mines, at the University of Nevada, the Department of Civil and Mineral Engineering of the University of Minnesota, and the former Postgraduate School of Mining at the University of Sheffield. It is not possible to mention everyone by name, but it would be remiss to omit special reference particularly to the late Dr. Ivor Hawkes, and, in the United States, to Drs. Charles Fairhurst, Malcolm Mellor, Barrie Sellers, P. K. Dutta, and Pierre Moussét Jones; and in Britain, Peter Attewell, Ian Farmer, P. K. Chakravarty, Jim Furby, S. A. F. Murrell, Yilmaz Fisekci, J. W. Harrison, D. R. Moore, J. A. Hooper, and R. Dhir.

To all these persons, and to those not named, and to all those listed in the chapter bibliographies and references, the author's thanks are duly expressed.

Contents

Contents

Contents

Contents

Contents

Contents

CHAPTER 1

Rock Excavation by Blasting—I

The Blasting Action of an Explosive Charge in Solid Material

THE essential feature of an explosion is the generation of a large quantity of energy in a very short space of time. The method of energy generation may be electrical, chemical, or nuclear in origin. If the explosion is confined, very high pressures may be produced.

In the detonation of chemical explosives used in industry, pressures of the order of 150 kbar are common, and it may be expected that, within the confined space around a blast charge, pressures may reach 10–100 kbar within 2500–6000 ms after detonation, depending upon the degree of confinement.

Detonation

The process of detonation in a high explosive can be represented as shown in Figs. 1.1 and 1.2. The detonation, which is a rapid, self-propagating, chemical reaction involving change of state, moves through the explosive at a velocity D, where

$$D = W_{CJ} + C_{CJ},$$

where D is the detonation velocity, C_{CJ} is the velocity of sound in the gaseous products at the Chapman–Jouguet plane, and W_{CJ} is the particle velocity of the gas at the CJ plane [1, 2].

The CJ condition specifies the minimum stable detonation velocity that can exist in the explosive column. The length of column between the CJ plane and the shock front is the detonation zone. This zone moves through the explosive column at the detonation velocity, which is often used as an index of the "strength" of the explosive.

The detonation velocity forms the basis upon which may be estimated the detonation pressure, the energy liberated in the chemical reaction, and the general effectiveness of the explosive [3, 4].

The pressure exerted by the explosive blast upon the confining medium is lower than the detonation pressure. The explosion pressure, that is the pressure of the gaseous products when these have expanded to the initial volume of the explosive, is about half the detonation pressure at the CJ plane. A further drop in pressure is experienced by the time the gases have expanded to fill the confined space of the blast, and it is this pressure which impacts upon the confining medium.

The fragmentation produced by the blast is affected considerably by the degree of confinement of the charge. A charge completely stemmed or loaded without any free air space is said to be perfectly loaded. The existence of free air space results in partial loading.

AG - B

1

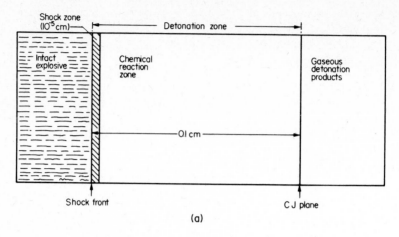

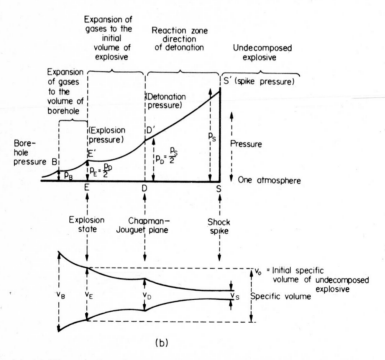

FIG. 1.1. (a) Diagram of steady-state one-dimensional detonation wave in a solid explosive charge. (b) Relation between various pressures from a detonating explosive.

The detonation pressure is therefore an important parameter in blasting theory since it determines the pressure exerted upon the interior of the blast cavity. Hino quotes, as a first approximation for practical purposes, the pressure P_r which exists at a distance r from the centre of a spherical charge of radius a is:

$$P_r = P_d(a/r)^n,$$

where P_d is the detonation pressure and n is a distance exponent [5].

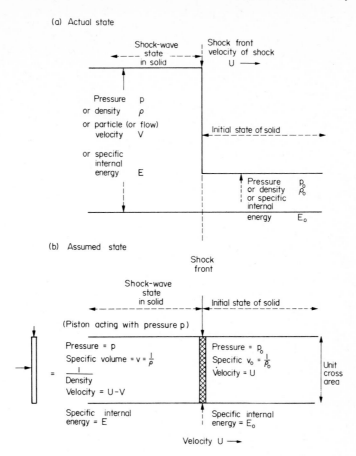

FIG. 1.2. Diagrammatic representation of a shock wave.

The results of detonating a confined explosive are twofold: (1) the propagation of stress waves, which may fracture the medium, and (2) the generation of gases at high temperature and pressure. These gases expand to fill any interstices that may pre-exist or that may be produced by the stress waves, and in so doing they break up the confining medium. Continuing expansion of the gases serves to move the broken fragments from the seat of the explosion.

Although explosives are commonly used as blasting devices in solid materials, the rapidity of the reactions has been a major obstacle to study of the processes of fracture during the actual explosion. Complete understanding of these processes does not yet exist, and there is no generally accepted theory of blasting in solid materials that will fully explain all the observed phenomena and also serve as a firm basis for design calculations. Consequently the degree of success or failure of blasting operations is still influenced, to a considerable extent, by the experience and intuition of the engineers concerned.

Nevertheless, in recent years notable advances in the application of theoretical concepts have been made. It is therefore of interest to review these in relation to some of the practical problems associated with blasting in solids.

Stress Waves Due to Blasting

The detonation of an explosive generates forces that impact upon any confining walls that surround the blast. Stress waves result, which propagate outwards in all directions through the confining medium. If the explosion pressure exceeds the strength of the confining solid, a zone immediately surrounding the explosive charge may be pulverized.

In conventional blasting with industrial explosives any pulverized zone is likely to be very small in extent. Surrounding it there is a transition zone in which non-elastic phenomena occur. These phenomena include the passage of shock waves, the occurrence of plastic flow, and the generation of radial cracks.

Shock Waves

At very high pressures of impact, shock waves may be produced which can be described by the Rankine–Hugoniot equation, derived from the equations for the conservation of mass, momentum, and energy, in the medium concerned. The essential feature of such a condition is that the compressibility of the medium is affected to the extent that the medium behaves hydrodynamically.

When chemical explosives are used in solids the extent to which shock waves may be produced in the confining medium is limited to the immediate vicinity of the explosive charge. The influence of shock waves is more significant, however, in nuclear blasting.

Plastic Flow

When a solid such as a hard rock is subjected to a sufficiently large strain, the stress–strain relationship ceases to be elastic and becomes non-linear. Some of the energy emanating from a blast in rock will therefore be transmitted in the form of strain waves through the confining medium.

Little work has yet been done on the study of strain waves in rock materials, but there is ample evidence of their passage in the form of residual effects. Several investigators, notably Bailey *et al.* [7] and Borg *et al.* [8], have described the results of microscopic and X-ray diffraction studies of rock materials that have been subjected to explosive attack. Phenomena have been noted, such as strain shadows and undulatory extinction in quartz, deformation lamellae in quartz, feldspars, and calcite, incipient fracturing within the crystal constituents, and marginal granulation at their boundaries.

The deformation of quartz and the quartz–coesite transition have been described by Boyd and England [9]. Fryer has identified coesite in samples of quartz sand subjected to explosive shock [10].

Radial Cracks

Immediately surrounding the crushed and transition zones that result from a blast, there exists in the confining medium a zone in which radial cracks occur. These cracks are the result of tangential tensile stresses, in the nature of "hoop stresses", that are produced in the confining medium as the dilatational stress wave expands outwards from the origin of the blast.

The expanding stress wave is compressional in a radial direction, in the immediate vicinity of the blast, and its amplitude is likely to be less than the compressive strength of the medium. The tensile strength of the medium being much smaller than its compressive strength, the tangential tensile stresses are large enough to produce radial fractures.

Broberg [11] explains the process of fracture as being attributed to stress concentrations at the edges of microcracks and discontinuities that are present before the passage of the stress wave. If the suddenly applied stress is substantially larger than the fracture stress, then the discontinuities may extend at an accelerated rate.

The acceleration will be stress-dependent, but the final velocity of the tip of the crack will be less than that of the stress wave, probably approaching the stress-wave velocity with increasing stress. Robert and Wells [12] estimate the final velocity of crack propagation to be about 40 % of the dilatational wave velocity and independent of stress, while Wells and Post [13] suggest that the limiting velocity of running cracks is about half the velocity of elastic shear waves.

Elastic Waves

The tangential stresses attenuate with distance until they reach a value below the tensile strength of the medium, after which, provided they do not interact with reflected stress waves in the elastic zone, they will pass through the medium without causing further fracture.

The hydrodynamic and non-linear zones surrounding the blast are of very limited extent compared with the elastic zone which surrounds them. The expanding compression stress waves and their associated tangential tensile stress waves pass into this elastic zone. Radial fractures are produced which may be extensions of the radial cracks generated in the non-linear zone or which may be initiated from microcracks and other discontinuities inherent in the confining medium.

Fracture Processes in Blasting

From a practical viewpoint, if nuclear explosives are not employed, the hydrodynamic and non-linear zones are only important in so far as (1) they act as a filter to determine what proportion of the blast energy passes into the elastic zone, and (2) they pre-condition the confining medium to failure in the elastic zone by the generation of incipient radial cracks, some of which may extend into the elastic zone.

It is in the elastic zone that commercial explosives perform the greater part of their useful function of breaking up the confining medium.

The quantity of energy contained in the stress waves is considerably smaller than the total energy released by detonation of the explosive. Fogelson *et al.* [15] quote observed figures of 9–18 % when blasting in granite, while Nicholls and Hooker [16] measured only 2–4 % in salt.

Langefors and Kihlstrom [17] estimate that only about 3 % of the total energy of the explosive is distributed by the stress waves within the volume of the confining medium that is broken by the blast. The major portion of the energy from the explosive charge is evidenced as heat, which goes to expand the gases generated by the detonation. These expanding gases are a prime agent in breaking up the confining medium, but in order for them to be most effective there must be fracture surfaces in the confining medium, against which they can exert pressure.

The final stage of the blasting process then follows, relatively slowly, during which time the expanding gases force open the radial cracks in the elastic zone. The process is aided if a free face or faces exist in the confining medium, into which the expanding gases, and fractured material, may yield and move.

Scabbing

There are occasions, however, when fracture at the free face may take the form of "scabbing", "slabbing", or "spalling", the mode of occurrence of which has been explored by several investigators, including Rinehart [18], Duvall and Atchison [19], and Hino [20]. The mechanism of this is illustrated in Fig. 1.3, in which a compressive stress wave incident upon a free face is reflected as a tensile pulse.

The reflected tension increases from zero at the free surface as it travels back into the confining medium. A fracture develops when the reflected stress pulse exceeds the tensile strength of the medium, and a detached slab forms which moves forward, impelled by the particle velocity in the incident stress pulse. A new free surface is thus formed, from which the now-attenuated tension again builds up from zero. The process is repeated until attenuation brings the reflected tension pulse below the critical value for fracture.

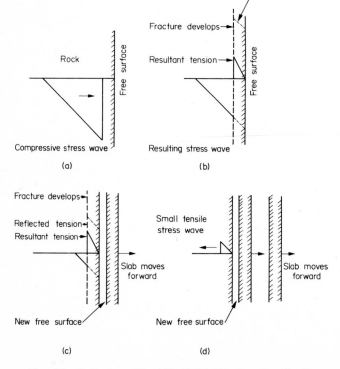

FIG. 1.3. Mechanism of "scabbing", "slabbing", or "spalling".

Many explosives engineers regard "scabbing", "slabbing", or "spalling" as of minor importance in conventional industrial blasting. For example, Langefors maintains that in hard material such as granite, scabbing only becomes important if the charge exceeds 1 kg/m^3, whereas in general quarry blasts the consumption of explosives is more likely to be around 0.15–0.60 kg/m^3. Consequently, if scabbing occurs many engineers would consider that the blast had been overcharged with explosive.

On the other hand, Hino used the scabbing principle to form "crater cuts", in which charges of around 5 kg/m^3 generate reflected stress pulses in rock to produce a conical crater extending from the free face at the base towards the charge at the apex. Reflected stress waves in this case are thus used to fracture the medium from without.

With lesser-weight charges it is probably more usual for the confining medium to be fractured by the blast from within, i.e. extending from the charge towards the free face. This must not be taken to imply that tension pulses reflected from free surfaces are not important in ordinary blasting technique. The reverse is undoubtedly the case. Rinehart [21] states that the interplay among the several waves that are reflected and interreflected from free surfaces will generate, within the medium, the substantial discontinuities in particle velocity that are necessary if it is to be fragmented.

Field and Ladegaard-Pedersen [22] suggest that reflected stress waves play a major role in determining which fractures develop from the blast origin and in which directions they will propagate. It is possible that stress-wave interactions can explain the differences in the "breakout angle" that are apparent when comparing the effects of concentrated and extended charges in rock blasting. Control of the stress-wave interactions, by suitable choice of relative positions as between charge and free surfaces, and timing of neighbouring charges to provide suitable free surfaces in planned succession and orientation, can therefore have great practical importance.

The "Equivalent Cavity"

Kutter and Fairhurst have investigated, both analytically and experimentally, the respective roles of the stress wave and gas pressure in the fragmentation of a confining solid by an explosive charge [23]. The fracture pattern generated by the stress wave was observed to be comprised of a zone of multiple-packed radial fractures immediately surrounding the blast cavity, surrounded by a ring of more widely spaced radial cracks.

The extent of this radially fractured zone was found to depend, not only upon the tensile strength and wave velocity of the confining medium, but also very largely on the degree of energy absorption in the confining mass. Their experiments showed that, for a given size of borehole, an increase in the size of the charge, once it had gone beyond an optimum size, did not produce a more extensively fractured zone. It only resulted in more crushing around the cavity walls.

The extent of the fractured zone was theoretically derived to approach six charge-hole diameters for a spherical charge, and nine hole diameters for a cylindrical charge. Kutter and Fairhurst maintain that it is in this expanded or "equivalent cavity" that the gas pressure begins to exert its effect and not, as might be assumed, the volume of the original charge-hole. This concept of the "equivalent cavity" thus leads to a larger stressed region, and consequently to more extensive crack propagation by gas expansion than is commonly assumed when using high explosives.

Kutter and Fairhurst also emphasize the important role played by the radial fractures in determining the stresses around the "equivalent cavity". For example, the stresses around a pressurized cylindrical cavity, with pressurized radial cracks of a length equal to two hole radii, are stated to be nine times higher than those which would exist around the original pressurized blast-hole. Consequently the stresses around a cavity with wave-generated fractures will probably be amply sufficient to propagate some of the cracks, to break through to the free surfaces, and fragment the confining medium.

Formation of the "Break Angle"

The existence of a free surface adjacent to the confined explosive charge is an essential requirement for the formation of a crater. This applies to craters formed by reflective spalling as well as to those formed by internal pressure. So far as the latter are concerned, the nearer the perimeter of the equivalent cavity approaches the free surface the greater will be the tangential "hoop stresses". In the theoretical case, the tangential stress varies with the angle φ in Fig. 1.4a and reaches a maximum at the perimeter of the equivalent cavity, where φ is a maximum. The maximum tension thus becomes higher the closer the cavity approaches the free face. Tensile stresses are also generated at the free surface. These reach a maximum at the point A, but if the depth of burden is larger than the diameter of

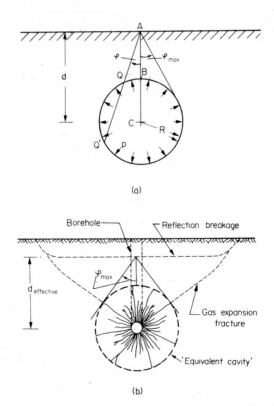

(a)

(b)

Fig. 1.4. Geometrical relations between equivalent cavity, depth of burden, and break angle (Kutter and Fairhurst [23]).

the equivalent cavity (which is usually the case in practice) the surface tensions are smaller than the hoop stresses around the cavity.

Kutter and Fairhurst then postulate a criterion for preferred fracture directions of those radial cracks which extend to form the break angle. The radial fractures within the equivalent cavity, which terminate in the zones of maximum tension (i.e. where φ is a maximum), will propagate at the lowest critical pressure, and thus determine the shape of the crater formed between the charge and the free surface (Fig. 1.4b).

This idealized process is, in practice, modified by the physical characteristics of the confining medium, which may be heterogeneous and non-isotropic. Blasting in rock is very largely conditioned by such geological characteristics as joints and cleavage, bedding planes, and shrinkage cracks. It is likely that these discontinuities will often determine preferential fracture directions.

Another major influence when blasting in rock strata is the *in situ* stress field, which in a real situation could be the resultant stress pattern generated by mining activities superimposed upon the tectonic stress field. Kutter and Fairhurst found that both the stress-wave and the gas-generated fractures propagated preferentially in the direction of the maximum principal stress, the effect being sometimes so marked, in a high-stress situation, as to drown the influence of geological discontinuities.

Kerkhof, in his studies of brittle fracture propagation in glass [24], showed that the direction of crack propagation is coincident with the direction of the maximum principal stress.

When considering the manner in which the radial cracks from a blast extend into the confining medium, and which of these are most likely to propagate to form the break angle at the free surface, Field and Ladegaard-Pedersen assume that the cracks grow outward at a velocity V, while the stress waves move at velocity C greater than V. The reflected C waves return from the free boundary to meet the outgrowing cracks.

Using the notation shown in Fig. 1.5, the blast origin is O and the free face AB, with depth of burden x.

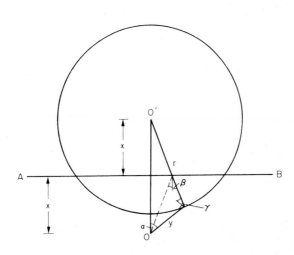

FIG. 1.5. Geometrical relations between free face, depth of burden, reflected stress wave, and break angle (Field and Ladegaard-Pedersen [22]).

where N is the critical depth, E is the strain energy factor, w_c is the critical weight of charge. E is a constant for a given explosive and a given confining medium.

If the weight of the explosive charge is increased until cratering of the confining medium is produced, the conditions pass beyond Livingston's strain energy range and enter what he calls the shock range, the upper limit of which is stated to coincide with the point of maximum efficiency of the utilization of explosives in blasting.

The weight of explosive producing this maximum blasting efficiency he terms the "optimum weight", and this is related to the upper limit of the shock range by the equation

$$d_o = \Delta E \sqrt[3]{w_o},$$

where d_o is the "optimum depth", Δ is the "depth ratio", and w_o is the "optimum weight" of charge.

The optimum weight of charge is defined as that weight of explosive, charged in a given shape, which produces maximum quantitative separation of the confining medium from the parent mass.

The depth ratio is

$$\frac{\text{depth of charge}}{\text{critical depth}}.$$

If the weight of charge exceeds the optimum, the conditions of the blast will enter the fragmentation range, in which the depth ratio decreases and the crater dimensions and proportions are progressively affected. The upper limit of the fragmentation range is characterized by extreme disintegration, the ejection of flyrock, and the production of dust and air blast. In these circumstances the quantity of energy dissipated into the atmosphere may exceed that utilized in breaking the confining medium, i.e. the numerical value of A may exceed that of B in the Livingston crater equation

$$V = ABC(N)^3,$$

where V is the crater volume, B is the rock fragmentation number, A is the energy utilization number, C is the stress distribution number, and N is the critical depth.

A expresses the effect of charge depth upon the crater volume for a given blast design (given depth ratio, explosive, confining medium, shape of charge). B is a variable function of the depth ratio. C expresses the effect on the crater volume of such factors as charge shape, geometry of free faces in the confining medium, geometry of mass to be broken by the blast, and the blast drill-hole and charge pattern.

At optimum weight, for a charge of a given shape, both A and C measure unity and B may be obtained experimentally.

Geometry of Blast Craters

Livingston describes the geometry of blast craters in terms of the three variables depth of burden d, radius of crater R, and the distance h measured between the apex of the crater and the centre of the charge C.

R/d is the radius–depth ratio.

$h/d = S_m$ is the similitude ratio.

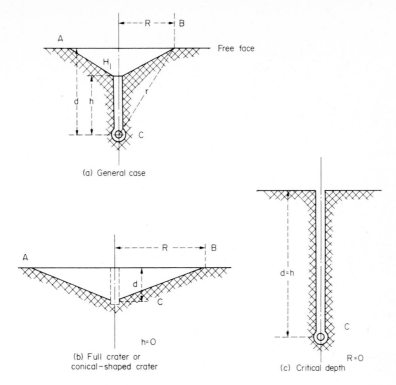

(a) General case

(b) Full crater or conical-shaped crater

(c) Critical depth

FIG. 1.8. Shapes of craters produced by spalling (Hino [5]).

When h = zero the crater is at its maximum depth and is termed a "full" crater.

For a given weight of charge, if d increases R tends to zero and h approaches the critical depth (Fig. 1.8).

Thus the crater geometry corresponding to a particular explosive in a given material may be represented by a set of curves such as those shown in Fig. 1.9.

Hino describes the formation of craters by the process of reflective "scabbing" in the following terms:

The peak pressure p_r produced in the confining medium at distance r from the centre of a spherical charge of radius a (corresponding to the reduced distance D where $D = r/a$) is

$$p_D(1/D)^n \quad \text{or} \quad p_r = p_D(a/r)^n,$$

where p_D is the detonation pressure and n is a distance exponent.

The outgoing shock wave is an exponential function of time and space, represented by

$$p_{rt} = p_D(a/r)^n e^{t/\theta}$$

where θ is a time constant, e is logarithm base, and t is the time.

This may be approximated to a linear triangular function in which the peak pressure p_m is reached over a time t, a shock wavelength L, and velocity U, where $L = Ut$.

This wave reflects at the free face AB of Fig. 1.10 as a tension wave, and if the resultant of the outgoing compression and the reflected tension exceeds the tensile strength S_t of the rock, then scabbing occurs. The separation results in the formation of another free face

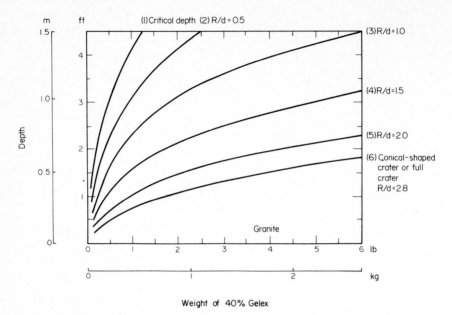

FIG. 1.9. Relationship between weight of charge and critical depth for 40% Gelex blasting in granite (Livingston [26] and Hino [5]).

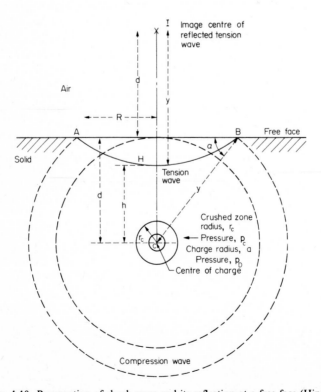

FIG. 1.10. Propagation of shock wave and its reflection at a free face (Hino [5]).

from which the process is repeated so long as the shock wave persists.

To produce a full crater of depth d_f the shock wave must have a duration $t_f = 2d_f/U$.

If the depth of the charge d is increased for a constant weight of explosive, the radius R of the crater becomes smaller, and either the number of scabs is reduced or else the thickness of the first slab is increased (depending upon the tensile strength of the confining medium) owing to the reduction of peak pressure at the first reflection.

The similitude ratio is then

$$S_m = h/d = \frac{d - d_f}{d} = 1 - \frac{d_f}{d}.$$

Since the reduced depth or "reduced burden" is $D_f = d_f/a$,

$$S_m = 1 - D_f a/d,$$

where a is the radius of a spherical charge.

Radius of Crushed Zone Around a Spherical Charge

The exponent n may be determined experimentally. Its value lies between 2 and 3 in close proximity to the blast origin, but approaches unity with increase in distance. From the data shown in Fig. 1.9 and assuming a loading density for the charge of 1.0 g/cm^2, confined in granite in which the pressure–distance characteristic is that shown in Fig. 1.11,

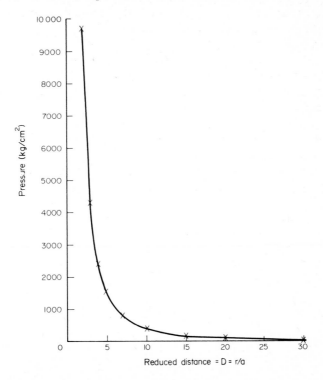

FIG. 1.11. Pressure – distance characteristic of a shock wave for 40 % Gelex blasting in granite (Livingston [26] and Hino [5]).

Hino derives a value $D = 4.76$, so that the radius of the crushed zone is smaller than 95.5 mm for a charge of 25 mm radius.

Crater Formation Using Nuclear Explosives

With very large charges of chemical explosives, or when nuclear explosives are employed, shock-wave phenomena assume much greater significance than they do in conventional quarry blasting.

In the process of crater formation above a buried nuclear charge, the Livingston concepts of optimum and critical depths apply on a larger scale. The apparent (i.e. visible) crater dimensions increase to a maximum at the optimum depth, then decrease until a depth of burst is reached at which no external crater is formed.

In nuclear blasting, however, a surface crater may form subsequent to the blast, as a result of progressive caving of overlying strata into a subterranean blast cavity. A chimney of rubble is formed above the shot. This ultimately may extend to the surface to form a subsidence crater. Hughes describes the mechanism [27].

Computation of Crater Geometry in Nuclear Blasting

Two basic approaches have so far been developed for predicting the crater dimensions from blasts of nuclear proportions. One approach involves computer calculations of the mound and cavity growth, used in conjunction with a free-fall, throw-out model, which gives reasonable estimate of the crater radius and ejecta boundary. The second approach involves empirical scaling relationships.

The Plowshare Division of the Lawrence Radiation Laboratory has developed two computer codes, SOC (spherical, one-coordinate) and Tensor (cylindrical, two-coordinate), which numerically describe the propagation of a stress wave of arbitrary amplitude through a medium.

These codes are Lagrangian finite-difference approximations of the momentum equations which describe the behaviour of a medium subject to a stress tensor in one (SOC) and two (Tensor) space dimensions [28].

A Lagrangian coordinate system (i.e. one that travels with the material) orientated with reference to the point of detonation of the explosive is established in the geologic media. The underground detonation of a nuclear device releases energy which imparts motion and alters the properties of the medium. The computer calculations are based upon a "feedback loop" analysis which numerically describes the effect of the stress-wave propagation in the medium.

The computation cycle is as follows:

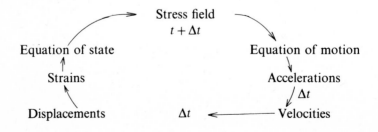

The equation of motion provides a functional relation between the applied pressure stress field and the resulting acceleration of each point in the medium. These accelerations, when allowed to act over a small time increment Δt, produce new velocities. The velocities, in turn, produce displacements which result in strains that establish a new stress field.

Cyclic repetition is accomplished as required. In order to ensure the stability of the iterative calculation procedure, the time increment Δt is selected in such a manner that it is less than the time necessary for a compressional wave to travel across the smallest zone.

The strain field in the code calculations is coupled to the stress field through the equation of state of the medium. Except for the vaporized region, this equation of state is determined experimentally. When the material is vaporized, the equation of state is expressed as a theoretical pressure–energy–density relationship, normalized to the Hugoniot and extending to the Thomas–Fermi–Dirac region at high energies. In the liquid and plastic states, the material is represented by a Hugoniot curve relating pressure and specific volume at the shock front.

The accuracy of the code calculations is limited by two main factors. One is the degree to which it is possible to describe completely the behaviour of the material under stress, especially failure mechanisms, and the other is the ability to describe the bulking of the fall-back material as it is deposited in the cavity void.

Empirical Scaling of Crater Geometry

This approach uses scaling laws which relate the crater dimensions for some reference energy yield to crater dimensions for any energy yield. The reference nuclear yield normally used is 1 kt of TNT.

A cube-root scaling law may be used in which

$$\text{crater radius } R_a = r_a (W/W_o)^{1/3}$$

and

$$\text{crater depth } D_a = d_a (W/W_o)^{1/3}$$

in which R_a and D_a are the predicted crater dimensions for explosive yield W, and r_a and d_a are the crater dimensions for some reference yield W_o.

This scaling law does not take into consideration the effects of gravity or the frictional forces in the rock. It appears to be valid for small-scale high-explosive experiments, but not for larger blasts or for nuclear cratering detonations in the optimum depth-of-burst region.

Empirical scaling laws are based on the premise that a relationship can be developed which will satisfactorily correlate the dimensional data obtained from craters resulting from explosive detonations of different energy yields. One approach which has been used is based upon the assumption that crater dimensions are proportional to depth of burst, i.e.

$$\frac{r_a}{R_a} = \frac{dob}{DOB} \quad \text{and} \quad \frac{d_a}{D_a} = \frac{dob}{DOB},$$

where DOB and dob are the depths of burst at which explosives at yields W and W_o, respectively, are detonated.

Applied Geotechnology

An analytical least-squares fit of crater dimensions resulting from the detonation of 256 high-explosive charges at various depths of burst in alluvium was used to develop a functional relationship between crater radius and crater depth and depth of burst, i.e.

$$R_a = f_r(DOB) \quad \text{and} \quad D_a = f_d(DOB).$$

These functions, f_r and f_d, were used in conjunction with the cratering dimensions from high-explosive detonations in alluvium, ranging in yield from 1160 kg to 0.5 million kg, to determine empirical scaling exponents, i.e.

$$\frac{R_a}{r_a} = \left[\frac{W}{W_o}\right]^x ; \quad \frac{D_a}{d_a} = \left[\frac{W}{W_o}\right]^y ; \quad \text{and} \quad \frac{DOB}{dob} = \left[\frac{W}{W_o}\right]^z .$$

The scaling exponents developed from this comparison of analytical techniques were:

$$x \simeq y \simeq z \simeq 1/3.4.$$

The empirical scaling law may then be expressed:

$$\frac{R_a}{r_a} = \left[\frac{W}{W_o}\right]^{1/3.4} ; \quad \frac{D_a}{d_a} = \left[\frac{W}{W_o}\right]^{1/3.4} ; \quad \frac{DOB}{dob} = \left[\frac{W}{W_o}\right]^{1/3.4} .$$

If a reference scale of 1 kt is used, these empirical scaling laws may be expressed as:

$$R_a = r_a W^{1/3.4},$$

$$D_a = d_a W^{1/3.4},$$

$$DOB = (dob) W^{1/3.4},$$

where the dimensions $r_{a.}$ $d_{a.}$ and *dob* apply to a crater produced by a yield of 1 kt and W is expressed in kilotonnes.

Single-charge Apparent Crater Shape

Detonations in the vicinity of optimum depth of burst produce craters, which are essentially hyperbolic in cross-section, with slope inclinations approximately equal to the angle of repose of the ejected crater material.

The following empirical relationships may be used in estimating the lip dimensions of craters resulting from detonations of single charges near optimum depth of burst:

	Lip height (H_{al})	Lip crest radius (R_{al})	Outer radius of lip (R_{eb})
Dry soil	$0.15 D_a$	$1.15 R_a$	$2.2 R_a$
Hard rock	$0.30 D_a$	$1.2 R_a$	$3.0 R_a$
Clay shale	$0.45 D_a$	$1.35 R_a$	$3.5 R_a$

Shape of Row-charge Crater

The simultaneous detonation of a series of buried charges of nuclear scale will form a relatively smooth, linear row crater. Experience so far indicates that the apparent crater resulting from a row-charge detonation is hyperbolic in cross-section normal to the linear direction. The same hyperbolic relationships apply as are those used to describe a single-charge crater.

The nuclear row-charge cratering experiment Buggy indicated that, for a row-charge array consisting of charges spaced at one apparent single-charge crater radius, the apparent width (W_a) and depth (D_{ar}) are enhanced approximately by 10% as compared with the corresponding single-crater dimensions.

Because the ejecta from the linear region of a row-charge are constrained to move laterally to the sides, rather than in all directions as in the case of a single charge, the apparent lips of a row crater are relatively higher than those which would result from a single charge. The lips at the ends of a row-crater are approximately the same as those for a single charge. The following relations may be used to estimate the side lip configuration for row craters:

	Lip height (H_{al})	Lip crest width (W_{al})	Lip edge width (W_{eb})
Dry soil	$0.40\,D_{ar}$	$1.5\ W_a$	$4\ W_a$
Hard rock	$0.65\,D_{ar}$	$1.5\ W_a$	$4\ W_a$
Clay shale	$0.60\,D_{ar}$	$1.5\ W_a$	$4\ W_a$

True Crater Size and Shape

The true crater is considered to include both the fall-back material and the underlying material which has experienced significant dislocation, and which would be recoverable with relative ease in a quarrying application.

The boundary of the true crater is not a sharp, well-defined surface. Consequently some judgement is required as to the boundary of the true crater in any particular case. A paraboidal shape is the geometric configuration that most accurately describes the true crater boundary between a horizontal plane through the zero point (effective centre of explosion energy) and the original ground surface.

Below the zero point the true crater boundary is that of the explosion cavity, and is hemispherical. The radius of this cavity in rock is approximately $45\ W^{1/3}$ ft, where W is in kilotonnes.

Blasting with Low Explosives

When blasting in solid media using explosives of low detonation velocity, or with explosives that deflagrate, the shock-wave analysis is irrelevant and the elastic-wave energy plays a less important role than does the energy contained in the expanding gases. Fracture and fragmentation of the confining medium is then a relatively slow process, and one that is largely controlled by shear forces. The theoretical treatment published in 1909 by A. W. and Z. W. Daw is still applicable to these conditions [29].

The Daws postulated that a shearing equation was applicable to the fracture surface such that

$$P = SWK$$

where P is the load, S is the periphery of pressure surface, K is the modulus of rupture in shear, and W is the rock burden.

They further postulated that the form of cavity produced by the application of an expanding pressure in ice is similar to that obtained by blasting in rock under similar conditions of free face and pressure surface.

Three interrelated empirical expressions were derived from trial blasts to include rock, charge, and spacing coefficients C_a, C_v, and e respectively.

More recently Clark and Saluga have duplicated the Daws' tests and have extended them into rock, suggesting that the modulus of rupture K is not always constant but is dependent on the ratio W/S, approaching a constant value when the ratio W/S is large [30].

Considering that the confining medium ruptures solely as a result of the explosion gas pressure acting upon the projected area of the shot-hole surface, Clark and Saluga develop from their experimental results

$$P = RS^a \ W^b$$

where P is the total load and R is the resistance to failure, and a and b are exponents, the value of which depends upon the nature of the confining medium. (In limestone a ranged from 0.8 to 0.85, and b ranged from 1.15 to 1.2.) As a and b approach unity, this expression approaches the Daws' rule $P = KS\,W$.

The results of the Daws' experiments on ice, when plotted in terms of the same dimensionless coefficients as used by Clark and Saluga, are shown in Fig. 1.12, from which values are obtained for $a = 0.915$ and $b = 1.085$.

Hence, for blasting in ice using low explosives,

$$P = RS^{0.915} \ W^{1.085}.$$

Clark and Saluga point out that the compressive strength of ice increases more rapidly than does its tensile strength, with reduction in temperature, and increases more rapidly with rise in temperature. That is, these properties approach one another in magnitude with rise in temperature. They suggest, therefore, that only at low temperatures will ice behave

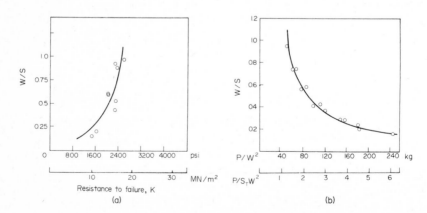

FIG. 1.12. (a) Influence of the ratio W/S over the value of resistance to failure K (Clark and Saluga [30]). (b) Relationship between dimensionless ratio W/S and $P/S_t \, W^2$ for ice (Clark and Saluga [30] after Daw and Daw [29]).

like rock, to obey the exponential law $P = RS^a W^b$. At higher temperatures ice will be more ductile and follow the Daws' rule $P = RS W$.

Clark and Saluga also suggest that the exponential law may be adapted to represent failure produced by high explosives by making appropriate changes to the values of the exponents. In actual field conditions W is the optimum burden for a given charge weight Q, and S is the perimeter of the projected pressure surface of the curved wall of a shot-hole of diameter d. Also m is the length of charge, so that $S = 2(m+d)$. For most commercial explosives $Q = C_v W^3$, where C_v is a charge coefficient (determined by trial blasts).

Coupling, Attenuation, and Transmission of Blast Energy Through the Confining Medium

Impedance

A proportion of the energy of the explosive charge is lost due to a mismatch between the characteristic impedance of the explosive and that of the confining medium (Tables 1.2, 1.3, and 1.4). The conditions of confinement and of geometrical contact between the explosive and the containing medium, the degree of pulverization of the wall in immediate contact with the explosive charge (this pulverization precedes the passage of the elastic-wave energy), all contribute towards the dissipation of explosive energy transmitted from the explosive to the confining medium.

TABLE 1.2.   *Characteristics of some typical explosives (Duvall and Atchison* [19])

Explosive	Specific gravity	Energy (Btu/lb)	Pressure (lbf/in²)	Detonation velocity (ft/s)	Characteristic impedance (lbf-s/in²)
Nitro-glycerine	1.6	650	2 939 000	26 250	47
50 % nitro-glycerine 2.3 % guncotton 41.3 % ammonium nitrate 5.5 % cellulose 0.9 % other	1.5	575	2 131 000	22 640	38
10 % nitro-glycerine 80 % ammonium nitrate 10 % cellulose	0.98	450	632 000	13 125	14
93 % ammonium nitrate 7 % carbon	1.0 to 0.8	436 to 430	679 000 to 429 000	13 900 to 12 000	15 to 10

TABLE 1.3.   *Compressional wave speeds and characteristic impedance for certain rocks* [31]

Rock	Velocity of compressional waves (ft/s)	Characteristic impedance (lbf-s/in²)
Granite	18 200	54
Marlstone	11 500	27
Sandstone	10 600	26
Chalk	9 100	22
Shale	6 400	15

TABLE 1.4. Summary of static and dynamic properties of the natural rocks, ceramic and plastic mixtures, and aluminium (Ricketts and Goldsmith [32])

Material	Density, ρ (g/cm³)	Rod-wave velocity, c_0 (in/s)	Specific acoustic impedance, ρc_0 (lb-s/ft³)	Compressive Young's modulus (10⁶ lb/in²)			Static strength of specimen (lb/in²)			
				Dynamic	Static (1)	Static (2)	Tensile (1)	Tensile (2)	Compressive (1)	Compressive (2)
Spessartite	2.89	244 000	114 000	16	8.9	2.8-(3)	2 500	2000-(3)	54 000	55 900-(3)
Basalt	2.72	194 000	85 200	9.6	6.5	6.7-(5) 7.0-(12)	1 400	595-(5) 1968-(12)	37 000	57 900-(5) 57 900-(12)
Diorite	2.83	144 000	65 800	5.5	5.1	4.65-(3) 3.56-(1) 3.2-(2)	800	1190-(3)	31 800	39 500-(3) 28 300-(1) 21 100-(2)
Leucogranite	2.59	135 000	56 500	4.2	4.45	4.2-(3)	450	460-(3)	30 000	27 000-(3)
Limestone	2.23	141 400	51 100	4.19	2.6	3.3-(14)	520	380-(14)	4 080	5 760-(14)
Sandstone	2.11	79 800	27 300	1.26	—	1.1-(14)	—	230-(14)	—	7 220-(14)
Scoria	1.49	131 000	31 600	2.4	1.1	0.66-(13)	625	360-(13)	4 190	2 800-(13)
Pumice	0.428	99 600	6 900	0.40	0.24	0.22-(11)	85	65-(11)	615	555-(11)
Diorite concrete	2.30	133 500	49 700	3.85	3.5	3.4-(1)	285	510-(1)	4 410	4 670-(1)
Scoria concrete	1.94	113 000	35 500	2.32	1.7	1.5-(1)	165	275-(11)	2 020	2 440-(1)
Concrete 7-day	2.17	143 500	50 300	4.18	2.9	3.1-(1, 2)	290	—	3 700	4 100-(1, 2)
Concrete 28-day	2.17	142 200	49 800	4.10	2.7	3.0-(1, 2)	330	—	4 700	4 700-(1, 2)
Lightweight aggregate concrete 7-day	1.34	123 300	26 500	1.90	1.55	1.53-(1, 2)	390	160-(1, 2)	4 640	6 160-(1, 2)
Lightweight aggregate concrete 28-day	1.34	123 300	26 500	1.90	1.55	—	390	—	—	—
Epoxy-aggregate concrete	2.04	108 500	35 600	2.33	2.04	1.72-(1)	—	65-(1)	6 800	7 050-(1)
Prestressed concrete	2.35	151 000	57 000	4.96	5.65	2.41-(2)	—	1690-(2)	10 120	10 440-(2)
Epoxy-aluminium powder	1.67	89 100	23 800	1.23	1.02	4.30-(1)	—	—	8 280	7 530-(1)
Cement paste 7-day	1.97	134 500	42 800	3.35	3.1	2.7-(1, 2)	530	—	11 500	10 500-(1, 2)
Cement paste 28-day	1.99	144 200	46 300	3.86	3.2	2.9-(1, 2)	350	—	12 700	10 500-(1, 2)
Epoxy	1.33	76 100	16 200	0.70	0.51	0.51-(1)	1 040	—	10 150	9 600-(1)
Aluminum 2024T-4	2.72	204 000	89 500	10.6	10.6	—	33 000	—	—	—

(1) Unshocked specimens.
(2) Shocked specimens. The number of previous shocks is indicated by the value in parentheses following the data, using averages for a particular bar.

The characteristic impedance of a medium is defined as density × stress-wave velocity. Nicholls and Atchison *et al.* state that the transfer of energy at the boundary of the shot-hole is determined by the ratio of the characteristic impedance of the explosive to that of the medium, a smaller impedance ratio resulting in more efficient transfer of energy to the medium [33–35].

Coupling

The transfer of energy between the explosive and the confining medium is determined by the conditions of coupling, and particularly by the conditions of contact and confinement. Sussa *et al.* [36] suggest the following relation between detonation pressure and the pressure imposed by an explosive in unconfined contact on rock:

$$P_t = \frac{2C\rho\, P_d}{D\rho_e + C\rho},$$

where P_d is the incident detonation pressure, P_t is the transmitted imposed pressure, ρ_e is the density of explosive, ρ is the density of rock, D is the detonation velocity, and C is the wave velocity, from which they calculated $P_d = 1.2 P_t$ for PETN/TNT—magnetite contact, and $P_d = 1.7 P_t$ for BeliteA 60%—magnetite contact.

With confined charges the explosive must be carefully tamped in place to improve the coupling. Conversely, deliberate "decoupling" techniques may be used, with the intention of limiting the transfer of energy and controlling the blast fragmentation. A "decoupling ratio" may then be quoted, this being radius of shot-hole/radius of charge.

Attenuation of Blast Energy

Noren [37] records observations of the effects of charge weight and distance on the peak pressure from single small explosive charges. The explosive used was DuPont 60% "Red Cross" "Extra". Comparative results using DuPont 60% RCX and DuPont "Pelletol" No. 1 are shown in Fig. 1.13. Noren's results are summarized in the formula

$$P = \frac{K}{D/W^{1/3}}\, 2.3,$$

where D is the distance, W is the weight, and K is the site constant (11 500 for the test conditions).

This is similar to Vortman's formula for high-explosive blasting in alluvium [38]:

$$P = \frac{3350}{D/W^{1/3}}\, 2.28.$$

Hino [5] relates the pressure produced by a spherical explosive charge in rock as

$$P_r = P_d(a/r)^n$$

where P_r is the pressure in the rock, P_d is the detonation pressure, a is the radius of spherical charge, r is the distance from centre of charge, and n is a distance exponent. n is stated to lie between 2 and 3 for small values of r, but approaches unity at greater distance.

23

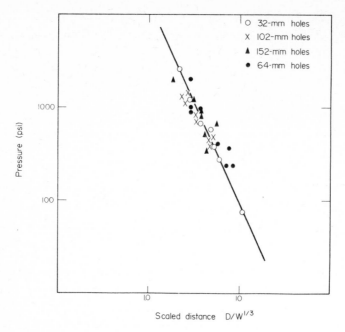

FIG. 1.13. Pressure vs. scaled distance (Pelletol No. 1) (Noren [37]).

Sussa and Ito [39] obtained a distance exponent for the peak radial stress at about − 2.4, but Coates [40] suggests that the attenuation of the peak stress is proportional to the inverse cube of the distance in the hydrodynamic state, and the inverse square of the distance in the elastic state.

Bass and Hawk [41] show a much lower decay constant (− 1.5) for a partially confined blast of TNT weighing 18 000 kg. Holtzer's data [42] from underground nuclear blasts suggest a decay factor about − 2.0 for alluvium and granite, but a lower exponent for tuff containing up to 25 % water.

Butkovich [43] predicts a variable decay factor reaching − 1.94 for pressures below 1 megabar. Various researches by the US Bureau of Mines quote decay exponents ranging from − 1.4 to − 2.1, depending upon the type of material and the distance from the blast.

As a general rule, when using commercial explosives in conventional-type blasting, a decay factor of − 2.3 is probably representative of near-spherical charges in hard rock, but this will vary in other materials. Noren suggests that the decay factor for dry sand will probably be much larger than 2.3, while for wet sand it will approach that given by Cook for water (− 1.13) [44].

From their observations of explosion-generated strain pulses in rock, Duvall and Petkof [45] quote a decay in strain according to the law

$$\varepsilon = (K/R)e^{-\delta R},$$

where ε is the strain, K is a linear function of the detonation pressure and the impedance, δ is a constant depending upon the rock type, and R is the scaled distance (distance/$\sqrt{\text{charge weight}}$).

24

δ was reported as having the values:

0.03	granite
0.048	sandstone
0.026	shale and chalk
0.08	marl

Rinehart [46] suggests that the amplitude of the peak pressure in the blast wave at distance r is equal to

$K(a/r)e^{-\delta' r}$ for spherical wave propagation and

$K(a/r)^{1/2}e^{-\delta' r}$ for cylindrical wave propagation, where a is the radius of the shot-hole and δ' is the attenuation coefficient.

Attenuation is not the same for all wavelengths. Auberger and Rinehart, measuring attenuation in rocks over the 100 kHz to 2 MHz range, found that attenuation peaks occurred at harmonic frequencies which could be related to the predominant grain size in the rocks. It appears that when the length of the energy pulse approaches the grain size of the rock the constituent crystals are set into resonance, which causes local increase in the attenuation/frequency relation.

Krishnamurthy and Balakrishna [47] have also determined attenuation characteristics for several rocks. Attenuation was seen to increase with frequency in fine-grained rocks, but with larger grain sizes in dolerite, limestone, and marble, attenuation was independent of frequency. Attewell and Brentnall [48] observed the attenuation exponent to vary linearly with frequency over a range of approximately 5–100 kHz, and they adduced that the attenuation is due to scattering at grain and pore boundaries (Fig. 1.14).

Ricketts and Goldsmith [32], investigating the dynamic properties of rocks and composite structural materials, using the Hopkinson-bar technique, determined the attenuation characteristics of pulses generated by impact velocities up to 254 m/s. The exponential attenuation coefficient for this level of impulse was small in limestone, but there was much greater dispersion in sandstone.

Dispersion and attenuation change with successive impacts in the same specimen, so that repeated shocks produce an increase in the pulse level and a decrease in the rod-wave velocity, indicative of weakening the material due to the stress-wave passage (Fig. 1.17). These observations are of significance in relation to blasting phenomena, as they help to explain the co-operative action of the stress wave, which weakens the rock, followed by the expanding gas pressure, which breaks it. They also help to explain the effectiveness of the multiple incidence and multiple reflections of the stress waves produced by rapid-sequence delay blasting.

Effects of Charge Shape on Coupling and Attenuation

Haas and Rinehart [49], experimenting with cylindrical charges of DuPont EL506A (PETN), found that the area of contact of a cylindrical charge had a much stronger effect on the peak stress produced in rock than did the length of the charge, i.e. a given quantity of explosive will transmit more energy into rock when the charge is short and squat than when it is long and thin. (In these experiments the charge rested is unconfined contact with the rock.)

An air gap between the end of the cylindrical charge and the surface of the rock greatly reduced the stress in the rock. For a 12.5 mm diameter charge and a 25 mm air gap the

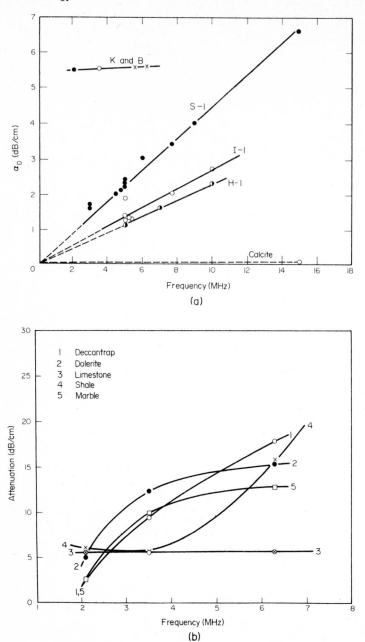

Fig. 1.14. Attenuation against frequency for several rocks (Krishnamurthy and Balakrishna [47]).

peak stress was approximately one-third that observed for direct contact of explosive and marble.

Langefors and Kihlstrom [17] recommend that, to obtain maximum fragmentation from a single shot, the blast-hole should be filled to the greatest possible loading density.

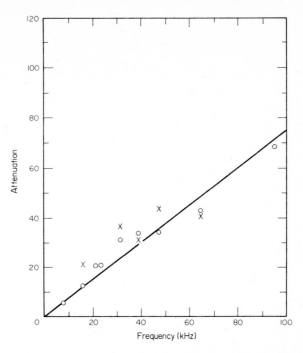

FIG. 1.15. Frequency response of Darly Dale Sandstone (Attewell and Brentnall [48]).

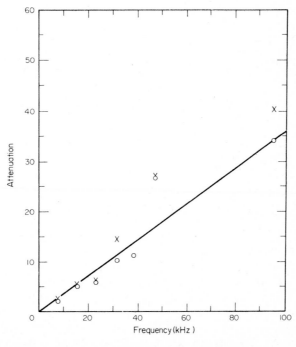

FIG. 1.16. Frequency response of calcareous sandstone (Attewell and Brentnall [48]).

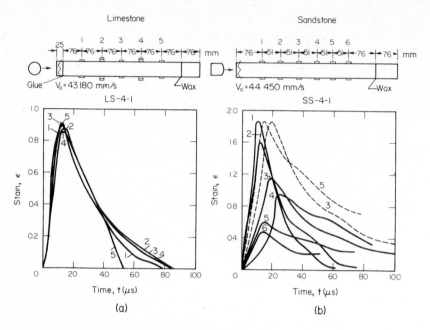

FIG. 1.17. Superposed pulses, with a common time origin, for limestone and sandstone (Ricketts and Goldsmith [32]).

Such a condition is intended to produce best possible coupling. Subsequent experiments by Personn *et al.* [50], exploring the effects of the larger diameter drill-holes that are now feasible using contemporary drilling techniques, quote the results shown in Fig. 1.18. In these experiments, for each borehole diameter a number of shots were fired with different depths of burden, up to the "critical burden" (where no surface fragmentation was visible). The dashed line on Fig. 1.18 gives the critical burden as a function of hole diameter, for a

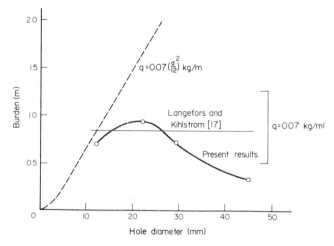

FIG. 1.18. Comparison of relationships between hole diameter and critical burden (Personn *et al.* [50]).

constant charge density of $0.07(d/12)^2$ kg/m bench height (d = hole diameter mm). The solid lines on the figure represent constant charge weight per unit bench height, $q = 0.07$ kg/m.

The critical depth of burden increases far more rapidly with hole diameter at a constant loading density than it does at a constant weight of charge.

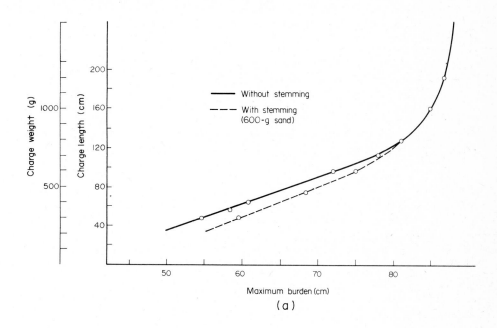

(a)

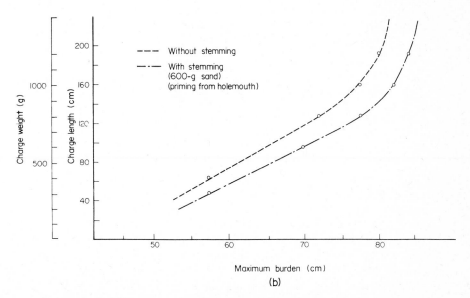

(b)

FIG. 1.19. Limit burden with and without stemming, and with forward and reverse initiation (Johnsson and Hofmeister [52]).

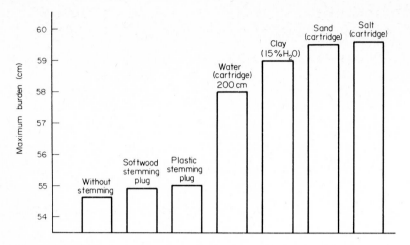

Fig. 1.20. Limit burden with various stemming materials (Johnsson and Hofmeister [52]).

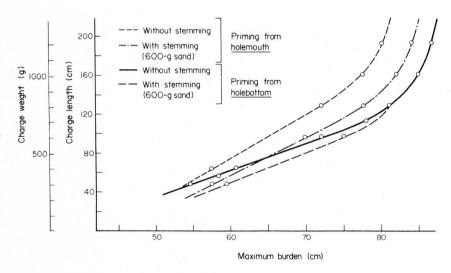

Fig. 1.21. Effects of stemmed and unstemmed blasts, with forward and reverse initiation (Johnsson and Hofmeister [52]).

Decking and Decoupling

The procedure of "air decking" of an extended charge in bench blasting, and decoupling a concentrated charge in crater blasting, is described by Melnikov [51]. Decoupling a crater blast is performed by drilling a larger hole than the diameter of the charge, so that a volume of air is left between the charge and the wall of the drill-hole. A "decked" charge consists of several separate charges distributed along the length of the hole, with intervening air gaps, the charges being fired simultaneously with detonating cord.

The object of the procedure is to minimize the extent of, or possibly eliminate altogether, the layer of pulverized wall rock immediately surrounding the perimeter of the blasted charge. This otherwise represents a waste of energy and it obstructs the passage of the stress wave into the confining medium.

The greater volume made available to the explosion gases slightly reduces the peak explosion pressure, but also, what is more important, it also extends the time during which the pressure builds up to its peak value. The effects of the explosion are thus extended to a larger volume of the confining medium than would be the case with a more intensive and shorter blast. This is estimated to give more widespread utilization of the explosive energy and more uniform fragmentation of the medium.

The technique is akin to that employed when blasting in coal-mines when the production of large-lump coal, rather than pulverized material, is desired. Low-density explosives, which contain a proportion of inert lightweight material disseminated throughout the charge, or the inclusion of air space between the charge and stemming, to give what is termed "cushion blasting", is the method then used.

Effects of Stemming the Charge

Stemming is used to confine the explosion gases and thus increase the amplitude of the peak explosion pressure. Johnsson and Hofmeister [52] describe comparative blasts in rock salt with variations in stemming. A charge of 300 g of explosive stemmed with sand or granular salt produced the same effect as 400 g of unstemmed charge (Fig. 1.19). The limit

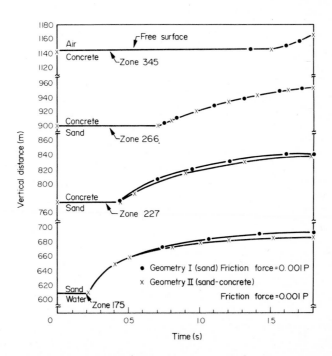

FIG. 1.22. Displacement of stemming of a large blast for a friction of 0.001P (Crowley *et al.* [53]).

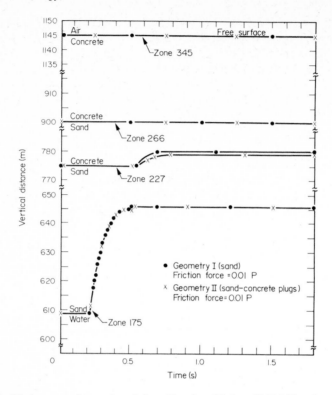

FIG. 1.23. Displacement of stemming of a large blast for a friction of $0.01P$ (Crowley *et al.* [53]).

burden attained with 300 g charges, using various stemming materials, was as shown in Fig. 1.20.

A comparison of the effects of priming at the bottom of the hole, compared with priming at the collar (reverse initiation and forward initiation, respectively), gave the results shown in Fig. 1.21. With small charges the influence of the stemming was more important than the method of initiation. For longer charges reverse initiation of an unstemmed charge produces higher blasting efficiency than does a stemmed charge with forward initiation. It appears that the damming action of the detonation gases with reverse initiation effects better confinement of the blast than does stemming on a charge initiated from the mouth of the shot-hole.

With underground nuclear blasts, stemming of the charge-hole and of other possible vent paths is required to contain the explosion. Crowley *et al.* [53] have described some of the containment problems associated with typical stemming configurations in nuclear blasting. The dynamic frictional force between the stemming material and the surrounding media were judged to be far more important than the static cohesive strength normally assumed, 0.07 MPa (10 psi) to 0.48 MPa (70 psi), for "mush concrete" plug/surrounding media contact.

The calculated displacement of the stemming plug with time after detonation is shown in Figs. 1.22 and 1.23 for various frictional force values, estimated as a function of the static pressure P. Figure 1.23 indicates no physical displacement of the top 1800 ft of stemming material when a value for the friction force of $0.01P$ is assumed.

References and Bibliography: Chapter 1

1. FISCHER, R. L., D'ANDREA, D. V., and FOGELSON, D. E., *Effects of Explosive Properties on Free-surface Displacement Pulses and Crater Dimensions*. US Bur. Min., Rep. Invest. No. 7407, 1970.
2. SADWIN, L. D., STRESSAU, R. H., PORTER, S. J., and SAVITT, J., Nonideal detonation of ammonium nitrate–fuel oil mixtures, *Proc. 3rd Symp. on Detonation, Princeton, NJ, 1960*, pp. 309–25.
3. BRIDGEMAN, P. W., *The Physics of High Pressure*, B. Bell & Sons, London, 1949.
4. COURANT, R., and FRIEDRICHS, K. O., *Supersonic Flow and Shock Waves*, Interscience, New York, 1948.
5. HINO, K., *Theory and Practice of Blasting*, Nippon Kayaku Co. Ltd., Asa, Yamaguchi-Ken, Japan, 1959.
6. ANDERSON, G. D., *The Equation of State of Ice and Composite Frozen Soil Material*, CRREL RR 257, 1968.
7. BAILEY, S. W., BELL, R. A., and PENG, C., Plastic deformation of quartz in nature, *Bull. Geol. Soc. Am.* **69**, 1443–66 (1958).
8. BORG, I., FRIEDMAN, M., HANDIN, J., and HIGGS, D. V., *Experimental Deformation of St. Peter Sand*, Geol. Soc. Am. Memo 79, ch. VI, pp. 133–91, 1960.
9. BOYD, F. R., and ENGLAND, J. L., The quartz–coesite transition, *J. Geophys. Res.* **65**, 749–56 (1960).
10. FRYER, C. C., Shock deformation of quartz sand, *Int. J. Rock Mech. Min. Sci.* **3**, 81–88 (1966).
11. BROBERG, K. B., Some aspects of the mechanism of scabbing, *Int. Symp. on Stress Wave Propagation in Materials, Penn. State Univ., 1960*, pp. 231–8, Interscience, New York.
12. ROBERT, D. K., and WELLS, A. A., The velocity of brittle fracture, *Engineering* **178**, 820–1 (1954).
13. WELLS, A. A., and POST, D., The dynamic stress distribution surrounding a running crack: a photoelastic analysis, *Engineering* **192**, 705–29 (1968).
14. DOBRIN, M. B., SIMON, R. F., and LAWRENCE, D. L., Rayleigh waves from small explosions, *Trans. Am. Geophys. Un.* **32**, 822–32 (1961).
15. FOGELSON, D. E., DUVALL, W. I., and ATCHISON, T. C., *Strain Energy in Explosion-generated Strain Pulses*, US Bur. Min., Rep. Invest. No. 5514, 1959.
16. NICHOLLS, H. R., and HOOKER, V. E., *Comparative Studies of Explosives in Salt*, US Bur. Min., Rep. Invest. No. 6041, 1962.
17. LANGEFORS, U., and KIHLSTROM, B., *Rock Blasting*, 2nd edn., Wiley, New York, 1967.
18. RINEHART, J. S., Fracturing under impulse loading, *3rd Am. Symp. on Mining Res., Univ. of Missouri, Rolla, 1958*, pp. 46–83.
19. DUVALL, W.I., and ATCHISON, T. C., *Rock Breaking by Explosives*, US Bur. Min., Rep. Invest. No. 5336, 1957.
20. HINO, K., *Theory and Practice of Blasting*, Nippon Kayaku Co. Ltd., Asa, Yamaguchi-Ken, Japan, 1959.
21. RINEHART, J. S., The role of stress waves in the comminution of brittle, rock-like, materials, *Int. Symp. on Stress Wave Propagation in Materials, Penn. State Univ. (1960)*, Interscience, New York.
22. FIELD, J. E., and LADEGAARD-PEDERSEN, A., The importance of the reflected stress wave in blasting, *Int. J. Rock Mech. Min. Sci.* **8**, 213–26 (1971).
23. KUTTER, H. K., and FAIRHURST, C., On the fracture process in blasting, *Int. J. Rock Mech. Min. Sci.* **8**, 181–202 (1971).
24. KERKHOF, F., Neue Ergebnisse der Ultraschallfraktographie, *Symposium sur la Résistance Méchanique du Verre, Florence, 1961*; and Fraktographische Untersuchungen von mechanischen Impulsen in Platten. *Proc. 7th Int. Congr. on High Speed Photog., Zürich, 1965*, p. 345.
25. PERSONN, P. A., LADEGAARD-PEDERSEN, A., and KIHLSTROM, B., The influence of borehole diameter on the blasting capacity of an extended explosive charge, *Int. J. Rock Mech. Min. Sci.*, **6**, 277 (1969).
26. LIVINGSTON, C. W., Fundamentals of rock failure, *Colo. Sch. Mines Q.* **51** (3) 1–11 (1956).
27. HUGHES, B. C., *Nuclear Construction Engineering Technology*, NCG Technical Rep. No. 2 Lawrence Radiation Laboratory, 1968.
28. CHERRY, J. T., Computer calculations of explosion-generated craters, *Int. J. Rock Mech. Min. Sci.* **4**, 1–224 (1967).
29. DAW, A. W., and DAW, Z. W., *The Principles of Rock Blasting*, E. & F. N. Spon, London, 1909.
30. CLARK, L. D., and SALUGA, S. S., Blasting mechanics, *Trans. AIME New York* **232**, 78–90 (1964).
31. LEET, L. D., *Vibrations from Blasting Rock*, Harvard Univ. Press, 1960.
32. RICKETTS, T. E., and GOLDSMITH, W., Dynamic properties of rocks and composite structural materials, *Int. J. Rock Mech. Min. Sci.* **7**, 315–55 (1970).
33. NICHOLLS, H. R., and HOOKER, V. E., *Comparative Studies of Explosives in Salt*, US Bur. Min. Rep. Invest. No. 6041, 1962.
34. ATCHISON, T. C., and PUGLIESE, J. M., *Comparative Studies of Explosives in Granite*, US Bur. Min., Rep. Invest. No. 6434, 1964.
35. ATCHISON, T. C., and PUGLIESE, J. M., *Comparative Studies of Explosives in Limestone*, US Bur. Min., Rep. Invest. No. 6395, 1964.
36. SUSSA, K., COATES, D. F., and ITO, I., Coupling and stress waves in close proximity to surface explosions, *Int. J. Rock Mech. Min. Sci.* **4**, 229–43 (1967).
37. NOREN, C. H., Pressure–time measurements in rock, in *Status of Rock Mechanics, 9th Symp. on Rock Mech., Golden, Colo.*, pp. 285–96, AIME, New York, 1968.

38. VORTMAN, J. J., Cratering experiments with large high explosive charges, *Geophysics* **28** (June 1963).
39. SUSSA, K., and ITO, I., Dynamic stresses induced within rock in blasting with one free face, *7th Symp. on Rock Mech., Penn. State Univ., 1965*.
40. COATES, D. F., Rock mechanics applied to the design of underground installations to resist ground shock from nuclear blasts, *5th Symp. on Rock Mech., Univ. of Minnesota, 1962*.
41. BASS, R. C., and HAWK, H. L., *Close-in Shock Studies*, Project 1 : 4 Ferris Wheel Series POR 3005, Sandia Corpn., 1965.
42. HOLTZER, F., Measurements and calculations of peak shock waves parameters from underground nuclear detonations, *J. Geophys. Res.* **70** (4) (1965).
43. BUTKOVICH, T. R., Calculation of the shock wave from an underground nuclear explosion in granite, in *Peaceful Applications of Nuclear Explosives*, Plowshare, 1965.
44. COOK, M. A., *The Science of High Explosives*, Reinhold, New York, 1958.
45. DUVALL, W. I. and PETKOF, B., *Spherical Propagation of Explosion-generated Strain Pulses in Rock*, US Bur. Min., Rep. Invest. No. 5483, 1959.
46. RINEHART, J. S., The role of stress waves in the comminution of brittle, rock-like, materials, *Int. Symp. on Stress Wave Propagation in Materials, Penn. State Univ., 1960*, pp. 247–69, Interscience, New York.
47. KRISHNAMURTHY, M., and BALAKRISHNA, S., *Geophysics* **22**, 268–74 (1957).
48. ATTEWELL, P. B., and BRENTNALL, D., Internal friction—some consideration of the frequency response of rocks, *Int. J. Rock Mech. Min. Sci.* **1**, 231–54 (1964).
49. HAAS, C. J., and RINEHART, J. S., Coupling between unconfined cylindrical explosive charges and rock, *Int. J. Rock. Mech. Min. Sci.* **2**, 13–24 (1965).
50. PERSONN, A., LADEGAARD-PEDERSEN, A., and KIHLSTROM, B., The influence of borehole diameter on the rock blasting capacity of an expended explosive charge, *Int. J. Rock Mech. Min. Sci.* **6**, 277–84 (1969).
51. MELNIKOV, N. V., Influence of charge design on the results of blasting, *Int. Symp. on Mining Res., Univ. of Missouri, Rolla, 1961*, pp. 147–55, Pergamon, New York, 1962.
52. JOHNSSON, G., and HOFMEISTER, W., The influence of stemming on the efficiency of blasting, using 36 mm shot-holes, *Int. Symp. on Mining Res., Univ. of Missouri, Rolla, 1961*, **1**, 91–102, Pergamon, New York, 1962.
53. CROWLEY, B. K., GLENN, H. D., KAHN, J. S., and THOMSEN, J. M., Analysis of Jorum containment, *Int. J. Rock Mech. Min. Sci.* **8**, pp. 77–96 (1971).

CHAPTER 2

Rock Excavation by Blasting—II

Explosives, Blasting Agents, and Explosive Substitutes

EXPLOSIVES are commonly classified as being either "low" or "high" explosives.

Low Explosives

Low explosives have a relatively slow-burning action or deflagration. They consist of chemical mixtures of sulphur, charcoal, and ammonium or sodium nitrate, and are sold commercially as black powders. These explosives have a wide field of application, where a slow heaving action is required, as, for example, in the production of large rock fragmentation or in log splitting. They are also used in core-type fuse (safety fuse) and igniters.

Black powders are relatively insensitive to shock, but ignite instantly at about 572°F (300°C) from flame, hot wire, friction, or electric spark. Their rate of deflagration varies with the conditions of confinement. When unconfined they burn at a rate measureable in seconds per metre, but when confined black powders will burn at from about 180 to 600 m/s depending upon the constituent grain size, the finer powders burning more rapidly. While they are burning, large volumes of smoke and toxic fumes are given off.

High Explosives

High explosives detonate at a high velocity. A prototype is nitroglycerine, which is itself an explosive liquid, but unstable and dangerous to handle in that form unless it has been desensitized by mixing with some other substance such as acetone.

Dynamites

Dynamites consist of nitroglycerine contained in an absorbent and a proportion of chemical ingredients with the aim of providing stability to facilitate handling, control the strength, and limit the fumes produced on detonation.

Nitroglycerine freezes at 12.3°C, so that in winter conditions explosives containing this substance as a constituent require careful thawing before being charged in the shot-hole. Low-freezing explosives are therefore preferable in cold climates. In these the nitroglycerine is replaced by an equivalent amount of nitrated ethylene glycol.

A wide range of commercial dynamites are available. Straight dynamites are designated in a series of grades, depending upon the proportion of nitroglycerine included in their composition, with "strengths" ranging from 30% to 60% NG. High-strength, straight, dynamites are characterized by a high detonation velocity, giving a rapid shattering action on the confining medium. They are water-resistant, generate a large volume of toxic fumes, are relatively sensitive to shock and friction, highly inflammable, and expensive.

Ammonia Dynamites

By replacing some of the nitroglycerine by ammonium nitrate, a range of explosives is produced with a slightly lower detonation velocity but at considerably less cost. They comprise probably the most widely used type of high explosive for general purposes. The ammonia dynamites are less sensitive to shock and friction than the straight dynamites, but they have a lower water-resistance (which is countered, to some extent, by the incorporation of waterproof cartridge wrappers).

The ammonia dynamites are available in standard small and large diameter sizes, and some grades are also bag-packed for pouring into the shot-hole.

Blasting Gelatines

Gelatine dynamites have, as their main base, a nitrated mixture of cellulose and glycerine, forming a nitrocotton–nitroglycerine gel. They provide a range of dense, plastic, cohesive explosives with excellent water-resistance. Graded in strength from 20% to 100% equivalents to straight dynamites, they have a high detonation velocity while their plasticity facilitates their being loaded to a high charge density, and thus give effective coupling.

Some grades are well suited to submarine work. The high velocity gelatines being relatively costly, for general purposes a range of semi-gelatine dynamites are available. These combine the economy of the ammonia dynamites with the plasticity, cohesiveness, and water-resistance of blasting gelatine.

Blasting Powders

A range of explosive mixtures, with ammonium nitrate as the main base, are termed blasting powders. These include the granular "permissible explosives", which are tested and approved for use in dangerous situations, such as in mines, where the external atmosphere may contain explosive gases and dusts.

The chances of flame from the detonated charge igniting gas and dust in the external atmosphere are reduced, so far as the explosive itself is concerned, by including cooling salts and inert materials in the explosive mixture, to reduce the flame temperature, and, in some cases, also to produce a blanket of CO_2 interposed between the gases from the explosive charge and the external atmosphere.

Also important in explosive mixtures intended for underground blasting is the incorporation of oxidizing agents to provide an oxygen balance, which reduces the toxicity of the fumes produced.

The use of "permissible explosives", either granular or gelatinous in type, is not in itself an adequate safeguard against the possibility of generating an external explosion by blasting in dangerous atmospheres. Concomitant with their use are certain modes of procedure, involving detailed attention to adequate and firm stemming, the prohibition of naked lights and fuses (rendering electrical detonating techniques obligatory), the provision of adequate ventilation and dust-control technology, and the prohibition of blasting altogether if dangerous gases and dusts are known to exist in proximity to the charge-holes.

Blasting Agents

Blasting agents are chemical compounds or mixtures which in themselves are non-initiating but which are capable of exploding when initiated by another high-explosive charge or primer. Several explosive manufacturers supply blasting agents under various trade names, but in general terms they may be classed as being either nitro-carbo-nitrates containing no high-explosive ingredients, or mixtures containing some (non-nitroglycerine) high-explosive agents.

AN–FO Mixes

Representative of the former class are the ammonium nitrate–fuel oil mixtures. The outstanding features of commercial blasting agents are their relatively safe-handling properties and, particularly with the AN–FO mixes, relatively low cost. Since they contain no nitroglycerine, they are excellently suited for application at low temperatures.

The prilled ammonium nitrate compositions can be applied in the "free-running" form, i.e., poured into the blast-hole instead of being loaded in canisters or cylindrical cartridges. This gives effective coupling between the explosive and the charge-hole wall, which is conducive to good blasting performance.

Water Gel or Slurry Explosives

Water gel explosives are derived from blasting agents, but they also contain from 10 % to 30 % water, which, together with other ingredients, combine with the blasting agent to form a gelatinous slurry. Included in the formulation are materials such as TNT, to act as a high-explosive sensitizer, and sometimes also powdered aluminium. The resulting mixture is fluid and water-resistant, so that it can be poured into a waterlogged hole to displace the water and form a dense, effectively coupled charge. The material can be safely bulk-loaded from trucks under severe climatic and environmental conditions, although it tends to lose its fluidity at low temperatures.

The loss of fluidity of early type water-gel explosives was due to lack of control of the chemical constituents in the aqueous slurry medium. Crystallization of the constituents was apparent even at summer temperature, while in winter they became so hard that they could not slump to fill the blast-hole.

They also had a limited water-resistance. Water flowing through a blast-hole could leach out the soluble salts, leaving the residual solids segregated in the toe of the hole. Improvements in technology have now resulted in gels that will withstand storage temperatures up to 45° C and also remain plastic in temperatures as low as −20° C. The strength of the gel is increased by adding TNT and aluminium, the optimum concentration being around 13–16 % of pigment grade aluminium flake.

Igniting and Detonating the Explosive Charge

Charges of low explosive, such as black powder, are ignited by match, squib or time fuse, or by safety fuse. (The fuse is itself a core of black powder.) High explosives are detonated by a small charge of more sensitive explosive, usually lead azide detonated in a small blasting cap or by detonating cord. (This is a core of PETN in a textile-covered asphalt tube.)

Safety fuse burns at the rate of from 90 to 150 s/m depending upon the ambient atmospheric and climatic conditions. Detonating cord transmits a detonating shock wave at the rate of about 6000 m/s.

Blasting caps may be non-electrical, in which case they are initiated by an igniter, fuse, or detonating cord, or they may be electrical, initiated by a hot bridge-wire. They may be instantaneous in operation or they may incorporate time delays, in intervals usually ranging from 5 to 35 ms.

Blasting agents such as the AN–FO mixtures and the water-gel explosives are not self-initiating, so that the charge needs to be initiated by a primer, which is detonated by conventional means. The primer may be a small charge of commercial dynamite but it may also be one of a number of proprietary non-nitroglycerine compounds which include sensitizing agents.

Military Explosives

Military explosive composition C_3 is a plastic demolition explosive containing 88.3 % RDX and 11.7 % of a non-explosive plasticizer. The composition is plastic at temperatures ranging from 0° C to 90° C. It tends to become brittle and loses some sensitivity below 0° C, becomes hard at −29° C, and exudes its oil at temperatures above 40° C.

For low-temperature work, composition C_4 may be preferable. Its composition is:

RDX	91 %
Polyisobutylene	2.10 %
Motor oil	1.60 %
Di(2-ethylhexyl) sebacate	5.30 %

This composition is still plastic at −57° C and does not exude oil when stored at 77° C.

The compositions C_3 and C_4 have about the same sensitivity to impact as TNT with a comparable degree of safety in transportation and handling. Their rate of detonation (7625 m/s for C_3) is 111 % that of cast TNT, with a comparable strength 115 % and 126 % that of TNT.

Liquid Explosives

Liquid nitroglycerine was occasionally used as a high explosive in the mid-nineteenth century, but its extreme sensitivity and instability rendered it too dangerous for widespread use. Within recent years desensitized nitroglycerine has been used for special purposes, notably to break rock with the intention of increasing its permeability, to support *in situ* leaching of oil-shale [4].

The procedure adopted was to introduce and displace the liquid into a permeable zone of rock, through a borehole, and then to detonate this so as to propagate an *in situ* explosion of the permeated strata.

The results of such a process, called the "Stratablast" process, have been reported [5]. The process uses a multiple component system of hypergolic fluids that explode when combined in the rock formation. Two alternatives are available. One uses rocket-type fuels, altered to behave as liquid explosives, and the other uses a heavy slurry of metallized ammonium nitrate.

In the oil-shale experiments a charge of 216 l of NGI was poured through a hose into the injection shot-hole. The NGI was displaced into the rock by a hydrostatic head of water, the NGI–water interface being monitored by hydrometer. Offset holes were sampled during the injection process to observe the NGI migration pattern.

The samples indicated that the liquid explosive migrated a distance of 7 m during the injection period. Air at a gauge pressure of 103 kPa (15 psi) was applied to displace the last 23 l of NGI.

A detonating device, in the form of a shell 0.6 m long by 50 mm diameter, containing seven 0.15 kg primers connected to an electric blasting cap, was constructed. The hole was tamped by crushed rock, pea gravel, and sand, and the charge was detonated electrically from a surface service truck [6].

LOX

Explosive compositions utilizing liquid oxygen have sometimes been employed. Strictly speaking, however, these are not liquid explosives, since the fuel, which is comprised of carbon in the form of lampblack, granular carbon, or charcoal, is applied in cartridge form.

The fuel cartridges are dipped into the liquid oxygen immediately before being lowered into the charge-hole, and the cartridges are fired by ordinary detonator. A minimum of delay after loading is imperative, since LOX rapidly loses its power through evaporation of oxygen. Confined detonation velocities of 3700–5800 m/s are attained.

Explosive Substitutes

Blasting in coal-mines presents the hazard of the ignition of atmospheric gas/air and dust/air mixtures by the flame from the blast charge. Permissible explosives are designed to minimize the risk but, in themselves, cannot eliminate it altogether. An alternative is to use explosive substitutes, which emit no flame. Such methods of breaking rock have some applications in other fields also, for example, when breaking relatively weak ground in urban areas, where blast noise and ground vibrations, and the hazard presented by flying fragments, may be objectionable.

A number of substitutes for explosives have been devised through the years. Of these, only the Cardox and the compressed-air systems are of significant importance today.

The Cardox System

The Cardox device consists essentially of a steel tube, closed at one end by a firing head and at the other by a steel shear disc. The tube is filled with liquid CO_2 at a pressure around 14 MPa (2000 psi). Immersed in the CO_2 is a heater unit which, when ignited by an electric fuse-head, raises the internal pressure in the shell, partly by hydrostatic compression, of the liquid CO_2, and partly by heat exchange to the gasified CO_2.

When the internal pressure reaches the rupture pressure of the shear disc, the disc bursts, allowing CO_2 to discharge through the ports. In practice the shell is inserted into a borehole drilled in the material to be blasted, with the discharge ports at the back of the borehole. The sudden discharge of CO_2 ruptures the borehole and the expansion of the gases provides the blasting action.

Air-blasting System

The air-blasting system consists essentially of a steel tube, similar to that used in the Cardox system, closed at one end by a steel disc or discharge valve and at the other end connected by means of a hose and pipeline to a high-pressure air compressor.

A firing valve in the pressure line controls the flow of air into the shell. When the valve is in the open position, air flows into the shell. When the pressure reaches a critical value the shear disc ruptures, or the discharge valve opens, allowing the compressed air to discharge through the portholes. In practice the shell is inserted into a borehole, with the discharge ports at the back of the hole. The jet of compressed air released from the discharge ports drives into the rock around the head and expands to produce the blast.

The Blasting Action of Cardox and Compressed-air Discharges

The blasting action of the explosive substitutes is substantially one of gas expansion. Any body stress waves are likely to be of little significance, although the surface Rayleigh waves can be of an order comparable with those produced by some chemical explosives [8].

The initial jet of gas or liquid impacts upon the sides of the borehole and penetrates into the confining medium to a depth determined by the initial velocity and the mass flow. Expansion of the gases forces open any interstices that may exist within the fabric of the confining medium to generate fracture surfaces that may yield to the gas pressure. The breaking action is much slower than that produced by chemical explosives.

Experiments to determine the pressure–time characteristics of blasting by Cardox and compressed air shells were conducted by Davies and Hawkes [8].

The recorded pressures in the vicinity of the shell discharge ports were defined by the relationship

$$P = KD^{1.3}$$

where P is the gas pressure and D is the distance from the discharge ports.

The value of the constant K was approximately 3500 when P was measured in psi and D in inches.

For any particular type of Cardox shell the wave velocity remained substantially constant at a value determined by the nature of the confining medium and the gas pressure.

For example, the C74 Cardox shell gave a blast-wave velocity of 67 m/s in shale, while the weaker shell discharging at 33 MPa gave 46 m/s in the same material. In coal, the blast wave from a 63.5-mm air-blast shell was measured at 23 m/s.

The outstanding feature of this work by Davies and Hawkes is that it demonstrates how a comparatively low gas pressure can effectively break up a confining medium, if that medium contains discontinuities or incipient fractures upon which the expanding gas can exert its force. During an air-blast the speed of the gas discharge jet was measured at 490 m/s to 67 m/s, falling very rapidly from 67 to 23 m/s as the expanding gases moved into the opening fractures. The jet pressure from a Cardox or air shell falls to values of from 1.4 to 0.7 MPa within a short distance from the discharge ports.

Mellor and Sellman describe tests of compressed-air blasting in frozen silt and gravel [10]. The system was that described by McAnerney *et al.* [11]. Air-blast shells were fired in vertical boreholes 76 mm in diameter and from 1 to 1.5 m deep, in frozen gravel. Firing pressures of 42–62 MPa were applied.

Comparative tests with chemical explosives in frozen gravel showed that an air-blast shell discharge of 5000 ml from an initial pressure of 62 MPa produced a result equivalent to that obtained with 2.3 kg of 60% dynamite. The conditions were judged to be favourable to air blasting in that below a depth of 1 m the gravel was unsaturated and permeable to air. This facilitated injection of the compressed air and improved the heaving action on the upper layers of gravel.

References and Bibliography: Chapter 2

1. ANON., *Blaster's Handbook*, E. I. Du Pont de Nemours & Co. Inc., Wilmington, Del., 1969.
2. LANGEFORS, U., and KIHLSTROM, B., *Rock Blasting*, Wiley, New York, 1963.
3. GREGORY, C. E., *Explosives for Engineers*, Trans. Tech. Pub. Clausthal, 1973.
4. MILLER, J. S., and HOWELL, W. D., Explosive fracturing tested in oil shale, *Colo. Sch. Mines Q.*, **62** (3) 63–73 (1967).
5. BRANDON, C. W., Method of explosively fracturing a productive oil and gas formation, US Patent 3,066,733 4 Dec. 1962; and Method of fracturing a productive oil and gas formation, Am. Petroleum Inst., Apr. 1965.
6. ANON., Space age explosives may revive well-shooting, *Oil Gas J.* **64** (38) 6 (19 Sept. 1966).
7. DICK, R. A., *Factors in Selecting and Applying Commercial Explosives and Blasting Agents*, US Bur. Min., Inf. Circ. No. 8405, 1968.
8. DAVIES, B., and HAWKES, I., The mechanics of blasting strata using the Cardox and airblasting systems, *Colliery Engng. London*, Nov. 1964, pp. 461–7.
9. DAVIES, B., FARMER, I. W., and ATTEWELL, P. B., Ground vibrations from shallow, sub-surface blasts, *The Engineer London*, Mar. 1964, pp. 553–9.
10. MELLOR, M., and SELLMAN, P. V., *Experimental Blasting in Frozen Ground*, US Army CRREL Rep. SR 153, Nov. 1970.
11. MCANERNEY, J., HAWKES, I., and QUINN, W., *Blasting Frozen Ground with Compressed Air, Proc. 3rd Canadian Conf. on Permafrost, 1969*, Tech. Memo. 96, Assoc. Committee on Geotechnical Research, NRC, Canada.
12. ROBINSON, R. V., Water gel explosives—three generations, *Proc. CIM 71st AGM, Montreal, April, 1969*.

CHAPTER 3

Rock Excavation by Blasting—III

Outline of Blasting Technology

The Evolution of Blasting Technology

Although the explosive mixture, black powder, had been known since the thirteenth century in Europe, and possibly longer in Asia, its effective application to rock excavation was not feasible until William Bickford invented the method of ignition by safety fuse in 1831. The subsequent history of explosives technology is dominated by the contributions of Alfred Nobel and included the introduction of the fulminate of mercury detonator, which did for guncotton and nitroglycerine what Bickford's safety fuse had done previously for black powder, turned the explosive into an engineering tool with wide applications and great potential. Nobel's inventions of dynamite, which made nitroglycerine safe to use, followed by blasting gelatine and a whole range of low explosives, extended the frontiers of blasting technology throughout the late years of the nineteenth century and on into the mid-1930s.

This was about the time that blasting agents such as the AN–FO mixes were introduced, followed in turn by the metallized slurries. The contribution made by explosives to rock excavation in mining and civil engineering is very great indeed. Explosives did much to ease the arduous tasks involved, and they made an effective contribution to the betterment of working efficiency and the improvement of conditions of labour, particularly in tunnelling. It is only in recent years that the applicability of explosives in such work has been challenged by the introduction of the continuous techniques that new and improved mechanical excavators have made possible.

At the same time, the advent of nuclear explosives has presented us with a new blasting tool of immense power and significance. While international agreements limit the surface testing and use of such explosives for civil applications, enough work was done in the early exploration of nuclear blasting to demonstrate its potential ability for bulk rock excavation and earth movement. The advent and subsequent prohibition of nuclear explosives for civil uses provided engineers with the incentive to apply conventional explosives, using large-scale charges free from the dangers of radioactive fall-out. As early as 1885, underwater obstacles to shipping in the form of rocks and reefs in the East River channel off Long Island Sound, New York, had been attacked by blasting, including a charge of 131 061 kg of dynamite, to remove Flood Rock.

Among the largest recorded non-nuclear explosive blasts was that used in 1958 to demolish the Ripple Rock, an underwater mountain the peaks of which lay only some 3 m below low-water line in Seymour Narrows, British Columbia. This rock was approached by a shaft sunk on the shore from which a tunnel was extended to reach a base in the rocks under the seabed. From this base level the rock was honeycombed by a series of vertical raises intersected by horizontal tunnels or drifts into which no less than 1 250 000 kg of a high-strength blasting agent were packed and fired by detonating cord.

In Russia, too, large explosive charges have been used to excavate canals and to construct rock-filled embankments. At the Vakhsh River a dam 60 m high was constructed by firing a charge weighing 1800 tonnes, which smashed and displaced 1.55 million m^3 of limestone rock, followed by 30 000 m^3 of clay ejected from the opposite bank.

The major explosives manufacturers all maintain comprehensive sales and service organizations to assist prospective users in the selection of suitable explosives and the adoption of appropriate techniques of applying them in priming and main charges, together with exploders, detonators, and firing circuits, for all possible forms of application. The technology of rock blasting may conveniently be considered in relation to (a) tunnel blasting, (b) quarry bench blasting, (c) surface cratering and trenching, and (d) pre-splitting and smoothwall blasting.

Tunnel Blasting

By tunnel blasting is meant the construction of an excavation the cross-section of which advances as a complete face, over the final perimeter, in the direction of its longitudinal section. Vertical and inclined shafts are thus included in this class of operation. In practice, a tunnel of large cross-section may be advanced in two or more operations, which comprise the creation of a tunnel heading (either blasted or bored), followed by enlargement of the heading to the final finished perimeter by a process of parallel cylinder blasting, or by bench blasting.

The operation of tunnel blasting includes two phases: first, the creation of a free face or faces, in effect an advance excavation, into which the rock broken by the main blast charges can yield, and, secondly, the placement and firing of the main charges. The first phase is termed the "cut" and the second phase is the "blast". In hard-rock mining both cut and blast phases are usually conducted by drilling and blasting, in the process of which explosive charges are placed and fired in holes drilled into the rock face. When mining weaker rocks, however, the cut may be excavated by rock-cutting machines.

There are various types of cut, which may be classified as follows.

Angled Cuts

In which the cut holes are inclined at angles to form, respectively, a fan cut (Fig. 3.1), a wedge, or vee cut (Fig. 3.2), or a pyramid cut (Fig. 3.3).

Parallel-hole Cuts

Limitations on the depth of advance of angled hole cuts, such as fans, vees, or pyramid cuts, as a percentage of the length of drill-hole, are imposed by the geometry of the face in relation to the rock burden and hole spacing. These must be so related as to produce suitable break angles to form the cut. An increased ratio of depth of advance to depth of drill-hole can be obtained by adopting various forms of parallel-hole cuts in which the cut holes are drilled perpendicularly into the face.

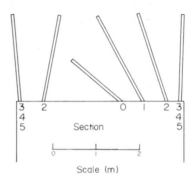

End view

Section

Scale (m)

FIG. 3.1. Fan cut.

Burn Cut

This is the oldest type of parallel cut. One or more holes, of the same diameter as the drill-holes, are left uncharged. Blasting then takes place from the other holes towards these central uncharged holes (Fig. 3.4). If the depth of cut is increased too far, in an attempt to increase the rate of advance, the charge concentration produced by an ordinary burn cut may be so high as to produce plastic deformation at the back of the cut. The rock then sticks or "freezes" instead of being fragmented.

Various adaptations of the burn-cut principle are used, sometimes employing larger diameter cut holes than the ordinary blast-holes, and loosely charging some of the holes (Fig. 3.5).

Cylinder Cuts

In cylinder cuts, holes are drilled perpendicularly into the face at suitably spaced intervals around a cylindrical locus. These holes are fired in rapid sequence towards an empty hole, so that the cut is enlarged cylindrically along its whole length. The charge concentration being uniform along the whole length of the hole, the possibility of rock "freezing" is removed, and the length of advance per blast is limited only by the degree of deviation in drilling the hole.

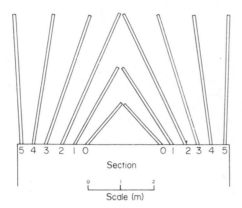

End view

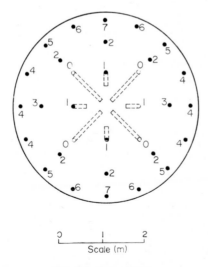

Section

Scale (m)

FIG. 3.2. Wedge or vee cut.

Scale (m)

FIG. 3.3. Pyramid cut.

Fragmentation and Throw-in Parallel Cut Blasting

Since the direction of break is towards the central cylindrical axis, along the full depth of cut, the broken rock is thrown with great force against the opposite wall of the cut, as

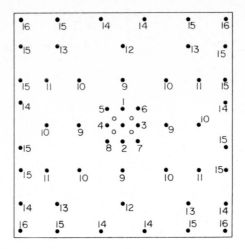

• Charged holes
○ Uncharged holes

FIG. 3.4. Tunnel blasting with burn cut.

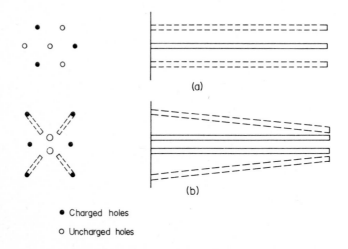

• Charged holes

○ Uncharged holes

FIG. 3.5. (a) Parallel burn cut. (b) Large hole burn cut with pyramid rakers.

each sequential hole is fired. This impact contributes to improved fragmentation. For maximum effectiveness the resulting debris should be cleaned from the cut by the expanding blast gases so as to leave clear space for the breakage from each hole in turn. With increased depth of cut the required distance of throw to effect complete clearance increases. By using short delay detonators each successive charge-blast may be made to follow its predecessor so closely that momentum of the broken mass can be maintained, almost in the nature of a fluid flow of rock fragments, during the overall period of the total blast.

Mechanical Cutting

The use of mechanical cutting, boring, and ripping machines is feasible in weak rock and also in ice and frozen ground. The machines may be used to cut slots horizontally at floor level (undercut), at roof level (overcut), at intermediate heights (middle cut), or vertically into the face (shear cuts), to depths averaging about 3 m.

The rock so exposed for that depth of advance is blasted by comparatively light charges of explosive, breaking into the free face or faces that are created by the slots. In contemporary tunnelling technology the use of explosives is largely being reduced by the use of mechanical heading machines which excavate the whole of the tunnel cross-section.

Rotary Raise, Shaft, and Tunnel Borers

Present tendencies are to try to extend the use of rotary-boring techniques, even to the hardest rocks, by increasing the size and power of machines, by improvements in technology, and by better understanding of the rock mechanics problems that are involved. Boring of shafts 2–4 m in diameter is now commonplace in ore mining, and larger machines are constantly being tested. Vertical and inclined shafts of greater diameter can be constructed by boring a small central shaft and then blasting into this on the cylinder-cut principle.

Quarry Bench Blasting

In bench blasting the charge-holes are drilled vertically downwards, or else inclined, in rows parallel to the free face (Fig. 3.6). Each row when fired together then forms a new free face for the succeeding row of blast-holes. In each row the charge may be concentrated at

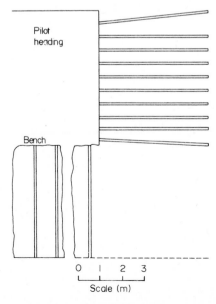

Fig. 3.6. Heading and bench blasting in a large tunnel.

the base of the hole, termed bottom charge, or distributed along the length of the hole, termed column charge, or combine both bottom and column charges. Each row of holes may be charged and fired separately, or else in multiple rows using delay detonators. Short delay detonators usually give better fragmentation.

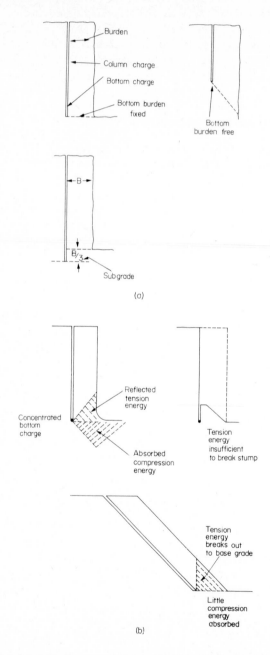

FIG. 3.7. (a) Quarry bench blasting—nomenclature. (b) Use of inclined benches to prevent toe "stumps".

Inclined holes give a reduced "back break" and also reduce the tendency of unblasted base "stumps" being left after the blast. The latter condition is liable to occur in hard rock if the bottom charge produces insufficient tension energy at the toe of the hole. With vertical holes drilled only to bench depth a large proportion of the charge energy is dissipated as compression pulses that are absorbed into the bedrock, and the remaining tension energy component may not be large enough to break out to a level base. Thus, unblasted stumps are left at the toe. This condition may be corrected by taking the blast holes to a depth below bench base level for a proportion of the burden distance termed "subgrade". Alternatively, by drilling the bench-holes at an angle of 45 degrees and taking them down to base level, most of the charge energy can be utilized in the form of tension pulses reflected from the bench rock, to break out to a level base (Fig. 3.7).

Crater Blasting and Trenching

Crater-cut blasting makes use of the reflected tension wave to fracture the rock by surface "slabbing". The shock wave and the reflected elastic-wave principles in blasting have become more relevant with the advent of nuclear explosives and the use of chemical explosives on an ultra-large scale, while smaller charges of chemical explosives are sometimes used to produce an initial crater cut for open-pit excavations on virgin ground.

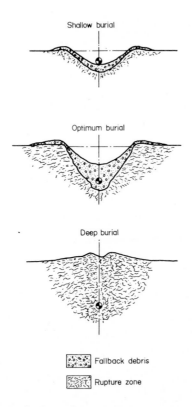

Shallow burial

Optimum burial

Deep burial

Fallback debris

Rupture zone

FIG. 3.8. Change in shape of crater with increasing depth of charge.

In craters produced by explosives at shallow depth the true crater boundary is characterized by an easily discernible change from the loose, disassociated materials in the bottom portion of the fall-back material to the fractured and slightly displaced rupture zone material. As the depth of burst is increased below an optimum value the true crater boundary becomes increasingly diffuse (Fig. 3.8).

Linear cuts or trenches may be constructed by the simultaneous detonation of charges buried in a row, so as to produce a linear crater. Safety considerations may require that the cut be excavated by a series of incremental linear craters which are subsequently interconnected, rather than by a single trench excavated at one blast. Trenching on a large scale, such as in canal construction, by means of conventional charges of chemical explosives, is a much slower process. Wide trenches must be opened up either by an angled cut or else by ditch blasting. Narrow trenches for pipelines, utility services, and drains, etc., may be excavated in rock conveniently and economically by blasting. By using short-delay blasting the preliminary drilling may be completed over a wide area in advance of the excavation (Fig. 3.9).

Pre-splitting and Smoothwall Blasting

Conventional tunnel blasting has certain disadvantages. These include the occurrence of "overbreak", i.e., the excavation of an area of cross-section larger than that which is desired. This is due to the inability of explosives, when used as described, to give precise control over the finished dimensions of the blast.

The structural design of the finished tunnel sets the maximum limits on dimensions, but to ensure that these are achieved must inevitably involve some damage to the rock walls beyond those limits. A rough, uneven, contour is the result and this must be made good in the subsequent construction of supports and linings. The cost of the operation is thus increased, both by causing excessive rock fracture and subsequently filling this in to the final perimeter. Also, the strength of the rock walls is weakened to a variable and unknown extent beyond the break limits.

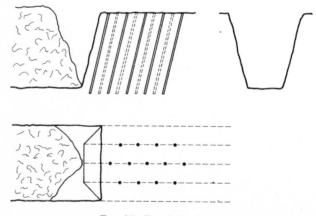

FIG. 3.9. Trench blasting.

Smoothwall Blasting

By drilling a series of holes at closely spaced intervals around the perimeter of the excavation, the blast may be trimmed to the finished dimensions with a minimum of overbreak. Under the ideal conditions for smoothwall blasting a pilot heading is carried in advance of the trim operations so that the rock may be broken to a free face unconstrained by broken debris. The main blast round should be so arranged that the rock is broken towards free surfaces giving throw towards the centre of the cleared cross-section. The final sequence of trim holes should be fired simultaneously (or as near simultaneously as possible), with a low explosive density in the holes. As a general rule this should be about one-tenth of the full-load concentration in holes spaced at from 10 to 20 equivalent diameters.

Cushion blasting of the trim-holes, or the use of "decked" charges, are among the methods that are sometimes used to give smoothwall control. The perimeter holes are commonly drilled on a burden-spacing ratio of about 1.5 to 1. In cushion blasting the trim-holes, fired after the main blast has been cleared, are loaded with light charges distributed along the length of the hole at intervals, the hole being completely stemmed and fired. The stemming cushions the explosive shock wave from the finished rock wall. A "decked" charge is an incremental charge, spaced at intervals, sometimes separated by spacers. The air space gives effective decoupling and damps the shock wave transmitted to the rock.

Pre-splitting

Paine *et al.* [9] further developed the smoothwall blasting technique by instantaneously firing a closely spaced series of holes around the desired perimeter before any of the main round was blasted. The effect of this procedure is to connect the holes by a tension crack. This crack decouples the main blast from the perimeter rock walls, which are thereby left intact and unbroken. There is a complete absence of overbreak, the object being to obtain a smooth perimeter to a controlled profile.

In an experimental and theoretical examination of the pre-splitting technique. Ato [10] shows that the cracking phenomenon in pre-splitting is mainly dependent on stress waves propagated from the pre-split blast, and not upon wedge action by the gases produced. His experiments show, too, that the velocity of outward crack propagation from the holes is less than the stress-wave velocity, although incipient cracks, which later expand and coalesce to form the complete break, are generated by the passage of the stress wave.

The pre-split hole diameter should be smaller than the main blast-holes so as to avoid the extra radial cracking of the walls that would be produced by large charges. Continuously loaded, moderately coupled charges are recommended. These may take the form of continuous columns 20–25 mm in diameter, packed into cardboard tubes and loaded into blastholes 30–40 mm in diameter, in underground work. In surface bench blasting, the boundary limit holes usually range from 50 to 125 mm in diameter, spaced 0.5–1 m apart. As with smoothwall blasting, air decking is sometimes applied, but this is time consuming and troublesome, and is not recommended when AN–FO mixtures are used. In weak and relatively soft rock, alternate advance pre-split holes may be left uncharged. The uncharged holes then form relief holes which guide the shear fracture along the desired separation plane.

Selection of Explosives

The choice of explosive that is best suited for a particular blasting operation is usually a matter of economics in the final assessment. However, there are many factors involved and it is often found that certain factors have a predominating influence over others and so reduce the range of selection.

The relevant properties of the material to be blasted include its density, porosity, and strength. The latter term requires qualification. If the medium to be blasted is hard, strong, and brittle, the blasting operation will require the use of a "strong" explosive, i.e. a high explosive, and the tensile strength of the confining medium will be the most important strength parameter. But if the confining medium is weak and yields at comparatively low loads, so that blasting by a low explosive becomes feasible, then the tensile strength of the medium will probably be less important than its shear strength.

As yet there is no universally acceptable scale of strength that can be applied without exception to materials that may have to be blasted. For convenience in testing and classification the uniaxial compressive strength and the effective moduli of elasticity are often applied as a combined strength parameter. (The dynamic moduli are relevant in this context.) In heterogeneous rock strata the geological characteristics, such as structure, joints, and bedding, will all be important as to their effects on the physical processes of fracture. The utilization of explosive energy in relation to the fracture process will be very dependent upon the characteristics of energy absorption and the attenuation characteristics of the medium concerned.

Strength of Explosives

The term "strength" as applied to an explosive has, by convention, been applied by explosives manufacturers to describe various grades of explosives. It is not always indicative of the energy that is available for blasting. Two empirical tests, used to classify explosives on a comparative basis, are the ballistic mortar test and the Trauzl lead-block test.

Ballistic Mortar Test

This measures the ability of a measured charge, when detonated, to deflect a heavy steel mortar. The "strength" of the explosive is then expressed as a percentage of the strength of TNT, as demonstrated by the same test.

Trauzl Lead-block Test

This test measures the distension produced by detonating 20 g of the explosive, using standard procedure, in a lead cylinder.

Submarine Testing of Explosives

Methods of underwater evaluation of explosives performance are described by Sadwin *et al.* [11]. In such tests the underwater shock energy is presumed to be related to that which causes shattering and spalling of a solid confining medium, while the bubble energy is considered to be analogous to that which produces the heaving or expansive action of a confined explosive charge. Atchison *et al.* [12] of the US Bureau of Mines compared the effects of explosives detonated in a solid-rock medium with the same explosives detonated under water. The two methods of testing show general agreement, but there are some significant differences in performance due to differences in explosive/rock and explosive/water impedance ratios.

Industrial Strength Ratings

The term "strength" was originally applied to dynamites in terms of the percentage of nitroglycerine in their composition, the remainder being inert filler. Sixty per cent dynamite contained 60% nitroglycerine by weight, and was judged to be three times stronger than 20% dynamite. The subsequent development of various high explosive compositions complicated the issue, however, because of the addition of the other active compounds. Consequently, today a 60% straight dynamite which contains 60% nitroglycerine is only about one and a half times stronger than a 20% straight dynamite, because of the energy supplied by sodium nitrate and carbon in the 20% grade explosive.

Also, different grades of explosive, all rated with the same percentage nitroglycerine figure, will produce very different results when blasting in the same medium if they have different detonation velocities. Ratings of strength that are commonly used in industrial blasting practice today are "weight strength" and "cartridge strength", which compare explosives on a weight basis and on a basis of volume, respectively. "Bulk strength" is an alternative term meaning the same as cartridge strength.

Strength is commonly expressed as a percentage, with straight nitroglycerine dynamite taken as the standard, for both the weight and the cartridge strength ratings. By definition, the weight strength and the cartridge strength of an explosive are equal when the specific gravity of the explosive is 1.4—the density of straight dynamite. The USBM nomograph by Dick [13] can be used to correlate the two ratings, some industrial explosives being identified by weight strength and some by cartridge strength (Fig. 3.10).

Criteria for Selecting an Explosive

Various criteria may be applied when selecting an explosive which is to be used for a rock-blasting operation.

Detonation Velocity

Detonation velocity may be expressed either as a confined or as an unconfined value. The confined value is the more significant one for use in the present context. However, most manufacturers measure the detonation velocity in an unconfined column of explosive 30-mm diameter, although some measurements are made within the confinement of a metal pipe or at a different diameter.

The detonation velocity of an explosive is dependent upon the density of the explosive, the ingredients of the explosive, the particle size of the ingredients, the charge diameter,

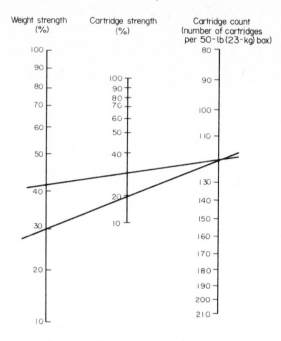

FIG. 3.10. Nomograph for comparing weight strength and cartridge strength (Dick [13]).

and the degree of confinement. Since decreased particle size, increased charge diameter, and increased confinement all tend to increase the detonation velocity, it is important to know the conditions under which the manufacturer has made his velocity determination. In choosing an explosive on the basis of detonation velocity it is particularly important to know whether the confined, or the unconfined, velocity is the one specified. Unconfined velocities are generally 70–80 % of the confined values. The confined detonation velocity of commercial explosives ranges from about 1500 to 7500 m/s.

Density

The density of an explosive can be expressed in terms of specific gravity or as cartridge count (representing the number of 30-mm by 200-mm cartridges in a 23-kg box).

The specific gravity of commercial explosives ranges from 0.6 to 1.7, with corresponding cartridge counts of 232 to 83. For free-running explosives the density is often specified as the weight of explosive per unit length of charge in a given size of drill-hole. With few exceptions the denser explosives give the higher detonation velocities and pressures.

For difficult blasting conditions, or where fine fragmentation is required, a dense explosive is desirable. In easily fragmented rock, or where coarse fragmentation is required, a low-density explosive will suffice. The density of an explosive is important when working under wet conditions. An explosive with a specific gravity less than 1.0, or a cartridge count greater than 140, will not sink in water.

Detonation Pressure

Detonation pressure is not usually given in the technical data sheets or in the blasters' handbooks published by the explosives manufacturers. Nevertheless, it is an important parameter in blasting technology. The relationship between density, detonation velocity, and detonation pressure is complex. The following approximation may be used:

$$P = 4.18 \times 10^{-7} DC^2/(1 + 80D),$$

where P is the detonation pressure (kbar), D is the specific gravity, and C is the detonation velocity (ft/s).

A high detonation pressure is preferable when blasting hard, dense rock, whereas in softer rock a lower pressure may be better. The detonation pressures of commercial explosives range from about 10 to over 240 kbar (Table 3.1).

TABLE 3.1. *Detonation pressure of typical explosives*

	(kbars)
Composition B	240
Pentolite	240
Power primer	125
Slurry gel	80
TNT-slurry	70
60% dynamite	50
40% dynamite	40
Metallized slurry	33
AN–FO	15
Thermal primer	10

Water Resistance

In dry work, water resistance is of no consequence. If water is standing in the drill-hole and the time occupied by loading and firing is short, an explosive with a water resistance rating of "good" will be adequate. If the exposure to water is prolonged, or if water is percolating into the borehole, "very good" to "excellent" water resistance is required. In general, blasting gelatines offer the best water resistance, and the higher density dynamites have fair to good water resistance, while low-density dynamites have little or none. The emission of brown nitrogen-oxide fumes from a blast often means that the explosive has deteriorated due to contact with water, and indicates that a change should be made in the choice of explosive.

Fume Class

In underground work, or in confined spaces where ventilation may not be good, the fume rating of an explosive is an important factor. The rating that is quoted by a manufacturer for a cartridged explosive assumes that the explosive will be detonated in its cartridge. Removing the explosive from its wrapper will upset the oxygen balance.

Choice of Explosive

The optimum choice of explosive for a particular operation depends upon relating the properties of the explosive to those of the material to be blasted so as to achieve the desired result in terms of fragmentation and displacement of the medium concerned. Within this context there may be overriding environmental conditions, such as the presence of water and the conditions of ventilation, or the need to minimize vibration and noise levels, which will affect the choice of explosive.

In addition to safety considerations, other factors that must be taken into account are the effect of the choice of explosive upon the operations as a whole, including drilling, firing, loading, and handling the debris. It might be, for example, that when blasting in hard rock the use of a more expensive, high velocity, high density, explosive will increase blasting costs, but at the same time might result in lower overall operating costs. Such a situation could very well exist on an underground mining or tunnelling project.

On a surface blasting or on an open-pit project, involving large charges of explosive, the cheaper AN–FO mixes associated with bulk-loading techniques might be preferable. The advent of metallized slurries has made such an approach technologically feasible, even in hard ground. Non-metallized grades of AN–FO are now used mainly as top loads, or in easier breaking formations, where fragmentation is assisted by well-developed joint planes or by sedimentary laminations. Where protective packing is needed to give AN–FO mixes added water resistance, the use of metallized powder additives gives added density as well as energy, so that the packaged product can break rock effectively despite the decoupling effect of the package material.

Determination of Charge Weight Required in Rock Blasting

Successful blasting depends upon the optimum use of the available explosive energy to perform the work involved in fragmenting the rock and moving the rock fragments. The operation consists of using properties that are inherent in the explosive to control properties that are inherent in the rock mass, and the result greatly depends upon the spacial distribution of the explosive charge within the rock mass in relation to the physical properties (strength, anisotropy, heterogeneity) of the rock mass and the orientation of the excavation concerned.

Explosive properties of major importance include the total available energy, bulk density, the rate of energy release, and the pressure–time characteristics of the gases produced on blasting. The important rock properties include density, compressibility, porosity, energy-absorption characteristics, and mechanical strength. The strength of a rock mass is a parameter that, as yet, cannot be defined in precise terms. Much research still needs to be done to determine the relevance of mechanical tests which measure the effects of material cohesion, internal friction, effective moduli of elasticity, the characteristics of fracture and flow, and the overall response of the rock material to dynamic loads. It is not surprising, therefore, that, in practice, rock-blasting problems are generally approached by empirical methods.

Single Charges

One of the earliest approaches was that undertaken by A. W. and Z. W. Daw, who more than 50 years ago proposed the relationship for single charges,

$$W = CV^3,$$

where W is the weight of explosive charge required, V is the weight of rock broken, and C is an empirical blasting coefficient.

This may be considered as the basic law in the theory of blast design. It is usually more convenient when expressed as the cube-root law

$$B = C\sqrt[3]{W},$$

where B is the burden to be fractured.

The blasting coefficient C is determined by the rock properties, the blast design geometry, the explosive characteristics, and the charge geometry. It is therefore determined by trial blasts.

Crater Methods of Blast Design

Before an extensive blasting programme is instituted, a number of experiments may be carried out to observe the effects of particular types of explosive on the ground concerned.

Livingston Formulae

One such procedure uses the Livingston crater formulae, based on the concepts of "energy factor" and "critical depth". To determine these parameters, charges of the same weight are fired at various depths in the material concerned and the volume broken by each blast is measured. The critical depth N is the depth of charge when surface rock failure is first apparent.

The energy factor E is calculated from

$$N = E\sqrt[3]{W}$$

and, if $\Delta =$ ratio B/N,

$$B = \Delta E\sqrt[3]{W}.$$

Therefore for any given burden B the required charge weight W may be determined.

Values for E range from 1.8 to 4.6, the larger magnitudes appertaining to the harder brittle rocks. The depth ratio coefficient ranges between 0.45 and 1.0. Livingston suggested that a transition in the mode of failure, from shear in the softer rocks to tensile in the harder rocks, takes place when E is around 3.3 [15].

Design of Crater Cuts—Hino's Method

The results of experimental crater tests may be plotted as shown in Fig. 3.11. This shows the relation between the depth of burden (i.e. the distance d_f between the free face and the centre of the charge or group of charges) and the weight W of charge required to produce a "full" crater (i.e. a crater whose apex reaches the centre of gravity of the charge).

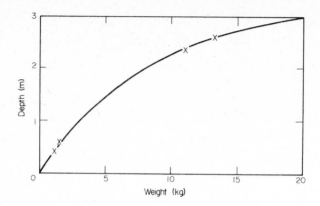

FIG. 3.11. Experimental crater curve (Hino).

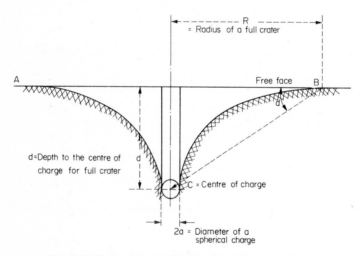

FIG. 3.12. Reduced depth·of burden for a full crater (Hino).

Hino maintains that a practical working assumption is

$$R_f/d_f = 1.3 \quad \text{and} \quad W = 4/3\pi a^3 \delta,$$

where R_f is the radius of a full crater, d_f is the distance (depth of burden from free face to centre of charge), a is the radius of a spherical charge, and δ is the loading density.

If d is the depth of charge, the reduced depth of burden $D_f = d/a$ for a full crater (Fig. 3.12) [16].

Blasting to a Free Face. Determination of Charge Weight

Langefors and Kihlstrom [4] quote a basic equation for determining the quantity of explosive required in a single blast-hole, with free burden, as

$$W = k_2 B^2 + k_3 B^3 + k_4 B^4,$$

where W is the weight of bottom charge required, B is the burden, k_2 and k_3 are coefficients depending on the elastoplastic properties of the rock, and k_4 is a coefficient depending upon the weight of rock to be excavated.

The first term in the equation corresponds to the energy involved in processes of plastic flow, the second term to the energy involved in purely elastic processes, and the third term to the energy involved in the remainder of the fracture process in blasting.

For general purposes, when blasting in Swedish bedrock the equation resolves to

$$W = 0.10B^2 + 0.40B^3 + 0.004B^4.$$

Darrell W. Hansen [17], extending Langefors and Kihlstrom's methods to problems of blasting at the Morrow Point Dam and Power Plant, quotes

$$W = 0.052(k/B + 1.5)B^2 + 0.4C(k/B + 1.5)B^3,$$

where W is the total charge weight (lb) in a single hole with free burden, k is the height of free face (yd), B is the burden (yd), and C is a rock constant determined by trial blasts.

Values of the rock constant ranged from 0.4 in biotite schist (compressive strength 22 MPa (3170 psi)) to 1.0 in quartzite (compressive strength 198 MPa (28 820 psi)). The computed charge was corrected for the influence of the degree of fixation of the drill-hole, the strength of explosive, geometry of the drill pattern, and the effect of other charges in the same delay order, by the expression

$$W_c = f/s.E/B.0.8W,$$

where f is the fixation factor, s is the explosive factor, E is the hole spacing, and B is the burden.

Values of f recorded by Langefors and Kihlstrom ranged from 1.00 in vertical holes with maximum fixation to 0.75 when there was free breakage at the bottom of the drill-hole, while s ranged in value from 0.9 for 30% dynamite to 1.3 for 60% dynamite.

For tunnel blasting the calculated charge weight, as thus determined, may need further modification, depending upon the conditions of confinement and fixation. Experimental blasts are usually necessary before firm procedures can be established. Much depends upon the drilling pattern adopted.

Distribution of the Charge

In underground work, and in bench blasting, as a general rule more explosive is needed at the back of the hole than is required near its mouth, and the charge should be well tamped for a length above the intended back of the cut equal at least to the depth of burden. The holes should extend and be similarly charged to a depth of about one-third to one-quarter the burden beyond the intended back of the cut.

The main length of the hole is then filled up to about B from the collar (where B is the depth of burden), using a lower density of charge (about 0.4–0.5 the base-charge density). The remainder of the hole to the collar should then be stemmed, dry sand being convenient for this in vertical holes, or by means of cartridges of stemming-clay, sand, or water in horizontal and near-horizontal holes.

When AN–FO and slurries are used for bench blasting in wet conditions the charge at the bottom of the hole may with advantage consist of waterproof, gelatinous based, explosive, with AN–FO forming the upper part of the charge where the hole is dry. The depth of water may necessitate a greater length of gelatinous explosive, however, so as to extend above the water level. An alternative would be to use the more recently introduced slurry explosives and water gels, which give a higher loading density than AN–FO.

Initiation of the Blast

Priming cartridges of Composition B or of Pentolite are commonly used with AN–FO, these being threaded and positioned along the detonating cord at intervals of about 20-hole diameters in order to insure against what otherwise might be a poorly sustained propagation of detonation throughout the column. Sometimes it may be better to initiate the blast either at the bottom (reverse initiation) or at the top of the charge (forward initiation), using a primer or a stepped primer–booster combination that will cause the entire column to detonate at the steady-state velocity of AN–FO.

The blasting performance of AN–FO is very dependent upon the priming system. After initiation the detonation velocity of an AN–FO charge passes through a transient phase before reaching a steady state. The initial velocity may be somewhere in the range between 600 and 6000 m/s, depending upon the transfer characteristics of the shock wave from the primer to the receiver. For good transfer performance or "overdrive" not only must the primer itself have a high detonation pressure and display a rapid rise to steady-state detonation velocity, but it must fill the borehole in order that the shock wave be transmitted over the entire cross-section of the AN–FO receiver. In large holes of 150 mm in diameter or more, the primer should have a length-to-diameter ratio of not less than one, while its length in smaller holes should exceed the diameter, in some cases by several times. Present-day trends in quarry blasting include the use of larger charges in holes of much larger diameter than were formerly used. This requires that a close watch be kept on the economics of priming systems, for primers such as Pentolite and Composition B are expensive. There is, therefore, a move towards two-stage priming in large holes, using a cast primer in water gel or slurry, plus a booster to give effective "overdrive" to the main AN–FO charge.

Detonation Systems

High explosives are detonated either electrically or by means of detonating cord. Electrical systems are often more convenient to use with small cartridge charges, or in underground operations such as tunnel blasting, where multiple delay firing may be employed on a complex blasting pattern.

The trend towards the use of AN–FO mixtures loaded through pneumatic systems underground, and the increasing use of electrical machinery and power systems, can render electrical detonators unsafe, since they may be exploded inadvertently by electrostatic discharge or electrical ground leakage. In surface blasting, added hazards exist in the possibility of lightning discharge and of radio interference. Furthermore, electric cap wires are subject to mechanical damage and are liable to corrosion by AN–FO mixtures. There is, therefore, a general trend towards the use of detonating fuse wherever possible, and initiation by this means is the general rule in open-pit work.

Designing Blast Patterns

Empirical formulae for the design of blast patterns in quarry blasting have been described by Ash [18]. More recently, J. M. Pugliese [19] of the US Bureau of Mines has compared Ash's formulae with the observed results of blasting practice in a number of limestone quarries. His observations apply to blasting in vertical holes ranging over 40–265 mm in diameter and from 2 m to 80 m deep. Vutukari and Bhandari [20] also quote relationships between hole diameter, burden, and spacing, as obtained from a statistical analysis of mining practice in several countries. The following parameters are suggested.

Burden

If the burden B is more than an optimum value then the powder factor increases and fragmentation is poor. If it is too small then the drilling costs are increased. A good practical value is

$$B = 0.024d + 0.85$$

where d is the blast hole diameter (mm) and B is the burden (m).

Pugliese suggests

$$B = K_B d/12,$$

where K_B is the burden ratio, equal to 30 in average conditions, in rock with density around 2.7 g/cc. If the rock density is higher than this, then K_B should be reduced, and vice versa.

Spacing Between Holes

Vutukari and Bhandari suggest

$$s = 0.9B + 0.91,$$

where s is the spacing (m).

Pugliese quotes $s = K_s B$, where K_s is the spacing ratio. For simultaneous detonation of charges in the same row, K_s lies between 1.8 and 2.0.

The optimum spacing is related to the ratio hole length/burden (H/B) and s approximates to $\sqrt{BH}$ if H lies between $2B$ and $4B$ and approximates to $2B$ if H is $4B$ or more.

The ratio H/B has an influence on the spacing ratio K_s, which should be reduced to less than 1.8 if H/B is less than 3. K_s should be around 1.1 for sequence delay charges in the same row.

Optimum Length of Hole

For bench blasting in limestone $H = K_h B$, where K_h is the hole length ratio.

K_h ranges from 1.5 to 4.0 with a general average around 2.6. If K is less than 1.0, the blast is liable to be overcharged, and this results in overbreak and cratering. When a single primer is employed and K_h is less than 4.0, the toe of the hole is liable to remain unbroken, but if multiple priming is adopted the hole may safely be made deeper. K_h may then exceed 4.0.

The Blast Pattern

Interaction between adjacent delay holes is an important factor in the design of the blast pattern. The basic requirement is that a free face must exist in front of each hole when it is blasted. This requires a systematic progression in the delay sequence from the first to the last hole, or group of holes. The delay interval must be such that each blast occurs leaving sufficient time between blasts for the fracture process to be completed before the next group is initiated.

The time factor must provide for (1) the travel of compressive and tensile waves through the burden. (Incident and reflected waves between the charge and the face occupy about 4 ms for 9 m of burden); (2) adjustment of the ambient stress field. (The generation of primary radial cracks and the effects of reflections from the free face occupies from 10 to 20 ms after initiation, depending upon the rock and the type of explosive); (3) movement of the rock debris, which must be accelerated to a high velocity by the expanding explosion gases (30–50 ms are necessary for this).

If the delay time is too short then a cratering-type blast will result. Therefore the minimum delay time should be not less than 10–20 ms, or 1.5–3.0 ms/m of burden. The maximum delay time should be that required for optimum movement of the debris and approaches 30–50 ms or 4.5–7.5 ms/m of burden. In patterns comprised of multiple rows of holes, a staggered arrangement of holes in successive rows is often preferred since it promotes uniform fragmentation throughout the rock mass.

Blast Design Procedure

The following procedure is suggested when designing a quarry-blasting pattern:
  (i) Draw a plan to scale showing the original rock face and the proposed new face with the desired direction of rock movement. Major geological features such as joints, cleavage, and faulting should be indicated and their possible effects considered.
  (ii) Bearing in mind the type of explosive to be used and the rock characteristics, particularly the rock density, estimate the burden.
  (iii) Select a pattern of drill holes that is estimated to be able to provide the desired rock movement as indicated in step (i). Take into account any possible ground vibration effects that may limit the maximum permissible charge weight or the number of holes to be fired per delay interval.
  (iv) Select the hole spacing distance.
  (v) Mark the sequence of charge initiation on the sketch plan, selecting an appropriate number of rows and the number of holes per row, as required to conform with steps (iii) and (iv).
  (vi) Check the hole length in relation to the rock burden.
  (vii) Calculate the desired subgrade length and the collar stemming distance.

Placing the Charge in Underground Blasting

Small charges in cartridge form are usually loaded by hand and tamped with the aid of a wooden rod. By this method, to achieve an overall loading density comparable with that of the explosive itself requires careful and painstaking tamping with the rod after each cartridge has been inserted. The cartridge dimensions must be less than those of the hole in order that there is adequate clearance to pass the explosive down the hole. General recommendations are:

Hole diameter		Cartridge diameter	
(in)	(mm)	(in)	(mm)
$1\frac{1}{4}$	32	$1\frac{1}{8}$	28
$1\frac{1}{2}$	40	$1\frac{1}{4}$	32
2	50	$1\frac{3}{4}$	44

The packing density and efficiency of coupling are sometimes improved by using cartridges having a perforated casing, which collapses under the pressure of the tamping rod. With certain types of explosive (not powders) the cartridge cases may be slit lengthwise before being inserted. In any event, hand loading is slow and the loading density that is commonly achieved compares unfavourably with that which is obtained by other methods.

Pneumatic Cartridge Loaders

Various explosives manufacturers offer pneumatic cartridge loaders by means of which the explosive cartridges are forced by compressed air down a pipe inserted in the blast-hole, to be ejected from the pipe under pressure. The pipe is withdrawn as the explosive feeds to the back of the hole, and with some loaders the air pressure also controls the packing density.

The loading tube may be of antistatic polythene or of aluminium alloy and brass. The metallic tubes can be used in smaller diameter holes than those possible for polythene tubes, but safety considerations limit the type of explosive that may be loaded with their aid. Gelatinous AN–dynamite with 35 % nitroglycerine is about the safe limit in dry holes. Plastic tubes have the disadvantage that they tend to become sticky internally due to the adherence of some dynamite. The tubes should therefore be thoroughly cleaned by flushing them with water after use. However, when they are used carefully, plastic tubes are safe with all civil explosives in dry holes.

In down-holes of large diameter, free running AN–FO mixtures can be poured down a funnel into the hole. For horizontal and up-holes a pressure-loading system is required, and this is usually pneumatic in operation. Pneumatic loading systems, employing anti-static tubes of plastic, are also convenient to use in underground situations such as long-hole blasting in circumstances of restricted space at the collar of the hole, since they may be spirally coiled and then fed into the hole in a continuous length.

The AN–FO mix is contained in a small tank, either carried by hand to the loading point or else loaded on to a sack barrow, giving a loading capacity of from 30 to 70 l. The operating pressure is from 1 to 2 atm, the mix being fed by gravity to the base of the tank and then being blown by an ejector valve into the loading tube. When loading horizontal and up-holes the muzzle of the loading hose is forced against the side of the hole, either by a spring or by air pressure, a filter allowing the air to exhaust while confining the explosive mix.

Placing the Charge in Surface Blasting

Trenching

Small charges, as used when trenching, are loaded by hand. Cartridged explosives are commonly used for this, with reverse initiation from the base charge. A low-charge density is required here, so that cushion blasting, decking, and decoupling, by using, say, 22 mm cartridges in 40 mm diameter holes, are the techniques employed. Decking the charge is done by placing wooden spacers between successive charges along the length of the hole.

Perimeter Holes in Smoothwall Blasting

Similar techniques of decking and decoupling are used in the trimming holes for smoothwall blasting. Some manufacturers supply special stick charges for this, contained in thin tubes, screwed together to form a continuous length for insertion into the hole. Alternatively, cartridges of conventional blasting gelatine are slit longitudinally into quarters, and these are attached to a thin rod which is then inserted into the hole.

Bench and Open-pit Blasting

Holes 100–200 mm in diameter are commonly used for bench blasting and open-pit work. Cartridge explosives such as the gelatine dynamites that are used for bottom charges in wet holes must be lowered down the hole by hand, on a rope, using a self-release hook. The cylindrical cartridge, being smaller in diameter than the hole, a granular free-flowing explosive such as Pentolite is then poured in to fill the annulus round the charge, and thus give better coupling and blast performance.

Water-resistant gelatinous slurries may be contained in plastic bags, which, when lowered down the hole and released, expand laterally to fill the hole completely and thus give good coupling. The remainder of the hole may be hand-loaded by pouring free-flowing AN–FO mixes or else slurries shucked from sacks at the mouth of the hole. In large blast programmes the quantities of explosive consumed are considerable. A reduction of the time occupied in loading and a higher and more uniform charge density may be obtained by adopting bulk-loading techniques.

Bulk-loading Techniques

When large quantities of AN–FO or slurry are used it is generally more economical to load in bulk than to use packaged charges. Two methods may be used. In one, the AN–FO is blended in a stationary mixing plant, remote from the blast site, to which it is transported by hopper-truck. This truck has a vee-shaped hopper at its base, fitted with a screw discharge which feeds the AN–FO into an air lock. The AN–FO is then picked up by a 1 atm air stream, which carries the explosive through about 15 m of 75-mm hose to the charge hole. These trucks have a capacity ranging from 4500 to 9000 kg AN–FO and discharge rates of 110 to 220 kg/h.

The second, more recent, development is the AN–FO mix-truck system. These trucks carry dry ammonium nitrate prills and fuel oil in separate tanks for on-site blending and discharge into the borehole. One type employs pneumatic feed, oil being injected into the stream of prills at the entry point to the delivery hose. Another type uses mechanical screw feed in which the prills emerging from the hopper are picked up by a second screw operating inside a swinging boom. Oil is metered into this second screw and the blended mixture is then discharged vertically about the borehole. Such trucks have a capacity of about 9000 kg and a discharge rate of from 140 to 300 kg/min.

Pneumatic feed has the advantage of greater flexibility than the screw, and several holes can be loaded from one set-up. The loading mechanism is operated by the blaster, by remote control, from the end of the delivery hose. In wet situations, plastic borehole liners are used, tied at the bottom end and lowered into the hole with a bag of slurry to serve as a weight. The hole must be de-watered and then the liner lowered into place before the water re-enters. De-watering can be done successfully by compressed air jet in holes up to 175 mm in diameter and 15 m deep. Larger or deeper holes must be pumped dry.

Slurry explosives are loaded either by mix-truck or by pump-truck. The slurry mix-truck is a truck-mounted continuous mix process plant which pumps its product direct into the boreholes through a hose. These trucks carry a liquor of ammonium nitrate, sodium nitrate, and oxidizing salts, in a tank maintained at 43–55°C. Solid fuels and gelling agents, separately or pre-mixed, are also carried in hoppers. The liquor from the solution tank and the dry ingredients from the hoppers are fed in measured quantities into a mixing chamber, and thence through 40- to 50-mm hose into the borehole. The slurry pump-truck delivery system uses slurry made in a stationary mixing plant remote from the pit. The pump-truck is then a transporter system employing pressurized tanks to facilitate unloading. From 4500 to 14 000 kg of slurry are carried per trip, and pumping rates range from around 130–300 kg/min.

Powder Factor

The performance of a blasting programme is sometimes assessed in terms of the "powder factor".

Powder factor = Volume of rock broken/Weight of explosive used.

The concept is implicit in A. W. and Z. W. Daw's required charge weight formula for single charges, in which the empirical blasting coefficient may be regarded as the apparent powder factor.

In ordinary quarry-bench blasting the consumption of explosive usually ranges from 0.15 to 0.60 kg/m³.

Another way of expressing powder factor is the weight of explosive used per tonne of rock broken. Bauer has observed a linear relation, in quarry-bench blasting, between the powder factor so expressed and the compressive strength of the rock being blasted [21].

Tunnel blasting involves a much greater consumption of explosives, from 4 to 10 times that which applies to quarry-bench blasting. Langefors and Kihlstrom quote an empirical rule for tunnels with cross-sectional area ranging from 4 m² to 100 m²,

$$q = 14/A + 0.8,$$

where q is the powder factor (kg/m³) and A is the area of cross-section being blasted.

Computer Analysis of Blast Design

Lang and Favreau have constructed a computer model to predict the effect of varying parameters, e.g. the ratio of spacing to burden, on blast performance. The procedure depends on the attachment of representative numerical values to rock properties (dynamic Young's modulus, Poisson's ratio, compressive and tensile strengths, longitudinal wave velocity, and density). The interrelation between explosive and rock, which is represented by "powder factor" in empirical blast design, is replaced by the term "energy density", which relates the energy of the explosive to the mass volume of material broken [15].

References and Bibliography: Chapter 3

1. MELNIKOV, N. V., Influence of explosive blast design on results of blasting, *Int. Symp. on Mining Res., Univ. of Missouri, Rolla, 1961*, **1**, 147–56, Pergamon, New York, 1962.
2. ARIEL, R. S., *Explosively Constructed Dam on the Vaksh River*, Trans. No. 789, Bureau of Reclamation, Denver, Colo., 1968.
3. ANON., *Blaster's Handbook*, 15th edn., E. I. Du Pont de Nemours & Co. Inc., Wilmington, Del., 1969.
4. LANGEFORS, U., and KIHLSTROM, B., *The Modern Technique of Rock Blasting*, 2nd edn., Wiley, New York, 1963.
5. GREGORY, C. E., *Explosives for Engineers* Univ. of Queensland, St. Lucia, 1966.
6. PFLEIDER, E. P. (ed.), *Surface Mining*, AIME, Maple Press, Pa., 1968.
7. GNIRK, P. F., and PFLEIDER, E. P., On the correlation between explosive crater formation and rock properties, *9th Symp. on Rock Mech., Golden, Colo.*, pp. 321–45, AIME, New York, 1968.
8. KOCHANOWSKY, B. J., New developments in drilling and blasting, *Engng. Min. J. New York*. **164** (12) (Dec. 1964).
9. PAINE, R. S., HOLMES, D. K., and CLARKE, H.E., Pre-split blasting at the Niagara power project, *Explos. Engr.* **39**, 72–93 (1961).
10. ATO, K., Phenomena involved in pre-splitting by blasting, PhD thesis, Stanford Univ., 1966.
11. SADWIN, L. D., COOLEY, C. M., PORTER, S. J., and STRESSAU, R. H., Underwater evaluation of the performance of explosives, *Int. Symp. on Mining Res. Univ. of Missouri, Rolla, 1961*, **1**, 125–34, Pergamon, New York, 1962.
12. ATCHISON, T. C., PORTER, S. J., and DUVALL, W. I., Comparison of two methods for evaluating explosive performance, *Int. Symp. on Mining Res., Univ. of Missouri, Rolla, 1961*, pp 135–46.
13. DICK, R. A., *Factors in Selecting and Applying Commercial Explosives and Blasting Agents*. U.S. Bur. Min., Inf. Circ. No. 8405, 1972.
14. CLARK, L. D., JONES, R. J., and HOWELL, R. C., Blasting mechanics, *Trans. AIME New York*, **250**, 349–354 (1971).
15. LANG, F. C., and FAVREAU, R. F., A modern approach to open-pit blast design and analysis, *Trans. CIM*, **75**, 87–95 (1972).
16. PEARSE, G. E., Rock blasting—some aspects on the theory and practice, *Mine Quarry Eng.*, **21**, 25–30 (1955).
17. HANSEN, D. W., Drilling and blasting techniques for Morrow Point power plant, *9th Symp. on Rock Mechs., Golden, Colo.*, pp. 347–60, AIME, New York, 1968.
18. ASH, R. L., The mechanics of rock breakage, *Pit Quarry* **50**, 98–118 (1963).
19. PUGLIESE, J. M., *Designing Blast Patterns using Empirical Formulae*, US Bur. Min., Inf. Circ. 8550, 1972.
20. VUTUKARI, V. S., and BHANDARI, S., Some aspects of design of open pits, *Colo. Sch. Mines Q.* **56**, 55–61 (1961).
21. BAUER, A., Trends in surface drilling and blasting, *Proc. CIM 70th. AGM, Vancouver, April, 1968*.
22. JANK, N. M., Principles of priming and boosting AN/FO with slurry explosives, *Proc. AIME AGM, New York*, preprint No. 68-F-7, Feb. 1968.

CHAPTER 4

Vibrations and Noise from Blasting and Other Engineering Processes

Engineering and the Urban Environment

Within recent years the expansion of urban development in proximity to mining and quarrying activities, together with civil engineering earthworks and associated activities concerned with urban redevelopment, have brought a corresponding increase in the occasions of conflict between the needs and desires of urban populations and some of the consequences of the mining and civil engineering operations concerned. Two such consequences are frequently found to be the occurrence of noise and vibrations.

To a very large extent these two phenomena are associated with one another, as, for example, in blasting and pile driving, and with railroad and heavy road traffic; and when individual complaints are made it is sometimes not easy to identify which is the major factor involved. There are cases on record when complaints have been made about the effects of vibrations from quarry blasting when what in fact was experienced was noise, the blast occurring in the early morning when all else was quiet. When the time of the blast was changed to mid-day the complaints ceased. Although the vibrations were no less, the noise was submerged into that of the urban road traffic and went relatively unnoticed.

There are occasions, however, when the effect of vibrations generated in the earth from such sources as quarry blasts or pile-drivers can be seen as consequential damage to neighbouring buildings and structures.

Criteria for Structural Damage Due to Ground Vibrations

Various criteria for structural damage due to blasting vibrations have been used, and a review of criteria was published by Duvall and Fogelson, of the US Bureau of Mines [1]. Other publications of note include contributions from Northwood *et al.* [2 and 3] discussing data obtained in Canada and Sweden, and Melvedev in the USSR [4].

One of the first attempts to establish a damage criterion was made by Rockwell [5] who pointed out the need for measuring the vibration level as a function of charge size and distance. His criterion may be considered to be semi-quantitative in that it simply designated a distance of 70–100 m from the blast as defining a safe limit.

Subsequent investigators proposed more objective criteria, but it was first necessary to establish a damage threshold in specific terms. Thoenen and Windes [6] defined the threshold of damage as the first occurrence of cracks in wall plaster, and three degrees of structural damage were proposed based on the extent to which wall plaster was affected.

These were:

(1) No damage—below the threshold.
(2) Minor damage—occurrence of fine cracks and further opening of old cracks in plaster.
(3) Major damage—serious cracking and collapse of plaster.

The plaster-damage threshold, which is eminently practical and easily identified in buildings and structural interiors, is still in widespread use, but circumstances may arise when it cannot be used. Damage to industrial plant and to laboratories may occur, for example, due to the interference of extraneous vibrations with the proper functioning of delicate control gear and laboratory equipment. In such cases it is necessary to establish, by direct measurements, what may be regarded as acceptable limits of vibration in the particular circumstances concerned.

Thoenen and Windes [6] established their damage threshold by vibrating the ceilings, floors, and wall-panels of houses by means of a mechanical shaker, which was adjustable to give disturbances of various frequencies and amplitudes. The resulting induced phenomena were measured by seismometers and correlated with the vibrations from quarry blasting. A number of investigations of this kind conducted between 1935 and 1942 enabled the US Bureau of Mines to specify a damage index based on acceleration and expressed as

$$I = \frac{4\pi f^2 A}{12g},$$

where I is the damage index, A the maximum displacement, f the frequency, and g the gravitational constant. Vibrations with peak accelerations below $0.1\,g$ were considered to be safe, between $0.1\,g$ and $1.0\,g$ required caution, and higher values had a high probability of causing damage, the threshold of minor damage being regarded as $1.0\,g$.

Crandell [7] recommended safe limits of vibration, basing his criterion on the peak energy in the disturbance. He termed his criterion the peak-energy ratio, defining this as the second power of the acceleration divided by the second power of the frequency or

$$\text{Energy ratio } ER = \frac{a^2}{f^2}$$
$$= 16\pi^4 f^2 u^2$$
$$= 4\pi^2 v^2,$$

where a is the peak acceleration (ft/s^2), u is the peak displacement (ft), v is the peak velocity (ft/s) and f is the frequency of vibration at peak amplitude (Hz).

Energy ratios below 3 were regarded as safe, between 3 and 6 required caution, and above 6 were dangerous.

Later investigators favoured a criterion based on the peak particle velocity rather than either displacement or acceleration. Langefors *et al.* [8] suggested four damage levels classified as minimal damage, slight damage, moderate damage, and serious damage, equivalent to particle velocities at the point of assessment of 70, 110, 160, and 230 mm/s respectively.

Edwards and Northwood [9, 10] described a series of controlled blasting tests using accelerometers and velocity gauges mounted inside several buildings and a seismograph on nearby ground. Explosive charges, buried at depths of from 5 to 10 m, were detonated at progressively closer distances to the buildings until damage occurred.

A considerable variation in the damage threshold was observed in connection with which there appeared to be some dependence on the predominant frequency, and they concluded that it is not realistic to assign a damage threshold in terms of displacement without reference to the frequency.

This study and that of Langefors *et al.*, reported slightly earlier in time, were conducted independently and in soils of different types. The conclusions from both studies indicated that particle velocity provided the best criterion, since it was least dependent on the soil type and the combined effect of the blasting charge and distance from the source.

For the purpose of relating damage to vibration energy, Edwards and Northwood defined three categories, as follows:

Threshold damage: widening of old cracks, the formation of new plaster cracks, and dislodgement of loose objects such as chimney bricks.

Minor damage: damage that does not affect the strength of the structure, such as broken windows, loosened or fallen plaster, and hairline cracks in masonry.

Major damage: damage that seriously weakens the structure, such as large cracks, shifted foundation and bearing walls, distortion of the superstructure caused by settlement, and walls out of plumb.

Edwards and Northwood proposed that a vibration level providing a peak particle velocity of 50 mm/s could be considered to be safe, 50–100 mm/s would require caution, and above 100 mm/s would present a high probability of damage. For single charges of explosive this study indicated that the damage threshold was given approximately by

$$\frac{W^{2/3}}{d} = 0.3,$$

where W is the weight of charge (lb), d the distance from the blast (ft), and the formula $W^{2/3}/d = 0.1$ was recommended as giving a safety factor of 3 when calculating the safe limit for normal blasting operations.

Duvall and Fogelson [11], in reviewing this and other work done prior to 1962, correlated the damage criteria, and concluded that a vibration with a peak particle velocity not exceeding 50 mm/s could be accepted as offering reasonable safety from the possibility of structural damage. The Edwards and Northwood criteria of minor and major damage were judged to correlate with particle velocities of 140 mm/s and 190 mm/s respectively.

A review of investigations in the USSR on the seismic effect of explosions, reported by Melvedev [4], quotes an empirical threshold criterion of structural damage due to ground vibrations to be a particle velocity of 65 mm/s.

The work of a number of investigators therefore suggests that, for a variety of site conditions involving frequencies of vibration ranging from about 2.5 to 400 Hz a peak particle velocity of 50 mm/s should not be exceeded if the vibrations from blasting are to be regarded as harmless in respect of structural damage. This recommendation has been further strengthened in a more recent report by Nicholls *et al.* [12].

Air-blast

Some of the energy near the source of a blast is transmitted to the air as pressure waves having a large range of frequencies extending into the audible range. Most of the energy in the air-blast, however, is carried at frequencies less than 100 Hz, and in a well-balaned and properly stemmed quarry-blast the energy dissipated as air-blast represents a very small proportion of the total available energy in the blast.

In general, when an explosive is detonated in the open air, or in a very shallow borehole, a stress wave forms which moves into the air at supersonic speed. Peak amplitudes measured in air-blast waves (Kringel [13]) have been observed to be extremely variable, depending on the explosive charge, the depth of cover, and—very critically—on weather conditions. Within a very short distance from the source the speed of the air-pressure wave falls to sonic velocity in air and its travel is then governed by air temperature, wind direction and speed, and the presence of obstructions such as buildings, vegetation, and ground contour.

Windes [14] reports that an air-blast pressure of around 70 mbar is necessary to cause window breakage, which may be regarded as the threshold criterion in respect of damage to structures from air-blast. To give some idea of the blast details required to produce such an effect, Edwards and Northwood report an air-blast pressure of 12 mbar at a surface distance of 24 m from 270 kg of explosive detonated at a depth of 9 m.

While the air-blast from conventional chemical explosive charges in surface blasting operations may therefore be regarded as relatively harmless, air-blast will be more significant in connection with the use of nuclear charges, or with massive charges of chemical explosive, such as might be used in large-scale cratering, earth moving, and demolition.

The air-blast experienced in confined spaces underground, in conventional mining and tunnelling practice is liable to be higher than that experienced from blasting above ground. For surface-blasts, the combined effects of charge weight and distance from the blast source may be estimated by Sach's cube-root scaling law [15]:

$$\frac{D_1}{D_o} = \sqrt[3]{\frac{W_1}{W_o}},$$

where D is the distance between the blast charge and the point of observation and W is weight of explosive charge required to produce a specified air-blast overpressure. For surface blasting, Nicholls *et al.* recommend 35 mbar air-blast overpressure as a safe working limit, which should not be exceeded if structural damage is to be avoided [12].

However, Hanna and Zabetakis [16] found that the cube-root scaling law for peak overpressures in still air was only valid over a distance of about one tunnel diameter from the charge, in underground blasting, for which Olsen and Fletcher quote:

$$P = 4.9 \times 10^3 \left(\frac{D}{\sqrt[3]{W}} \right)^{-2.15},$$

where P is the air-blast over pressure (psi), D is the distance from the charge (ft), and W is the charge weight (lb) (instantaneously detonated) [17].

Air-blast overpressures measured by Olsen and Fletcher, and resulting from production blasts with charge weights totalling 142 kg of AN–FO in forty holes fired in six sequential delay groups, and 136 kg of 60% ammonia–dynamite in a similar blast pattern, ranged from 49 to 122 mbar—much higher than the recommended damage threshold for surface blasting.

Blast Ground-wave Propagation Laws

Various investigators have studied ground vibrations from blasting and have developed theoretical treatments to explain the experimental data. If the energy released is taken to be directly proportional to the weight (B) of explosive fired, an equation of the simple form

$$A \propto B^{1/2}$$

is observed, where A is the amplitude of particle velocity, acceleration, or ground displacement caused by an explosive blast.

In more general terms, this is expressed as

$$A \propto W^{1/2},$$

where W is equal to the product of B and the weight strength of the explosive, expressed as an equivalent percentage of blasting gelatine.

In actual practice some of the energy of the explosion is lost due to a mismatch between the characteristic impedance of the explosive and that of the rock.

There may exist also poor geometrical coupling between the explosive and the rock, largely dependent upon the conditions of confinement of the explosive, and there is further dissipation of explosive energy by the pulverizing of the rock in immediate contact with the explosive charge. All these factors will influence the wave propagation characteristic.

Early exploration of these matters was made by Morris [18] and by Habberjam and Whetton [19]. Morris propounded that the amplitude of particle displacement is proportional to the square root of the weight of the charge and inversely proportional to the distance from the blast.

That is,

$$A = K\frac{W^{1/2}}{d},$$

where A is the maximum amplitude, W is the explosive charge weight, d is the distance from explosion, and K is the site factor.

This equation introduced the site factor K for the first time.

Habberjam and Whetton suggested a higher power for the charge weight in their formula

$$A \propto W^{0.805}.$$

Subsequent investigators have proposed further modifications of the propagation law. A general equation for conditions close to the explosive but beyond the zone of pulverization is

$$A = k_1 W^n$$

where k_1 is a constant and n is a constant exponent varying between 0.5 and 1.

As the wave propagates beyond the pulverized zone, under elastic conditions, its amplitude decreases inversely with distance while its total energy decreases inversely as the square of the distance. If we include a propagation exponent m to embrace any attenuation independent of frequency, another general equation might take the form

$$A = k_2 d^{-m},$$

where k_2 is a constant and m varies between 1 and 2.

Combining these two general equations gives us a wave propagation law in its simplest form:

$$A = KW^n d^{-m}$$

in which the maximum amplitude is a function of two variables distance and charge strength, and the site factor K.

In these equations W is expressed in pounds of blasting gelatine, D is measured in hundreds of feet, and A is the peak particle velocity in inches per second.

The exponents n and m must be determined for any particular conditions. They vary with the different components of velocity (radial, vertical, and transverse), but they are independent of other variables. Typical values for n range from 0.4 to 1.0 and a study of the literature (Davies *et al.* [20], Devine *et al.* [21, 22], Hudson *et al.* [23], Ito [24], Windes [14], and Nicholls *et al.* [12]) reveals that specific field determinations have given m as approximately $2n$.

Actual values of n and m quoted for specific investigations in various rock types are:

Rock	n	m
Shale [20]	0.84–0.90	− 1.60 to − 1.84
Sandy shale [25]	0.85–0.90	− 1.68 to − 2.44
Limestone [23]	0.67–0.84	− 1.28 to − 1.74
Massive limestone [26]	0.64–0.96	− 2.0
Laminated limestones and marls [26]	0.63–0.92	− 1.98 to − 2.44

Site Factors

The site factor constants included in these reported propagation laws range from 0.013 to 0.20. Other typical site factors (Table 4.1) are quoted by Thoenen and Windes [6] while Attewell [20] quotes figures of 0.05–0.15 as offering transmission properties representative of most of the strata likely to be encountered in seismic work. As, with the equipment currently available, there should be no difficulty in establishing the propagation laws for a given site in any particular investigation, there is little point in attempting to determine figures for general application.

Scaling Factor for Ground Vibrations

While, in theory, the laws of dimensional analysis suggest that, when comparing the effects of different charge weights of explosive, a cube-root scaling factor should be

TABLE 4.1. *Typical site factors (Thoenen and Windes [6])*

Blasting in	Structure sited on	Site factor K
Rock	Rock	0.05–0.1
Rock	Clay	0.1 − 0.2
Clay	Rock	0.1 − 0.2
Clay	Clay	0.2 − 0.3
Unconsolidated sand	Unconsolidated sand	0.4 − 0.5

applied [27], in practice, several investigators have observed a square-root scaling law to be operative.

Duvall *et al.*, from a statistical analysis of available field data, deduce that the square root of the charge weight $W^{1/2}$ is an appropriate scaling factor for ground blast-wave vibrations. The ground-wave propagation law then becomes

$$V = H \left(\frac{d}{W^{1/2}} \right)^{-\beta},$$

where V is the peak particle velocity, d is the charge to gauge distance, and W is the equivalent charge weight per delay (total charge for an instantaneous blast).

H and β are constants determined at a particular site by trial blasting. Plots of peak particle velocity against scaled distance, for a variety of charge weights and distances, are constructed on log–log coordinates, from which H is the particle velocity intercept of the regression line at unity scaled distance and β is the slope of the regression line. Nicholls *et al.* report average values of H and β from six sites to be as shown in Table 4.2.

TABLE 4.2. *Observed values of H and β in the scaling law v = HDᵝ, where D is the scaled distance (Nicholls et al. [12])*

Site	Particle velocity intercepts H			Average slopes β
	Radial	Vertical	Transverse	
1 (17 tests)	2.10–4.85	1.16–3.53	0.698–1.46	−1.24 to −1.66
2 (3 tests)	1.48–1.97	1.05–1.67	0.861–1.52	−0.99 to −1.45
3 (3 tests)	2.04–2.77	1.26–2.01	0.741–1.19	−1.17 to −1.46
4 (3 tests)	0.998–1.67	0.684–1.51	0.463–1.40	−1.37 to −1.65
5 (4 tests)	0.923–1.52	0.811–1.75	0.771–1.00	−1.17 to −1.34

Recording Ground Vibrations from Blasting

A review of equipment applied to fundamental studies of shock effects in rocks was given by Attewell [26]. Instrumentation and techniques employed prior to 1964, in the observation of the effects of quarry blasts, are discussed by Duvall [28]. Leet [30] describes the use of displacement seismographs and accelerographs operating by mechanical–optical lever systems. Northwood *et al.* [2] used a variety of instruments in the measurement of displacement, velocity, acceleration, and strain in the basement walls of buildings subjected to blast vibrations of varying intensity. A multi-recording oscillograph was used to give visual records on light-sensitive paper. Particle velocity was measured by a moving-coil transducer, and the displacement records were obtained simultaneously in two ways:

(a) by electrically integrating the output of a velocity transducer;
(b) by applying a differential transformer to measure the displacement of a 1-s pendulum system.

The outcome of this study was to confirm the advantages of using particle velocity as the damage criterion, and in his subsequent review Duvall concluded that the equipment available for such field measurements at that time was heavy, expensive, and inaccurate. He therefore proposed the development of a portable velocity seismograph and suggested minimum design specifications for two such instruments, which have been applied in subsequent work in the United States, the results of which are described by Nicholls *et al.* [12]. In Britain, Davies [29], Fisekci [31], Girayalp [33] and Attewell and Farmer [32] have applied vibration transducers to the measurement of ground vibrations from various types of explosive, from pile driving, from road and rail traffic, and from forge-hammer operations.

Field measurements were made employing vibration gauges of the Dawe 1403/1 type. This is a moving-coil gauge, which develops an output voltage proportional to the velocity under investigation.

Field Observation Procedure

The ground vibrations are resolved into three mutually perpendicular components— two in a horizontal plane, and one vertical. These are commonly referred to as the radial, transverse, and vertical components. The observations may be made either by mounting three gauges on three mutually perpendicular faces of an aluminium block or by mounting the gauges separately on aluminium plates. For observations on bedrock the superficial ground cover should be removed and the block or plates rigidly secured to firm ground, using rock-bolts. For observations in or on the overburden the gauges may be mounted in a box, buried in the soil, or pinned to the sides of an excavation in the soil. The transducer equipment, waterproofed for field application, is shown in Fig. 4.1 a – c.

The output voltages from the gauges are generally recorded either by oscilloscope and camera, or directly by oscillograph.

From the records so obtained, the elastic wave profiles are studied, wave propagation laws investigated, and the effects of blasting assessed in relation to various problems, e.g., the cumulative energy levels generated in millisecond-delay blasting, and damage thresholds for various charge limits and distances.

Using this criterion together with the wave propagation laws for the site, deduced from the observed records, limit curves may be drawn relating distance to maximum allowable explosive charge.

Alternatively, the results of a vibration survey may be presented in the form of limit lines drawn on a plan of a quarry and its environs to indicate the maximum extension of a quarry for a specific explosive charge and delay interval on the basis of a specific vibration amplitude at the point of interest.

To use the scaled distance as a ground vibration control in blasting the log–log plot of peak particle velocity against scaled distance (Fig. 4.2) may be used. From such graphs the critical scaled distance, at which the damage threshold velocity of 50 mm/s occurs, can be picked out. Table 4.3 shows such critical scaled distances for the various modes at the two sites concerned for the very limited number of observations recorded by Girayalp [33]. From which, considering vibrations in the radial mode at site A, due to charges exploded at a distance of 488 m (1600 ft) from the point of observation:

$$\text{Maximum permissible charge weight} = \left(\frac{1600}{16.37}\right)^2 \text{ lb or 4333 kg.}$$

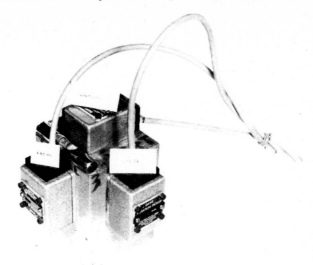

(a)

(b)   **(c)**

FIG. 4.1. Arrangement of vibration gauge.

For general application, Devine *et al.* proposed $20\,\text{ft}/(\text{lb})^{1/2}$ as the critical scaled distance, representing the damage threshold of 50 mm/s particle velocity. More recently, Nicholls *et al.* suggest that, in the absence of data obtained from instrumented blasts, a

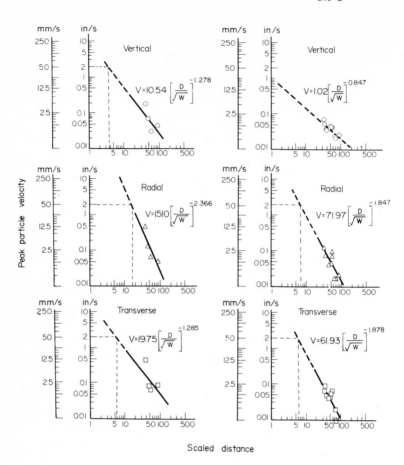

Fig. 4.2. Peak particle velocity against scaled distance (Girayalp [33]).

scaled distance of 50 ft/(lb)$^{1/2}$ should be used. This should ensure that vibration levels will not exceed 50 mm particle velocity. On this basis, at a distance of 152 m (500 ft) from the point of observation the maximum permissible charge weight per blast (or per delay interval in multi-delay blasting) is

$$\left(\frac{500}{50}\right)^2 \text{ lb} \quad \text{or} \quad 45 \text{ kg of explosive.}$$

Vibration Sources Other than Blasting

Earthquakes and blasting are not the only potential sources of vibrations that may cause damage to structures. Such things as pile-drivers, forge-hammers, rail and highway traffic, are frequent causes of complaint. In objective terms, unless the observer is very close to the vibration source, the vibration magnitude at the point of observation is likely to be less than the generally accepted threshold particle velocity of 50 mm/s. But none of

TABLE 4.3.   *Critical scaled distances
at two sites (Girayalp* [33])

Mode	Site A	Site B
Vertical	3.67	0.46
Radial	16.37	6.98
Transverse	5.94˙	6.22

the structural damage criteria so far mentioned take into account resonance effects, nor do they take into account any subjective assessment, which, from a human environmental aspect, is of paramount importance. Some case histories of investigations into these matters are therefore outlined in the following section.

Ground Vibrations Due to Pile Driving

In an urban redevelopment project the ground vibrations set up by a vibrator pile-driver were compared with those of a diesel hammer operating at 57 strokes per minute. The ground conditions at the site consisted of a surface layer of sand and gravel under which were several beds of clay on top of a thick bed of sand. Vibration gauges were set up at four stations, from 20 m to 100 m from the pile-drivers, about forty records being taken.

At distances greater than 70 m there was little difference in the magnitude of peak particle velocity produced by either machine. At lesser distances there was a marked difference, and at a distance of 23 m the diesel hammer produced a peak particle velocity of 2.5 mm/s compared with 0.75 mm/s produced by the vibrator pile-driver. In objective terms either value was much less than the suggested threshold of 50 mm/s. Subjectively, however, both machines were highly objectionable and caused much complaint, the diesel hammer on account of impact and noise, and the vibrator on account of resonance in neighbouring buildings.

At an industrial site the ground vibrations produced by the impact of forge-hammers were compared with those produced by heavy highway traffic. The general lay out of the site is shown in Fig. 4.3 together with a pictorial representation of typical records in the vertical, radial, and transverse modes. The maximum particle velocities at the recorder, due to a forge-hammer weighing 18 tonnes at distances of from 15 m to 21 m away, measured 1.0 mm/s on the radial and transverse modes, and 1.8 mm/s on the vertical mode. The peak vibration from the forge-hammer was almost instantaneous. The vibrations from highway traffic also peaked at 1 mm/s on the shear mode, and the peaks on all three modes were sustained over periods of 0.25 s. Nevertheless, from a subjective viewpoint the operation of the forge-hammers was regarded as objectionable while the highway traffic vibrations and noise were accepted without complaint.

The noise associated with an impact source or blast makes the vibrations appear to be much worse to an observer to whom the shock is unexpected, while vibrations from everyday, mundane, sources are often accepted without demur, or go unnoticed. During the excavation of foundations for an extension to a nuclear power station, it was feared that the effects of outside blasting would disturb the delicate relays and equipment in the control room within the existing building. Recordings of the vibrations at the control station, however, showed quite clearly that the careful closing of a panel door could cause vibrations to the contained relays at least as great as the effects of blasting outside the

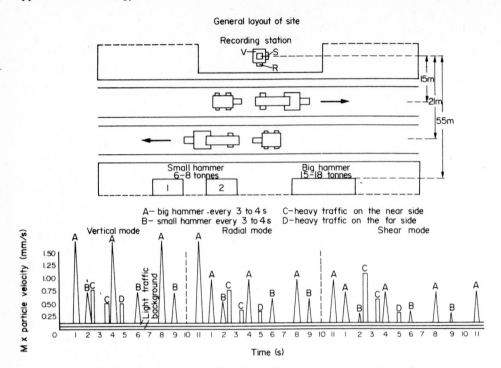

FIG. 4.3. Vibration from road traffic and an adjacent forge-hammer (Fisekci [31]).

power station and very much less than the effects of carelessly closing a door, or of carelessly dropping heavy objects on the floor within the station. For example, the vibration produced by 6 kg of explosive fired in the ground 25 m away from a relay was less than one-half that produced by carefully closing the door of the panel behind which the relay was contained.

A switch relay panel, and the main structure of the building containing it, were seen to be put into resonant vibration by the incoming ground wave from an exterior blast. It is well known that the surface waves that are associated with seismic shock are potentially the most destructive of vibrations so far as structural damage is concerned. They are also the most liable to generate resonance, so that the effect of the incoming surface wave from a minor seismic disturbance such as blasting may be magnified. Structural engineers should therefore give much attention to insulation from extraneous vibrations, to resonance effects, and to damping, where the housing of delicate equipment of critical importance is concerned.

Subjective Assessment of Ground Vibrations

The Reiher–Meister scale was introduced in 1931 as a multiple criterion method of subjective appraisal of the human response to ground vibrations generated by railroad traffic [34]. Six levels of response were defined:

(1) Imperceptible.
(2) Just perceptible.

(3) Clearly perceptible.
(4) Annoying.
(5) Unpleasant.
(6) Painful or harmful.

Some attempts were subsequently made to correlate these criteria with objective criteria of structural damage [35]. From these it would appear that subjective criterion (3) (clearly perceptible) corresponded to medium vibrations causing no structural damage, and subjective criterion (4) (annoying) to strong vibrations with the first appearance of cracks in wall plaster. Subjective criterion (5) (unpleasant) corresponded to heavy vibrations with damage to wall panels, and criterion (6) (harmful) to serious damage and structural collapse (Figs. 4.4 and 4.5). However, insufficient data have yet been forthcoming to enable such a detailed correlation to be made with much confidence. It is likely that in a quiet rural environment people will react strongly long before there is reason to fear structural damage to the buildings in which they are housed, whereas in a noisy urban environment complaint may not be forthcoming until signs of structural damage appear.

Goldman's investigation of the subjective response to vibrations, reported in 1948 [36], used three criteria, "perceptible", "unpleasant", and "intolerable". Nicholls *et al.*, in an analysis of Goldman's results over the predominant range of ground frequencies generated by blasting (6–40 Hz), show that blasting vibrations at the 50 mm/s particle velocity damage threshold would probably be subjectively assessed as highly unpleasant or intolerable.

This was borne out by the history of complaints from families located in the vicinity of the Salmon nuclear test, where 35% of the residents located in the zone where 50 mm/s particle velocity was exceeded filed complaints. From the Salmon test data it appears that a vibration level of 10 mm/s corresponded to around "just perceptible" on the subjective scale, for it resulted in only 8% of the residents filing complaints. A study of vibrations due to rail and road traffic, and forge-hammer operation, on the industrial site shown in Fig. 4.6 produced the results which are shown diagrammatically on Figs. 4.7 and 4.8. This puts the threshold of subjective perception of ground vibrations at a much lower level (around 0.50–1.8 mm/s particle velocity) [37].

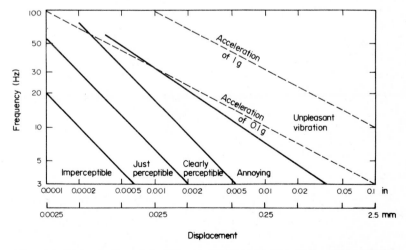

FIG. 4.4. Response of people to different vibration intensities (Reiher–Meister [34]).

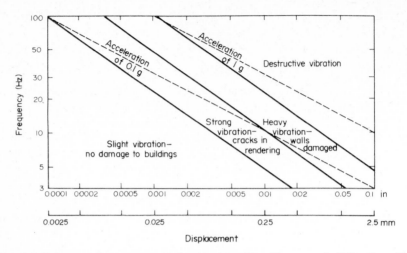

FIG. 4.5. Response of structures to vibration intensities (Ihara *et al.* [35]). *Note*: The displacement is plotted as to the main parameter, but two values of the acceleration (0.1*g* and 1.0*g*) are shown for comparison.

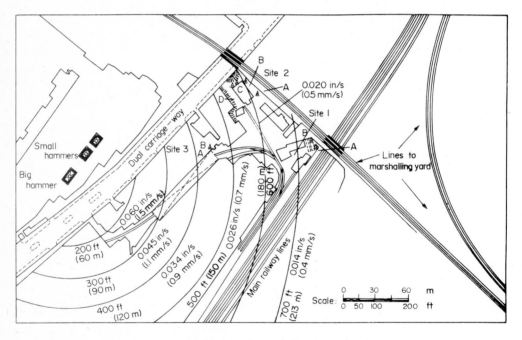

FIG. 4.6. Investigation of ground vibrations—site plan.

Minimizing the Effects of Ground Vibrations

When the engineer has arrived at a conclusion as to the maximum permissible charge weight which may be detonated at a given distance without causing damage to adjacent property, he then must reconcile his decision with the need to achieve an economic and efficient excavation procedure. In the end he may have to compromise between the

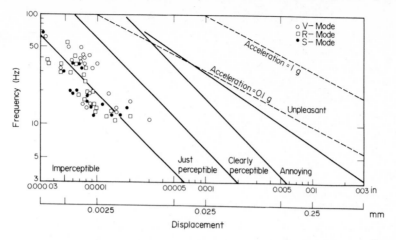

FIG. 4.7. Subjective analysis of ground vibrations due to railroad traffic—site 1.

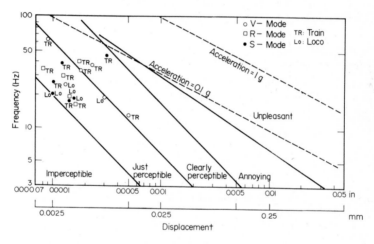

FIG. 4.8. Subjective analysis of ground vibrations due to railroad traffic—site 2.

possibility of causing structural damage, resulting in complaints and possible lawsuits, and the possibility of incurring extra costs by virtue of the procedures he employs. Such extra costs can arise either because by underestimating on his explosives or by not placing them in an optimum pattern he may need to undertake secondary blasting to make good his initial errors. Or by overestimating on his explosives he may cause overbreak beyond the limits of the desired perimeter. In general, the cost and inconvenience of returning to a rock face for secondary drilling and blasting exceed that of dressing loose rock and filling-in an overbreak to the final size, so that a contractor will probably aim for the latter alternative. But the limits are sometimes very finely drawn. For example, during the rock excavation associated with the Niagara Power Project several million square feet of rock face had to be prepared within a tolerance of 150 mm. No rock had to be left within the inner excavation limit, and any overbreak requiring extra concrete had to be at the contractor's expense [38].

In open-pits and quarry blasts the matter of overbreak is seldom of any significance, whereas an incomplete blast can interfere seriously with the excavation and materials handling procedures. The explosive charges will then be designed to ensure effective blasting, carrying the risk of generating ground vibrations that might damage neighbouring property, but employing techniques to minimize that risk as far as possible. To some extent, the techniques of cushion blasting, air decking, and decoupling may be employed, with the object of reducing the peak explosion pressure, lengthening the duration of the pressure build-up in the rock, and so reducing the intensity of impact of the initial stress wave.

Another alternative procedure, which is more frequently adopted, is to divide the total explosive charge into a number of increments that are detonated, not simultaneously, but at successive delay intervals.

The effective charge weight, which is operative in the wave propagation law, is then reduced from the total charge weight to the charge weight per delay. The interaction between successive stress waves will then determine the maximum stress pulse intensity that results in the rock. It has often been observed that increasing the number of delay intervals in a large round of charges produces a decrease in the amplitude of ground vibration, while at the same time it may aid fragmentation and movement of the broken rock.

Langefors and Kihlstrom make the points that the ground vibrations from a single charge acquire a maximum amplitude after one or two vibrations, but only about three vibrations exceed half the maximum vibration amplitude, so that all the others may be ignored. This means that if the delay interval is greater than $3T$, where T is the periodic time of vibration

$$\text{(i.e. } T = \frac{1}{f}, \text{ where } f = \text{frequency)},$$

it can safely be reckoned that the successive stress waves will not reinforce one another.

For blast-wave frequencies ranging from 6 to 40 Hz this would involve a delay interval of from 1/2 to 1/12 s, which is too long for most multidelay blasting projects, where delays ranging from about 5 to 400 ms (0.005–0.400 s) are employed. Therefore delay intervals should be selected so that the total time involved $= nz$, where z is the delay time interval, n is the number of delay intervals, and $nz = kT$.

Ideally k should approximate to a whole number, but not the ratio k/n, and should be more than 3, so that the whole round is completed in time equal to at least one interval of the periodic time of vibration.

Langefors and Kihlstrom argue that the time interval between delays should be of sufficient length to allow for several cycles of motion. Depending upon the site conditions and blasting pattern, this would probably give an interval of the order of 1/10 to 1.0 s. Duvall reports that delay intervals of around 9 ms are effective in limiting ground vibrations, and Oriade reports even shorter delay intervals to be important in relation to vibration and potential damage control. Oriade also quotes the usefulness of primacord as a means to this end. Primacord is often thought of as giving instantaneous detonation of the charges to which it is connected. Actually it does not, for its rate of detonation is around 3/20 ms per m of length. Hence a careful arrangement of the trunk lines in a blasting pattern, detonated by primacord, can produce a rapid delay sequence from what would be instantaneous if fired by electric detonators with no delays incorporated.

Protection Against Ground Vibrations from Other Sources

In quarry blasting and on similar projects it is possible to observe the characteristics of the ground vibration and then control the blasting procedure in terms of selecting delay time intervals and the number of delay intervals per blast to limit the vibration at a given point of observation. With other industrial vibration sources, and when natural seismic disturbances are involved, this may not be possible. The alternative procedure of incorporating protective construction principles in the structure itself must then be observed if damage is not to result.

So far as industrial operations are concerned, we must consider ground vibrations originating either from within the structural foundations or from without. In either case the structure may be designed to incorporate two basic principles—(a) decoupling and (b) damping.

The principles of foundation design embodying decoupling were investigated by Woods [39], and are illustrated in Figs. 4.9 and 4.10. Figure 4.9 shows a vibration source isolated from the surrounding foundations by a circular open trench, and Fig. 4.10 shows a critical location being protected from incoming ground waves by a straight, open trench. Woods determined the reduction factor in the amplitude of vertical displacement for various ratios of trench depth to incoming wavelength and various trench radii around an internal vibration source. The width of the trench does not appear to be critical and, in

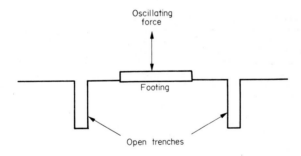

FIG. 4.9. Isolating vibration source by construction of circular trench (after Woods [39]).

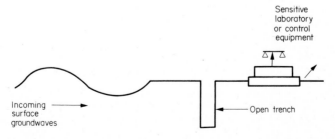

FIG. 4.10. Protecting sensitive zone from incoming ground vibrations (after Woods [39]).

practice, will depend upon soil properties and construction requirements. The depth of trench required usually ranges from 7 to 17 m, and this may present support problems in weak soils. In such circumstances the trench may be filled with porous, granular materials such as coke breeze, crushed pumice, and vermiculite, of high damping capacity, in which event the principle of complete decoupling is replaced by the interposition of a barrier where attenuation of the wave vibration is determined by the damping capacity of the material in the trench, whose width then becomes important.

Materials with a high damping capacity are also incorporated in the design of footings, both of the vibration source and of structures which are to be protected from vibrations [40]. Byrne describes such a form of construction in the foundations of the Union Carbide Building, which straddles the rail tracks of Grand Central Station, New York [41]. Insulating pads of steel and asbestos, contained in lead jackets, are interposed in the footings of the supporting columns of the building. Methods of analysis and design of dynamically loaded foundations, embodying limit design criteria of vibration amplitude, velocity, or acceleration, are described by Richart *et al.* [42].

References and Bibliography: Chapter 4

1. DUVALL, W. I., and FOGELSON, D. E., *Review of Criteria for Estimating Damage to Residences from Blasting Vibrations*, US Bur. Min., Rep. Invest. No. 5968, 1962.
2. NORTHWOOD, T. D., CRAWFORD, R., and EDWARDS, A. T., Blasting vibrations and building damage, *The Engineer, London* 25 (5600) 973–8 (May 1963).
3. NORTHWOOD, T. D., CRAWFORD, R., and EDWARDS, A. T., Further studies of blasting near buildings, *Ontario Hydro. Res. Q.* 15 (I) 1–10 (1963).
4. MELVEDEV, S. V., *Problems of Engineering Seismology*, Voprosi Inst. Seismology Pros., pp. 99–112, 1963.
5. ROCKWELL, E. H., Vibrations caused by quarry blasting and their effect on structures, *Rock Products* 30 58–61 (1927).
6. THOENEN, J. R., and WINDES, S. L., *Seismic Effects of Quarry Blasting*, US Bur. Min. Bull. 442, 1942.
7. CRANDELL, F. J., Ground vibrations due to blasting and its effect upon stressmeters, *Boston Soc. Civil Engng J.*, pp. 222–5 (1949).
8. LANGEFORS, U., KIHLSTROM, B., WESTENBERG, G., and JORN, B., Ground vibrations in blasting, *Water Power* 10 (9) 335–9, 360–6, and 421–4 (1959).
9. EDWARDS, A. T., and NORTHWOOD, R. D., Experimental studies of the effects of blasting on structures, *The Engineer, London* 210 (5662) 539–46 (1960).
10. EDWARDS, A. T., and NORTHWOOD, R. D., Studies of blasting near buildings, *Crushed Stone J.* 36 (10–23) (1961).
11. DUVALL, W. I., and FOGELSON, D. E., op. cit. (see 1).
12. NICHOLLS, H. R., JOHNSON, C. F., and DUVALL, W. I., *Blasting Vibrations and their Effects on Structures*, US Bur. Min. Bull. 656, 1971.
13. KRINGEL, J. R., Control of air blast effect resulting from blasting operations, *Min. Congr. J.* 46, 51–56 (Apr. 1960).
14. WINDES, S. L., *Damage from Air Blasts*, US Bur. Min., Rep. Invest. No. 3622, 1942.
15. SACHS, R. G., *The Dependence of Blast upon Ambient Pressure and Temperature*, Ballistics Res. Lab., Aberdeen Proving Ground, Maryland, Rep. 466, 1964.
16. HANNA, N. E., and ZABETAKIS, M. G., *Pressure Pulses Produced by Underground Blasts*, US Bur. Min., Rep. Invest. No. 7147, 1968.
17. OLSEN, J. J., and FLETCHER, L. R., *Air Blast Overpressure Levels from Confined Underground Production Blasts*, US Bur. Min., Rep. Invest. No. 7574, 1971.
18. MORRIS, G., Vibrations due to blasting and their effects on building structure, *The Engineer, London* 109, 394–5 (1950).
19. HABBERJAM, B. M., and WHETTON, J. T., On the relation between seismic amplitude and charge of explosive fired in routine blasting operations, *Geophysics,* 17 (1) 116–28 (1952).
20. DAVIES, B., FARMER, I. W., and ATTEWELL, P. B., Ground vibrations from shallow sub-surface blasts, *The Engineer, London* 217, 553–9 (1964).
21. DEVINE, J. F., Vibration levels from multiple holes per delay quarry blasts, *Earthquake Notes,* 33, 32–39 (1965).

22. DEVINE, J. F., BECK, R. M., MAYER, A. V. C., and DUVALL, W. I., *Vibration Levels Transmitted Across a Presplit Fracture Plane*, US Bur. Min., Rep. Invest. No. 6695, 1965.
23. HUDSON, D. E., ALFORD, J. L., and IWAN, W. D., Ground accelerations caused by large quarry blasts, *Seismol. Soc. Am. Bull.* **51**, 191–202 (Apr. 1961).
24. ITO, I., On the relationship between seismic ground amplitude and the quantity of explosives in blasting, *Mem. Fac. Eng. Kyoto Univ.* **15**, (11) 579 (1953).
25. DEVINE, J. F., and DUVALL, W. I., Effect of charge weight on vibration levels for millisecond delayed quarry blasts, *Earthquake Notes* **34**, (4) 17–24 (June 1963).
26. ATTEWELL, P. B., Recording and interpretation of shock effects in rock, *Min. Miner. Engng* **1**, 21–28 (Sept. 1964).
27. HENDRON, A. J., Ground vibrations and damage caused by blasting in rock, in *Rock Mechanics in Engineering Practice* (Stagg and Zeinkiewitz, eds.), pp. 217–23, Wiley, London, 1968.
28. DUVALL, W. I., *Design Requirements for Instrumentation to Record Vibrations Produced by Blasting*, US Bur. Min., Rep. Invest. No. 6487, 1964.
29. DAVIES, B., The time characteristics in compressed-air blasting, PhD thesis, Univ. of Sheffield, 1964.
30. LEET, L. D., *Vibrations from Blasting Rock*, Harvard Univ. Press, 1966.
31. FISEKCI, M. Y., A study of ground vibrations produced by blasting and mechanical impact sources, PhD thesis, Univ. of Sheffield, 1968.
32. ATTEWELL, P. B., and FARMER, I. W., Ground vibrations caused by blasting—their generation, form, and detection, *Quarry Managers J.* **64**, 191 (1964).
33. GIRAYALP, M., Subjective appraisal of ground vibrations, MEng thesis, Univ. of Sheffield, 1969.
34. REIHER, H., and MEISTER, F. J., Die Empfindlichtkeit ab Menschen gegen Erschütterungen, *JorschArb. Geb. Ingenieurwes.* **2**, 381–6 (1931)
35. IHARA, M., YOSHIMO, K., UEDA, C., KAJITA, Z., and MATSUD, A., Investigation on vibration and noise due to pile driving, *Railway Tech. Res. Q.* Rep. 4, No. 3, pp. 38–39, Tokyo (Feb. 1963).
36. GOLDMAN, D. E., *A Review of Subjective Responses to Vibratory Motion of the Human Body*, US Naval Med. Res. Inst. Rep. No. 1, Project NM 004001, Mar. 1948.
37. ROBERTS, A., Ground vibrations due to quarry blasting and other sources—an environmental factor in rock mechanics, *12th Symp. on Rock Mech., Rolla, Missouri*, pp. 427–56, AIME, New York, 1971.
38. STENHOUSE, D., Some applications of the pre-splitting technique in rock blasting, *Min. Miner. Engng*, pp. 453–44 (Dec. 1967).
39. WOODS, R. D., Screening of surface waves in soils. *J. Soil Mech. and Found. Eng. Proc. ASCE* **94**, (SM4) 951–79 (July 1968).
40. CROCKETT, A., and HAMMOND, B., *Reduction of Ground Vibrations into Structures*, Inst. CE, London, Structural and Building Engng Division, Paper No. 18, 1947.
41. BYRNE, D., Skyscraper sprouts through railroad terminal tracks, *Engng News Rec.* **161**, 34 (July 1958).
42. RICHART, F. E., HALL, J. R., and WOODS, R. D., *Vibration of Soils and Foundations*, Prentice-Hall, New Jersey, 1970.

CHAPTER 5

Excavation Technology—I

Small-bore Rock Drilling

TECHNIQUES of rock and earth excavation changed very little throughout recorded history until the advent of Brunel's tunnelling shield, designed to penetrate the London clay under the river Thames, followed by the application of compressed air as a source of power.

Countless generations of stone masons, miners, and labourers, totally oblivious of such concepts as critical stress, brittle fracture, and plastic yield, discovered for themselves that thrusting against hard rock with a blunt instrument is a process which consumes much energy for little reward, whereas a sharp pick will help to do the job more easily. Similarly, it does not take long to learn that strong, sharp, repetitive blows by a hammer are necessary to penetrate a hard and brittle rock when rock drilling, whereas a more plastic soil is best cut by the steady thrust of a clean spade.

At first sight it might appear, therefore, that the result of applying analytical design procedures to conventional rock drilling, boring, and cutting machinery is only to produce rational explanations for what has long been obvious. But while this is partly true, there are other considerations also to be borne in mind.

Ever-changing economics in the climate of advancing technology, and changes in social organization in general, put constant pressure on the conventional processes of rock drilling, tunnelling, and earth excavation.

These are fundamental engineering processes, essential for the supply of oil and natural gas, solid fuels, and mineral raw materials, the construction of urban subways and tunnels, and underground excavations of all kinds, for civil and military purposes. There is urgent need for us to be able to drill deeper and wider, to tunnel faster, and move greater volumes of material more rapidly and more cheaply than ever before.

There are limits to which conventional techniques may be scaled in size, in speed, or in their working range. We must find out what determines those limits so that the processes may be made more efficient and their limits extended. And in the search for greater efficiency and effectiveness new techniques must be explored and developed.

Basically there are four methods by which to attack rock materials: (a) by mechanical forces, (b) using stresses induced by thermal action, (c) by direct fusion and vaporization, and (d) by chemical means. Maurer has reviewed the general factors involved in all these methods, and he has outlined their incorporation in many new and novel drilling techniques [1].

Penetration of Rock by Mechanical Means

The majority of rock drilling and boring techniques employ principles of mechanical attack by means of which stresses are induced in the rock by impact, abrasion, wedge penetration or cutting, erosion, and sometimes by a combination of these methods.

Percussive Methods

Percussive drills reproduce mechanically the action of a hand-drill and hammer (Fig. 5.1). A reciprocating piston, usually actuated by compressed air but sometimes by hydraulic fluid, provides the necessary source of impact, and the energy transmission chain between energy source and rock embodies the transfer of that energy, and consequent energy losses, between several components. The train includes air/piston/anvil/drill-rod/rock-bit/rock interfaces.

There are three general methods of arrangement. (1) In churn drills, which are used mainly for drilling blast-holes in quarry benches, in soft to medium-hard rock, the whole of the piston–rod–bit combination is actuated by a reciprocating mechanism, using gravity impact. This arrangement has now largely been superseded by "down-the-hole" and by rotary drills, which have a much faster penetration rate. (2) In down-the-hole drilling the drill is positioned at the base of the hole to impact directly on to the bit. (3) Hammer-drills, used generally for making blast-holes up to about 75 mm in diameter when tunnelling, or in bench blasting, in which the energy transmission chain includes steel rods positioned between an anvil and the bit.

Although pneumatic drills were first used over one hundred years ago, a systematic study of the mechanics of energy transmission in percussive drilling was not attempted

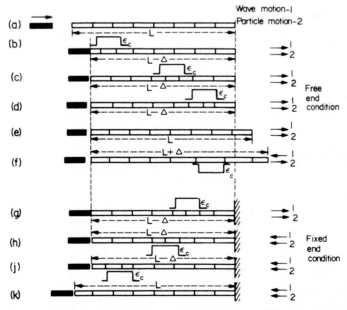

FIG. 5.1. Transmission of energy by percussive impact (Furby [12] after Fischer [3]).

until about 1930 [2]. However, progress was slow until methods of stress-wave analysis were developed and described by Fischer [3] and also by Arndt in 1959 [4]. Fischer used a graphical technique, while Arndt employed an analogue computer. Both these approaches came at a time when experimental methods employing bonded resistance strain gauges on the drill steel, coupled to cathode-ray oscilloscopes, were coming into general use, and subsequent progress in understanding was more rapid. There is now voluminous literature on rock-drilling technology. That on percussive drilling is reviewed, and the techniques analysed, both theoretically and experimentally, by Hustrulid and Fairhurst, by Lundberg, and by Dutta [5–7].

Mechanical Arrangements for Percussive Drilling

Percussive rock drills are manufactured in various sizes, and may be classified as jackhammers, drifters, and wagon drills. Jackhammers are relatively small, ranging in weight from around 10 to 30 kg, and are either hand-held or else supported by a compressed-air piston "air leg". Thrust is applied manually to drill small blast-holes around 25–40 mm in diameter.

Drifters are heavier machines, and when used underground in tunnelling are usually carried on a mobile rig, carrying arms or booms on which the drills are mounted. Thrust is applied mechanically, usually by a screw feed, which may be either manually operated or be driven by motor. When drilling down-holes, as in quarry-bench drilling, the weight of the drifter itself may provide the necessary thrust, but for holes 40–75 mm in diameter, in hard rock, additional thrust will be required. The machine is then mounted on a cradle and support mechanism, by means of which additional thrust may be applied and controlled from a mobile rig termed a wagon drill.

The mechanical principles of design of all the machines are basically similar, with some variations of detail. Compressed air actuates a piston and valve mechanism by means of which the piston, 50–100 mm in diameter, is caused to reciprocate and deliver hammer-blows at a rate varying from around 1500–3000 per minute, the stroke distance ranging up to 90 mm.

The hard-steel piston is a cylindrical annulus, cut internally to fit the rifled splines of a "twist bar" upon which it rides. The piston is turned to a smaller diameter at its forward end to form the piston shaft. This shaft is cut externally by straight splines which fit inside similar grooves on the "chuck", by means of which the hexagonally shaped drill-rod is attached to the drill.

The twist-bar is constrained by a ratchet-and-pawl mechanism so that it can rotate in one direction only. This causes the drill-rod and the chuck to partially rotate on the back stroke of the piston, while the twist-bar rotates slightly on the forward stroke. These combined movements cause the drill-bit to deliver successive impacts on to the rock on diameters inclined at an "indexing angle", which ranges from about 10 to 50 degrees, the smaller angles usually being applied in machines that are designed to attack the harder rocks.

In this manner a hole of circular section is cut. The piston may impact directly upon the end of the drill-rod held in the chuck, or else the impact may be transmitted through an interposed anvil. The latter arrangement is often employed in heavy-duty machines.

Energy Transmission in Percussive Drilling

When the drill-piston strikes the rod in drilling, a compressive strain pulse is generated which travels down the rod. This wave may be pictured as a rectangular pulse having a finite strain level ε and requiring a time interval t to pass any given point on the rod. Another compressive strain pulse passes in a similar manner back through the piston.

Considering the rod only, if the wavelength of the pulse is short, compared with the length of the rod, the initial impact will not be affected by reflections from the far end of the rod, so that the strain pulse travels down the rod at longitudinal sonic velocity c. During its passage the rod is compressed longitudinally and is constrained to be shorter than its original length by an amount Δ where $\Delta = \varepsilon ct$.

The conditions at the bit-end of the rod may be imagined in conditions of complete restraint, or "fixed end", and of no restraint ("free end"). If the pulse arrives at the bit-end of the rod under free conditions it is totally reflected and returns up the rod. However, in this reflected wave the particle motion remains in the incident direction, so that the reflected pulse is in tension.

When the reflected tension pulse arrives back at the piston-end of the rod, that end is caused to move away from the piston. The pulse is then reflected as a compression and the cycle is repeated. In the absence of other impacts the strain wave will oscillate up and down the rod, changing sign at each end with progressively decreasing energy until it dies out due to internal friction and damping.

If the bit-end of the rod is completely restrained, the incident particle motion in the initial pulse will be reflected as a compression pulse, and on arriving back at the piston-end the rod will move towards the piston.

The rod thus returns to its original length. In so doing a secondary impact is generated between the rod and the piston, and rebound energy is transmitted back to the piston, causing it to move away from the rod.

At the end of the complete cycle of wave incidence and reflection up and down the rod under fixed-end conditions there is no change in the length of the rod, but the stress generated at the contact interfaces between bit and rock, and between piston and rod, is double the incident stress value at each end. This contrasts with the free-end reflection in which the incident and reflected strains are superimposed, to double the resultant strain and zero the stress.

In practice the bit–rock contact conditions when drilling will range between the free-end and the fixed-end conditions. At either of these extremes no energy is transmitted to the rock and we may imagine the penetration process beginning with the free-end condition immediately before the bit makes contact with the rock, with restraint on the bit building up to reach the fixed-end condition at the instant when penetration ceases. Between these extremes, part of the incident energy is reflected in tension and part in compression. Only the compressional component transfers energy to the rock.

In practice the strain pulse is not rectangular nor is it totally reflected. Some energy must be transferred to the rock in order to penetrate it. The shape of the first incident pulse, i.e. the relationship between strain (and hence stress) and time, during the passage of the pulse at a given point on the rod, and the energy contained in the wave, are determined by the piston impact velocity and the combined geometry of the piston and the rod. Typical wave forms are shown in Fig. 5.2.

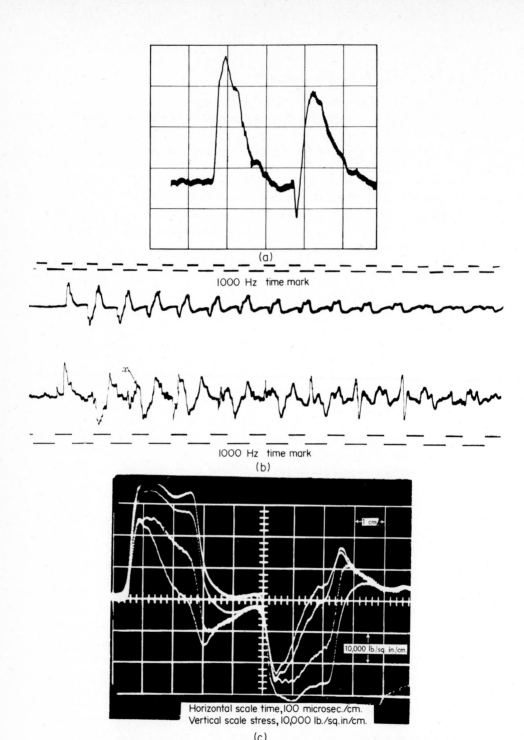

(a)

1000 Hz time mark

1000 Hz time mark

(b)

←1 cm→

10,000 lb./sq. in./cm

Horizontal scale time, 100 microsec./cm.
Vertical scale stress, 10,000 lb./sq.in./cm.

(c)

FIG. 5.2. Stress waves in percussive drill-rods. (a) Initial wave and first reflection at the bit–rock interface (Furby [12]). (b) Complete waveforms in a percussive drill-steel fitted with a chisel-bit and drilling into sandstone. Record A: air pressure 0.18 MN/m². Record B: air pressure 0.35 MN/m² (Chakravarty [9]). (c) Typical oscilloscope record of successive incident and reflected pulses (Fairhurst [18]).

The energy contained in the wave is given by

$$\text{Wave energy} = Ac/E \int_o^t \sigma^2 \, dt,$$

where A is the cross-sectional area of the rod, c is the longitudinal wave sonic velocity in the rod; E is Young's modulus for the rod material; σ is the amplitude of stress; and t is the time of duration of the wave.

Piston velocities are usually in the range from 6 to 12 m/s, and the longitudinal wave velocity in drill-steel is around 5200 m/s.

Energy is transmitted to the rod by the piston impact. The efficiency of this transfer of energy was observed by Arndt [4], and also by Chakravarty, and is nearly 100% provided that the drill-stem is not too short. The rod length should not be less than half the strain wavelength. For machines currently in use this means that the rods must be more than 1.2 mm long for optimum energy transfer, piston to rod. This is determined by the rebound energy transfer characteristics, rod to piston, when the strain waves return back up the rod.

Dutta [7] has published a method, and a computer program, for the determination of the stress waveforms produced by percussive drill-pistons of various size and shape, and it is now a relatively simple matter to produce any desired waveform at will (Figs. 5.3 and 5.4).

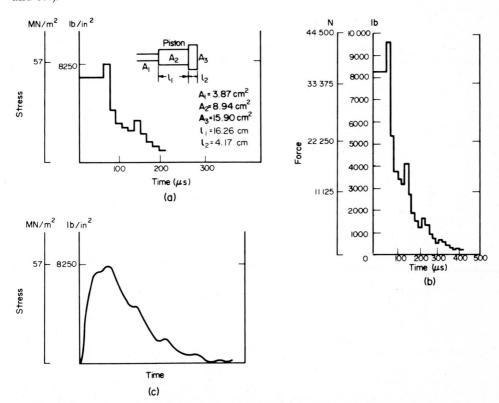

FIG. 5.3. Comparison of waveforms for a percussive rock-drill. (a) Calculated (Hawkes and Chakravarty [9]). (b) Theoretical. Calculated by computer (Dutta [20]). (c) Experimental (Hawkes and Chakravarty [9]).

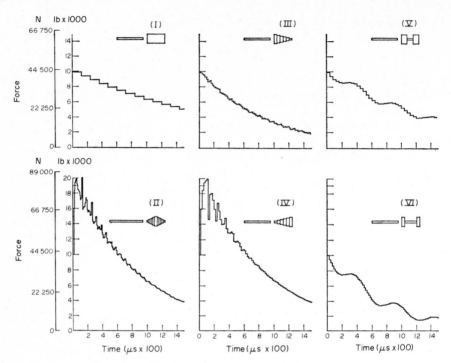

FIG. 5.4. Stress waveforms computed in detail for six pistons of different geometries. Drill-rod diameter 25 mm. Maximum piston diameter 150 mm, length 300 mm (Dutta [20]).

Energy Losses in the Drill-rods

Energy is lost in the drill-rods in three main ways: (1) transfer losses at connecting joints between successive rods; (2) flexural strain losses; and (3) energy trapped within the drill-stem by reflections at the bit–rock interface. Takaoka *et al.* [14] quote total energy losses of 25, 27, and 47% in taper, buttress, and upset joints respectively. The corresponding loss in an integral rod with no joints was 10%. Furby quotes an average figure of 10% transmission energy loss at each coupling of contemporary rolled-thread design (Fig. 5.5).

Under conditions of perfect impact, flexural strain would not exist. If, on impact, the axes of piston and rod were perfectly collinear, if the impacting surfaces were perfectly smooth and hemispherical, giving point contact, and if the rod was perfectly straight, only longitudinal waves would be generated. But in practice, under operating conditions, the axes of the piston and the rod are not collinear because of the necessary clearance between the chuck bush and the rod shank. Also, drill-rods are rarely perfectly straight and their bent condition is accentuated by the applied thrust, which causes the rod to bow, so that a bending moment is applied to the shank. Again, because of the clearance at the chuck bush, the mating faces of the piston and rod are not parallel at impact.

The situation is further complicated because the central hole in a drill-rod is never perfectly smooth, straight, circular, or truly central. Finally, because the percussive drilling process is one in which the rod is subjected to repeated impacts, the rod tends to vibrate at its natural frequency, and this aggravates the already bent condition of the rod.

Transverse particle motions in the drill-rod can make no contribution towards

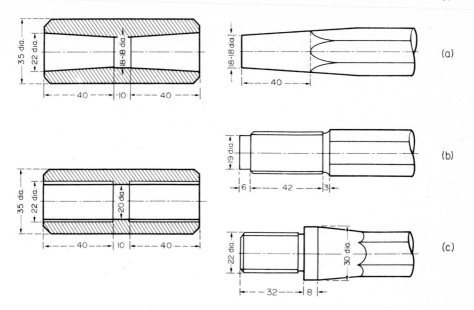

FIG. 5.5. Types of joint in percussive drill-rods: (a) taper joint; (b) buttress joint; (c) upset joint.
(Dimensions in mm)

longitudinal penetration, and they represent a direct loss of energy. The stresses which they generate in the steel only go to promote metal fatigue and premature failure of the drill-rod.

The flexural strain waves travel at a velocity around 3140 m/s and their waveform alters considerably as the wave progresses along the drill-rod (Fig. 5.6). It appears that under experimental conditions the shape of the initial strain pulse is related to the mass ratio piston-rod. The greatest strain levels rise in magnitude with increase in this ratio.

The maximum strain amplitude of the flexural wave is linearly dependent upon the piston velocity, no matter at what point along the drill-rod it is measured. It can be as high as 70% of the longitudinal wave amplitude, and while the maximum amplitude of flexural strain occurs within the first 150 mm of the rod length, the components of longitudinal and flexural strain are superimposed to produce a maximum resultant at about 280 mm from the piston–rod impact interface. This is thus a critical point on the rod at which metal fatigue may lead to fracture.

The characteristics of energy transfer from rod to bit and bit to rock are usually considered together since the loss during transmission from rod to bit is relatively small. Also, when considering the design of bits for improved penetration efficiency, that design includes the joint characteristics as well as the cutting-edge characteristics, and in a good design any energy loss at the joint will be compensated for by an improved energy transfer from bit to rock.

From theoretical considerations Lundberg maintains that energy transfer efficiencies as high as 90% are attainable. This figure must be compared with the estimates of earlier investigators, notably Hawkes and Chakravarty and Long, who reported losses of from 40 to 60% as shown by their experimental evidence [9, 15].

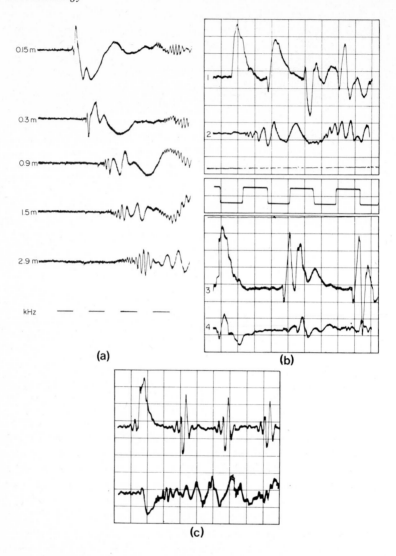

FIG. 5.6. (a) Oscillograms showing transverse strain waves at various distances from the struck-end of a drill-rod. (b) Oscillograms showing longitudinal strain waves (1) and (2) and flexural strain waves (3) and (4) in a drill-rod under operating conditions in the laboratory measurement stations (1) and (2) at 1.5 m, (3) and (4) at 300 mm from the struck-end of the rod. (c) Flexural (lower trace) and longitudinal (upper trace) strain waves at the bit-end of a drill-rod under operating conditions. Note the high bending strain in tension near the beginning of the flexural strain wave. (Furby [12].)

Rock Penetration by Percussive Drill-bit

Basic studies of the cutting action in percussive drilling were reported by Simon and also by Hartman. The process of rock failure and of crater formation, produced by the single impact of a vertical drop-hammer and chisel-bit, was pictured by Hartman as shown in

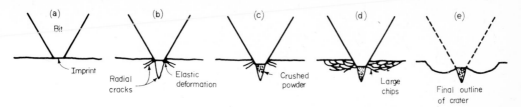

Fig. 5.7. Sequence of rock failure and crater formation when a blunt, wedge-shaped bit strikes an impact blow (Hartman [11]).

Fig. 5.7. He described the sequence of events as comprising:

(1) Crushing of surface irregularities as the bit first makes contact with the rock.
(2) Elastic deformation of the rock, produced as a result of the load applied by the bit. This is accompanied by the generation of subsurface cracks radiating from stress concentrations at the boundary of the cutting edge. Two major cracks propagate downwards to outline a vee-shaped wedge.
(3) Crushing of the central wedge of rock into fine fragments.
(4) Chipping-out of large fragments along a curved trajectory up to the surface adjacent to the crushed zone.
(5) Crumbling away of the crushed zone and displacement of the debris by the bit as it continues to penetrate.

Hartman further observed that if the impact energy was sufficiently great this sequence of events could be repeated more than once as the result of a single hammer-blow [11].

Basic energy relations demonstrate that the most efficient drilling is accompanied by the fewest and coarsest cuttings per unit volume, and faster penetration results when the greatest proportion of chipping to cutting takes place. Hence the advantage of a sharp bit. Chakravarty also observed that energy is transferred from the drill-steel to the rock several times as the result of a single blow from the piston, although the amount of energy transferred after the second incident wave is comparatively small. This progressive penetration by several increments per blow is attributable to the fact that only a part of the energy in the initial strain wave is transmitted to the rock, to be utilized in rock fracture. Most of the remainder is reflected back, away from the rock, and oscillates up and down the rod about 40 times between blows. Some is absorbed in elastic and plastic deformation in the drill components, rods, and rock.

The relative proportions of transmitted and reflected energy components are determined by the bit–rock contact conditions and the characteristics of the rock's reaction to impact. However, penetration depends not only on the amount of energy transmitted but also on the stress generated in the rock at the bit–rock contact. For any given rock there is a critical stress level below which penetration cannot be effected by impact.

If the transmitted stress level is sufficiently high to effect penetration, the amount of energy that will be absorbed in deformation processes without contributing to rock fracture is comparatively small. Simon considered it to be negligible, and Fairhurst has observed elastic rebound (other than that which occurs during the stepped penetration process), to be less than 3% of the total penetration.

Force – Displacement Characteristic

The instantaneous force F and the particle velocity du/dt developed by the longitudinal strain wave are:

$$F = (\sigma_i + \sigma_r)A + F_0,$$

$$du/dt = \frac{c}{E}(\sigma_i - \sigma_r) + V_0,$$

where F is the total force generated between bit and rock, F_0 is the force between bit and rock before the strain wave arrived, σ_i is the incident stress, σ_r is the reflected stress, A is the cross-sectional area, c is the longitudinal wave velocity in the drill-rods, E is Young's modulus of the drill-steel, u is the displacement, and V_0 is the bit velocity before arrival of strain wave.

The conditions at the bit–rock interface can be represented by a force–displacement characteristic, or a force–penetration curve, such as that shown in Fig. 5.8. This graph is typical for a brittle rock which exhibits both chipping and crushing during its penetration by the bit. The applied force builds up during increments of penetration when elastic deformation and crushing occur, and diminishes during increments of penetration when chipping occurs. As the load on the bit increases from 0 to F_{pi} during the first incident wave, the force–penetration characteristic builds up along the solid line to $U_{pi}F_{pi}$. With the passage of the first wave, as the load is removed an elastic expansion of the rock occurs

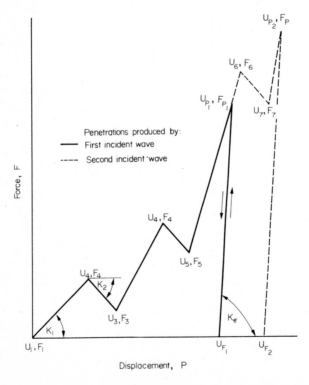

FIG. 5.8. Idealized force–penetration characteristic in percussive drillings (Hustrulid and Fairhurst [5]).

under the bit and the resultant penetration is less than the peak penetration by the amount of this elastic expansion.

When the load is reapplied by the second incident wave the elastic rebound portion of the curve is retraced to F_{pi} and then the original characteristic extended.

Hustrulid and Fairhurst's [5] analytical treatment provides equations for determination of energy transfer, bit force, penetration, and the reflected wave component, as functions of time, for any given incident stress wave and force–displacement characteristic, i.e. for any given machine and rock combination. Assuming that there is close bit–rock contact before the arrival of the first incident wave, they show that the bit and rock remain in contact for the first two incident waves, and then separate. Whether the bit regains contact with the rock before the next piston blow will depend upon the external thrust that is applied to the machine.

During the crushing phase of each increment of penetration the rock in contact with the bit is pulverized to a certain depth and the pulverized material is subsequently compacted as the force on the bit increases. The extent and shape of this zone are determined by the bit geometry (impacting wedge angle), the depth of penetration, and the angle of internal friction of the rock material.

The conditions of confinement between the wedge of pulverized rock material and the surrounding solid rock will have a profound influence on the succeeding chipping phase of penetration because it will cushion, and hence magnify, the force required to initiate chipping. Chipping is a momentary phenomenon associated with the sudden release of strain energy. Optimum conditions for such a release of energy require that the rock be able to relax, so that localized shear and tension fractures may open up in the process of brittle failure. Any confinement by the pulverized material will obstruct this relaxation. Hence the importance of clearing the cutting edge of debris by the use of air and water jets.

Effect of Bit Shape

The bit takes the essential form of a diametral wedge, as in a chisel-bit, or four radial wedges in a cruciform-bit. The angle of the wedge is an important factor. If it is broad and presents an obtuse angle to the rock, a small depth of penetration will generate a large area of contact, and the stresses developed at the bit–rock interface will be reduced accordingly. On the other hand, a sharp bit, which effects an acute-angled wedge penetration, generates higher contact stresses, with the maximum compressional component at a smaller angle to the surface of the rock, so that the available energy can be utilized more efficiently in chipping-off to the free surface (Fig. 5.9).

In practice a compromise must be effected between the higher initial penetration rate made possible by an acute-angled wedge design for the bit, and the subsequent deterioration in performance due to bit-wear as penetration proceeds. Percussive drill-bits are therefore usually made to present a wedge angle of from 100 to 200 degrees, the larger angles being used on bits drilling into hard abrasive rocks, so that wear can be kept to acceptable limits.

Effects of Thrust

To ensure that the bit and rock are in contact when the first incident wave arrives, at least a minimum value of thrust must be applied to the machine. Calculations based on

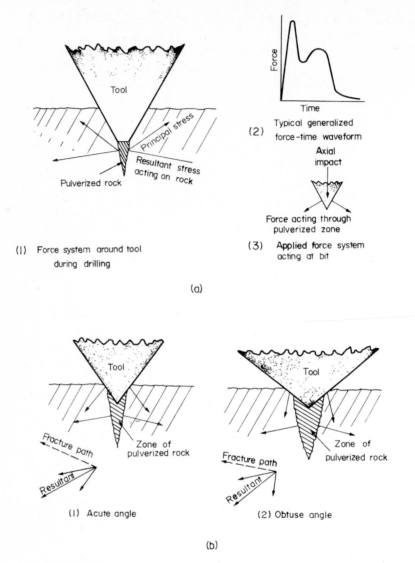

FIG. 5.9. (a) Simplified force system in percussive drilling; (b) effect of wedge angle in percussive drilling (Fairhurst [18]).

drill-steel momentum balance suggest that this minimum thrust requirement is

$$F_t = 2(I + \alpha)NW_p V_p,$$

where F_t is the minimum thrust required, α is the coefficient of restitution, N is the number of blows per minute, W_p is the piston weight, and V_p is the piston impact velocity.

In practice it is well known that to operate a percussive machine with too small a thrust gives poor penetration at high cost because energy is not transmitted to the rock efficiently. Dissipation of piston energy is then concentrated in the machine and in the drill-rods. Excessive maintenance costs on the machines and a short life of drill-steel, due to metal fatigue, are the inevitable result.

One of the reasons why heavy drifters and wagon drills are best fitted with motor-driven mechanical-feed mechanism is that the applied thrust can thus be increased and controlled. To what level that increased thrust should advantageously be raised is a matter of economic compromise between the benefits resulting from a higher penetration rate and the cost of increased flexural energy losses, fatigue due to bending in the drill-rods, and a shorter life of the drill-bits due to abrasion. The latter point is of considerable economic significance because, in order to prolong the life of the bit, its cutting edges are usually comprised of sintered tungsten carbide, which is expensive.

The optimum conditions for impact penetration would allow the cutting edges to retract clear of the crater under the bit, while the bit undergoes indexing. If the cutting edges do not clear the rim of the crater, then high torsional stresses will be generated between bit and rock during indexing. This leads to high abrasion losses, possible shear fracture, and chipping of the tungsten carbide.

When discussing his observed relationships between penetration, strain waveforms, thrust, and rate of piston impact, Furby [12] comments that, whereas a striking rate of around 1800 blows per minute is common in percussive drilling, rates as high as 6500 per minute might be used with advantage. He maintains that such striking rates would increase drill penetration by as much as 300 % without adversely affecting the service life of the rods.

In practice there is an optimum level of thrust below which penetration suffers due to inefficient energy transmission, and above which penetration falls off due to problems of indexing and bit flushing. Ultimately, if thrust continues to be increased at a given operating air pressure, a point will be reached at which the drill will stall. The optimum thrust and the point of stall increase with increase of air pressure supplied to the drill [24, 25].

Simon's thrust equation shows that impact rate and minimum thrust requirements are directly proportional to one another, so that the maximum benefit of increasing the impact rate can only be realized if complete indexing clearance can be achieved at the high values of thrust applied. The required indexing rate therefore becomes faster and the index angle smaller with increasing thrust. Ultimately a point is reached when it becomes impracticable to provide for clear indexing, and the bit must then be designed specifically to provide for rotational abrasion and torque. The rock-penetration process then changes from percussive drilling to rotary–percussive drilling (Fig. 5.10).

Down-the-hole Drilling

In conventional percussive drilling the energy transmitted from the piston to the cutting edge of the drill-bit diminishes as the depth of the hole increases. This is due partly to friction, damping, and flexural energy losses in the drill-rods, but mainly to energy transmission losses at their connecting joints. Also, since the impact energy, generated stress level and penetration rate of a given machine are all determined by the piston impact velocity, there is a temptation to try to improve drilling performance by increasing the air pressure at which the drills are operated.

The inevitable result of so doing is to increase the rate of fatigue and fracture in the rods and to shorten their working life. Operating heavy drifters and wagon drills at gauge pressures of 0.7 MPa (100 psi) or more can, in certain instances, produce unacceptable rod

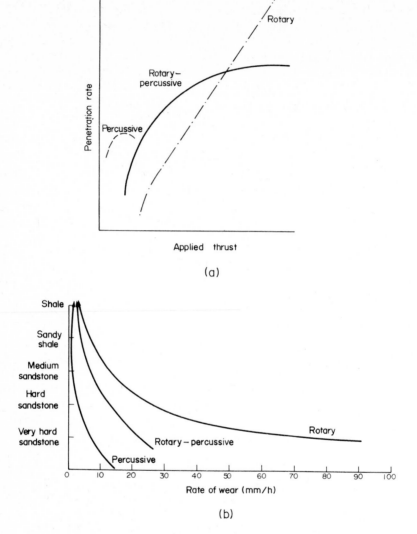

FIG. 5.10. (a) Typical thrust–penetration rate curves; (b) comparative wear rates (Fish [22]).

and bit failures. Relatively low air pressures, to a maximum of 0.5 MPa (80 psi), are therefore generally preferable on economic grounds.

These disadvantages are largely eliminated in "down-the-hole" drills, in which the impact mechanism is coupled to the drill-bit and follows this down the hole. The rods are then required to transmit torque only, and this is supplied by an independent motor at the mouth of the hole. In a down-the-hole hammer-drill the piston diameter is limited by the size of the hole. To achieve an energy input of useful practical penetrating power, a hole diameter of about 150 mm is the minimum needed. Since the rod fatigue and fracture problem is no longer of overriding significance, the operating air pressure may now be increased, and down-the-hole hammers generally use compressed air at 1.7 MPa (250 psi) and in some cases up to 2.8 MPa (400 psi).

However, high operating air pressures inevitably bring heavy abrasive wear and impact damage to the bits, with the result that another economic limit ultimately presents itself, and an alternative method of rock penetration must then be sought. One such alternative is to employ rotary rock drills.

Rotary Rock Drilling

While percussive drills are likely to remain a tool of major importance for boring small, relatively short blast-holes when excavating and tunnelling in hard rock, the method becomes unacceptably inefficient with increase of diameter and increase of depth to be drilled. Not only do the energy losses increase with increase in length of the hole but the difficulty of keeping the bit clear of debris becomes more intense with increase in the hole diameter. This is due to the fact that the fluid (air or water) velocity in the annulus between the drill-rod and the borehole wall decreases with increase of hole size. Ultimately a stage is reached when the rock debris cannot be effectively cleared from the hole.

The penetration rate then falls off due to "clogging" at the bit, until the drill eventually stalls.

Even for short-hole drilling the percussive technique is fundamentally inefficient, for the proportion of drilling time that is actually spent in rock penetration is quite small, being no more than around 1–3 ms per blow of the piston.

A much faster penetration rate is possible if the fracture process is made continuous, as it is aimed to do in rotary drilling provided the bit can be made strong enough to provide the necessary cutting edge and the cuttings are removed as fast as they are made.

Rotary Bits for Small Blast-holes

Rotary drills are used to make blast-holes up to about 50 mm in diameter in soft and medium-hard rocks, such as coal, shale, marls, some sandstones, lignite, salt, gypsum, potash, and weaker sediments in general. The evolution of bit design for such applications has largely followed empirical lines, and has resulted in a range of forms, as illustrated in Fig. 5.11. Larger holes, from 75 to 100 mm in diameter, are drilled in soft rocks by means of multiple bits which drill a pilot hole followed by reamers, sometimes in three stages.

The bit presents a continuous cutting edge or edges to the rock on a helical path, and on the analogy that this is similar to that presented by dragging a material on a straight path past the chisel edge of a cutting tool, the bits are termed "drag-bits". The cutting edges are comprised of tungsten carbide inserts.

After reviewing the characteristics of available drag-bits, Fish and Barker recommended that illustrated in Fig. 5.12 for use in carboniferous sediments. They found that both positive and negative rake-bits were liable to lose their cutting edge rapidly, due to chipping of the tungsten carbide. To provide the greatest possible support to the carbide insert a negative rake can be employed, although this reduces the effective cutting force for a given magnitude of thrust [24].

Performance Characteristics of Rotary Drag-bits

Several investigators have recorded the results of performance tests on rotary drag-bit drills. The results are usually portrayed in the form of a thrust–penetration curve, the

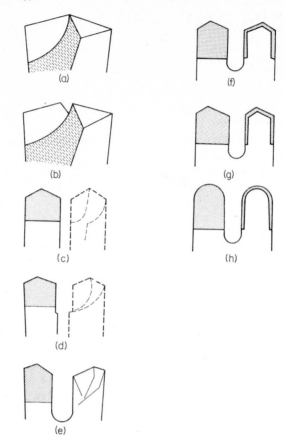

FIG. 5.11. Rotary drag-bits. (a) "D" or spade-bit; (b) notched centre-bit; (c) three-wing heavy-duty bit; (d) three-wing medium-duty bit; (e) U-bit; (f) concentric cutting edges; (g) eccentric cutting edges; (h) rounded cutting edges. (a) to (e) arranged as a series from hard rock to soft rock.

general form of which is linear between the limits of "clogging", which depends on debris clearance from the bit, and "chattering" when the thrust is too low to prevent excessive wear on the bit (Fig. 5.13).

Fish *et al.* [26] showed that the rate of bit wear is dependent not solely on the intrinsic abrasiveness of the rock, but also on the strength of the rock, because in a stronger rock a high level of thrust must be exerted to achieve penetration, and this increases the frictional force between bit and rock. Also, at higher thrusts the temperatures that are generated at the cutting edge are so high as to reduce the hardness of the sintered tungsten carbide and thus increase its susceptibility to wear.

The relationship between the machine torque and penetration shows less sensitivity to bit wear than does the thrust–penetration characteristic. This is attributed to the fact that the rotational shearing component is less sensitive than is the normal stress component to the increased thrust that wear on the bit makes necessary if penetration is to be maintained. The same effect is also displayed in the speed–thrust–penetration characteristics (Fig. 5.14). An increase in the speed of rotation of the drill reduces the thrust requirement, but at the same time it increases the rate of frictional wear on the bit. It also

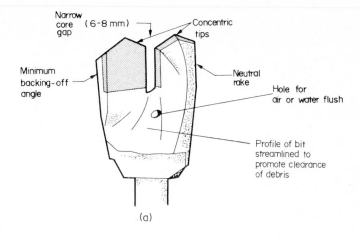

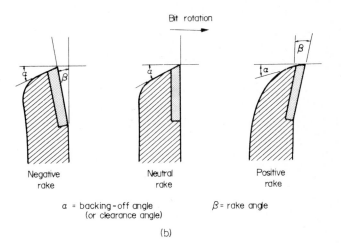

FIG. 5.12. (a) Desirable drag-bit features for soft to medium-hard rocks; (b) drag-bit rake and backing-off angles (Fish and Barker [24]).

reduces the maximum thrust that can be applied without stalling the drill. Thus the optimum drill speed is determined by the applied thrust and an economically acceptable bit-life.

Cutting Action in Drag-bit Drilling

The cutting action of a rotary drag-bit in rock is not an ideal continuous process, but is, to some extent, discontinuous. The sequence of events is pictured, both by Goodrich [27] and by Fish [24], in the following terms:

(1) Beginning the cycle immediately after the formation of a large fragment, elastic strain builds up due to angular deflection of the bit and torsional strain in the drill-rod (Fig. 5.15a).

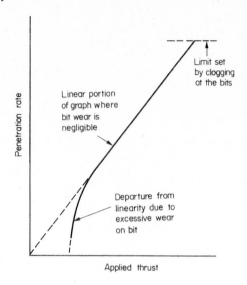

Linear portion
of graph where
bit wear is
negligible

Limit set
by clogging
at the bits

Departure from
linearity due to
excessive wear
on bit

Penetration rate

Applied thrust

Fig. 5.13. Basic thrust–penetration rate curve for rotary drag-bit drilling (Fish and Barker [24]).

(2) Strain energy is released, with consequent impact of the cutting edge against the clean rock surface, and comminution of rock fragments (Fig. 5.15b).
(3) Build-up of stress at the bit–rock contact, with further crushing and displacement of rock debris, until the cutting edge is effectively bearing on a step of unbroken rock which subsequently fails to create a large fragment or chip. This completes the cycle (Fig. 5.15c).

Experiments both by Fairhurst [18] and by Gray *et al.* [28] showed that the thrust and the torsional forces on a drag-bit undergo rapid oscillations due to the discontinuous nature of chip formation (Fig. 5.16). The thrust force goes through two or three oscillations as minor chips are formed and then it builds up to a higher peak just before the formation of a major chip, immediately after which the thrust falls, almost to zero force. The torsional force goes through similar oscillations, but at a lesser magnitude. The resultant cutting force thus oscillates between 30 and 90 degrees to the direction of rotation.

The nomenclature by which to describe rotary drag-bit drilling is shown in Fig. 5.17. During penetration a point on the bit advances along a helical path. The angle of inclination ω of this helix, for a point a at distance r from the centre of the bit, is

$$\arctan (p_v/2\pi r),$$

where p_v is the bit advance per revolution, i.e. penetration rate rotary speed.

Due to movement of the bit along the helix the effective clearance between bit and rock is reduced and the effective clearance angle $\varepsilon = \lambda - \omega$, where λ is the bit clearance angle.

For points near the centre of the bit the effective clearance is zero, so that the rock in this region is compressed under the bit. For this reason winged drag-bits with a central gap usually drill faster than spade- or centre-bits, because debris is more easily removed from the central gap.

There is some difference of opinion as to whether the major failure process in drag-bit drilling is by tension or by shear. In this connection a clear distinction should be made

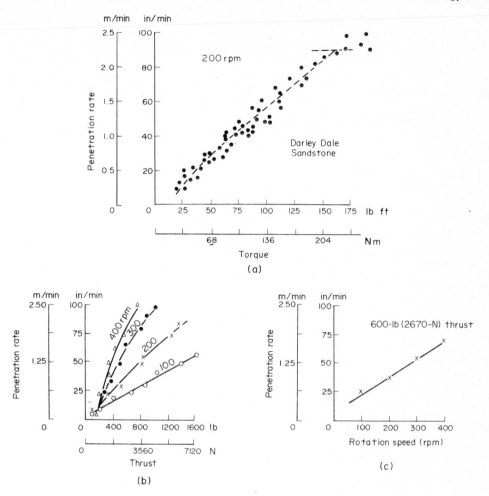

FIG. 5.14. Characteristic curves for rotary drag-bit drilling (Fish and Barker [24]).

between rotary drag-bit drilling and linear drag-bit operations such as rock ploughing and ripping. It is most unlikely that tensile failure is important in drag-bit drilling. The fact that the rock fragments are generally uniform in shape, even with widely different rocks, suggests that the line of fracture may be related to the plane of maximum shear stress. While the angle that this plane makes with the direction of the cutting force will be subject to variation, depending upon the internal friction characteristic of the rock, this variation will be small, even for a wide range of rock types. Also, tensile failure would be evidenced by the propagation of cracks in the direction of the cutting force, and this does not appear to happen (Fig. 5.18).

Gray *et al.* [28] observed that the depth of cut could advance beyond the tip of the cutting tool. Fractures were seen to be propagated below the cutting plane after initiation and then they curved upwards to the rock surface. This was taken to suggest that the initiation of failure was by tension fracture, which propagated into a combined tension–shear failure process.

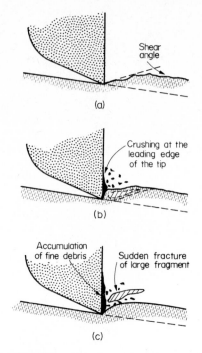

FIG. 5.15. Drag-bit cutting sequence (Fish and Barker [24]).

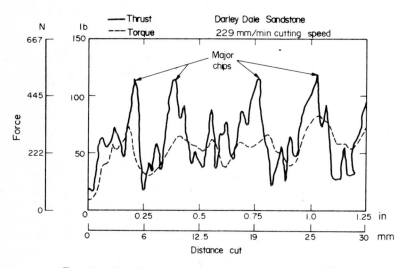

FIG. 5.16. Drag-bit force–displacement curves (Fairhurst [18]).

Rotary–Percussive Rock Drilling

The major disadvantage of percussive drills is their relatively low penetration rate, and the major demerit of rotary drills is the occurrence of excessive bit-wear at high values of thrust. On the other hand, the major advantages of the two systems are, in percussive

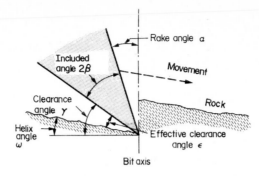

FIG. 5.17. Drag-bit nomenclature (Fairhurst [18]).

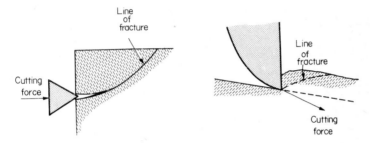

FIG. 5.18. The conditions postulated for linear drag-bit cutting in coal (Evans), and the conditions for major fracture in rotary drag-bit drilling (Gray *et al.* [28]).

drills the ability to penetrate hard rocks, and in rotary drills the feature of a near-continuous cutting process. The idea of designing a drilling machine which would combine the advantages of both systems and eliminate their major demerits has obvious attractions. Rotary–percussive rock drills have been designed with these objects in view.

Guppy [29] compares two typical machines which were introduced primarily for drilling underground blast-holes up to 50 mm in diameter. An early machine consisted essentially of a rotary air motor to drive the drill, combined with a percussive motor which delivered its impacts directly on to the drill-rod which passed right through the machine. The rotational drive was through gears which provided rotational speeds of 100, 200, and 300 rev/min.

An alternative arrangement had the percussive action superimposed on the rotary motion by two annular pistons which impacted on to a striker shaft, keyed to the main rotary shaft, and free to slide axially over it. In later machines hydraulic motors replaced the compressed-air units for rotation and thrust, while the percussive motor alone was actuated by compressed air. The interposition of the multistage gear-box was also eliminated in favour of a simpler single-speed gear, operating at 200 rev/min.

The Hausher machine, type DK7ES, applied an impact energy of 39J (29 ft-lb) at 6000 blows per min. Thrust was supplied by a feed screw arrangement driven by a 2-kw motor and automatically regulated by the penetration rate.

The design of bit that is employed with these machines reflects the compromise between percussive and rotary cutting requirements. The wedge angle is reduced from 70 to 80 degrees and is set asymmetrically to the central axis, with the bisector of the cutting wedge at an angle of 15–20 degrees to that axis.

Typical performance characteristics for such machines are reported by Inett [30], Guppy [29], Jahn [31], and Fairhurst [18]. The observed thrust–penetration relationship is exponential, and while at low values of thrust increasing the thrust gives a rapidly increasing penetration rate, a law of diminishing returns applies until eventually a point of stall is reached. At this point the torque required to maintain cutting is beyond the power of the machine to provide.

The penetration rate of the percussive machine is considerably exceeded by that which is made possible by utilizing the increased thrust at which the rotary–percussive machine operates. Rotary drilling is still superior in terms of maximum penetration rate, but rates of penetration that previously were attainable only at very high thrust on rotary drills are reached at much lower levels of thrust on the combined rotary–percussive machine.

Cutting Action in Rotary–Percussive Drilling

The rotary–percussive drilling process can be described as a continuous rotary cutting action upon which axial percussive blows are superimposed. The contact stress between bit and rock is thus momentarily increased, causing fracture at a lower value of rotary thrust and torque than that which is required in pure rotary drilling. Also, the rock directly beneath the tip of the bit, already elastically stressed by the rotary thrust, is rapidly subjected to high-intensity impact stress, leading to pulverization, fragmentation, and the development of subsurface microcracks, as in percussive drilling. Furthermore, since the bit is asymmetrical to the line of impact, the impact energy is focused towards the direction of rotation, thus adding to the stresses set up by the rotary action.

A clear distinction must be drawn between the role played by the rotary action in pure percussive drilling and that in rotary–percussive drilling. In the former case indexing takes place on the return (backwards) stroke of the piston. Any movement of rotation on this stroke will be relatively weak and does very little useful work towards rock fracture. Also, if indexing is not completely free of contact with the rock, at high values of thrust the machine begins to stall. The rotary action in rotary–percussive drills, on the other hand, makes a positive and powerful contribution towards rock fracture. Figure 5.19 is Patzold's interpretation of the characteristic cutting actions of the three drilling processes. Direct observation of the rock profile underneath the drill-bit in rotary–percussive

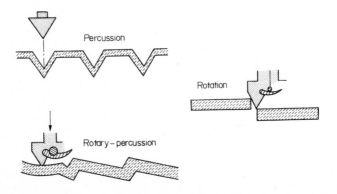

FIG. 5.19. Characteristic cutting actions in three rock-drilling systems (Patzold [32]).

drilling shows a relatively smooth work surface which must be determined greatly by the rotary cutting action which occurs between successive impact craters.

References and Bibliography: Chapter 5

1. MAURER, W. C., *Novel Drilling Techniques*, Pergamon Press, Oxford, 1968.
2. DONNELL, L. H., Longitudinal wave transmission and impact, *Trans. Am. Soc. Mech. Engrs* **52**, 153–67 (1930).
3. FISCHER, H. C., On longitudinal impact: I–III, *Appl. Sci. Res.* **A8**, 105–39, 278–308; **A9**, 9–42 (1959).
4. ARNDT, F. K., Untersuchung über die Energieübertragung beim Schlagvorgang im Blickfeld des schlagenden Bohrens, *Fortschr. Forsch. Geb. Bohr- u. -Schiebtechn.* **7**, 1–132 (1959).
5. HUSTRULID, W. A., and FAIRHURST, C., A theoretical and experimental study of the percussive drilling of rock, *Int. J. Rock Mech. Min. Sci.* **8**, 311–56 (1971).
6. LUNDBERG, B., Energy transmission in percussive rock destruction, *Int. J. Rock Mech. Min. Sci.* **10**, 381–436 (1973).
7. DUTTA, P. K., The determination of stress waveforms produced by percussive drill pistons of various geometric designs, *Int. J. Rock Mech. Min. Sci.* **5**, 501–18 (1968); and PhD thesis, Sheffield University, 1968.
8. FAIRHURST, C., Wave mechanics of percussive drilling, *Mine Quarry Engng*, pp. 122–78 (Mar./Apr. 1961).
9. HAWKES, I., and CHAKRAVARTY, P. K., Strain wave behaviour in percussive drill steels during drilling operations, *Mine Quarry Engng*, pp. 315–73 (July/Aug. 1961).
10. ROBERTS, A., HAWKES, I., and FURBY, J., Transmission of energy in percussive drilling, *Mine Quarry Engng*, pp. 447–58 (Oct. 1962).
11. HARTMAN, H. L., Basic studies of percussion drilling, *Trans. Am. Inst. Min. Engrs* **214**, 68–75 (1959).
12. FURBY, J., Strain wave behaviour in the percussive drilling process, *Trans. Inst. Min. Metall.* **73**, 393–413 (1964); and PhD thesis, University of Sheffield, 1964.
13. FURBY, J., Tests for rock drillability, *Mine Quarry Engng*, pp. 292–8 (July 1964).
14. TAKAOKA, S., HOGAMIZA, H., and MISAWA, A., On the reflection of elastic waves at a rod joint, *J. Min. Metall. Inst. Japan* **74** (834) 7–12 (1958).
15. LONG, V. H. F., An investigation aimed at improving the efficiency of percussive rock drilling, *J. S. African Inst. Min. Metall.* **66**, 276–96 (1966).
16. SIMON, R., Theory of rock drilling, *Proc. 6th Ann. Drilling Symp., Univ. of Minnesota (1956)*, pp. 1–14.
17. SIMON, R., Energy balance in rock drilling, *Proc. 1st Conf. on Drilling and Rock Mech., Univ. of Texas, January 1963*.
18. FAIRHURST, C., Discussion of J. Furby, Strain wave behaviour in the percussive drilling process, *Trans. Inst. Min. Metall.* **73**, 671–82 (1964).
19. SIMON, R., Transfer of the stress wave energy in the drill steel of a percussive drill to the rock, *Int. J. Rock Mech. Min. Sci.* **1**, 397–411 (1964).
20. DUTTA, P. K., A theory of percussive drill bit penetration, *Int. J. Rock Mech. Min. Sci.* **9**, 543–67 (1972).
21. FAIRHURST, C., and LACABANNE, W. D., Some principles and developments in hard rock drilling techniques, *6th Ann. Symp. on Drilling and Blasting, Univ. of Minnesota, 1956*, pp. 15–25.
22. FISH, B. G., Research in rock drilling and tunnelling, *Min. Elect. Mech. Engr*, pp. 1–13 (Feb. 1961).
23. McGREGOR, K., *The Drilling of Rock*, CR Books Ltd., London, 1967.
24. FISH, B. G., and BARKER, J. S., *Studies in Rotary Drilling*, National Coal Board, MRE Rep. No. 209, 1956.
25. FISH, B. G., *Studies in Rotary Drilling—The Basic Variables*, National Coal Board, MRE Rep. No. 2161, 1960.
26. FISH, B. G., GUPPY, G. A., and ROBIN, J. T., *Studies in Rotary Drilling—The Abrasive Wear Effect*, National Coal Board, MRE Rep. No. 2115, 1958.
27. GOODRICH, R. H., High pressure rotary drilling machines, *Proc. 2nd Ann. Symp. on Min. Res., Univ. of Missouri, 1956*, p. 25.
28. GRAY, K. E., ARMSTRONG, F., and GATLIN, C., Two dimensional study of rock breakage in drag bit drilling at atmospheric pressure, *J. Petroleum Tech.*, p. 93 (June 1962).
29. GUPPY, G. A., A laboratory study of two percussive–rotary drilling machines, *Mine Quarry Engng*, pp. 22–29 (1960).
30. INETT, E. W., A survey of rotary–percussive drilling, *Mine Quarry Engng*, pp. 2–8, 62–68, 106–10 (Jan., Feb., March 1957).
31. JAHN, R., Das Drehschlagbohren, *Glückauf*, pp. 1086–93 (11 Sept. 1954).
32. PATZOLD, F., La Rotation vibrée, *Rev. d'Ind. Minérale* **602**, 1058 (1953).
33. HARRISON, J. R., The response of rock to impact and abrasion, MEng thesis, Univ. of Sheffield, 1968.
34. HAWKES, I., and BURKS, J., Noise and vibration in percussive drill rods, *Int. J. Rock Mech. Min. Sci.* **16**, 363–76 (1979).

CHAPTER 6

Excavation Technology—II

Rotary Roller-bit Drilling

Roller-bit Drilling

The main limitation on rotary drilling with drag-bits is imposed by the hardness and abrasiveness of the rock being drilled. In all but the softer and weaker rocks the rate of bit-wear is rapid. This means that the diameter of the hole being drilled decreases with depth, and, in consequence in order to produce a hole of a required minimum size at depth, it must be started with a considerably greater diameter at its mouth.

While there are considerable advantages to be gained by blasting in wide holes, this is only so if the holes are of the same diameter throughout. A hole which tapers from a wide mouth to a narrow diameter at depth is wasteful of expenditure on drilling. It is also restrictive on the maximum weight of charge that can be placed, and hence is also inefficient for blasting purposes.

Roller-bits

The technique of deep-hole drilling was revolutionized by the introduction of the three-cone roller-bit, which is used extensively when drilling for oil and natural gas, where depths of penetration ranging from 5000 m to 7000 m are not unusual. These bits consist of a forging with a male-threaded portion to fit the hollow drill-rod. On the forging are mounted three steel cones that are carried, each apex-inward, on ball and roller bearings (Figs. 6.1B).

Each cone has three or four rows of hard-faced teeth. When the drill is rotated the cones roll against the rock interface, with the teeth on adjacent cones tracking on different radii, to attack the rock. However, all the cones carry a peripheral row of teeth that track at the full circumference of the hole. The effect is that although every point on the cutting structure contacts the rock only four times for every three bit-revolutions, all the cones have cutting elements working to remove the strongest part of the hole, which is the periphery at the rock face. Also, one cone carries a row of teeth centrally placed to form the "spear point" of the bit.

Application of Roller-bit Drills

The design of roller-bits has evolved empirically to suit different rock conditions. Bits for use on soft rocks have large teeth that are widely spaced and sharply pointed. The large

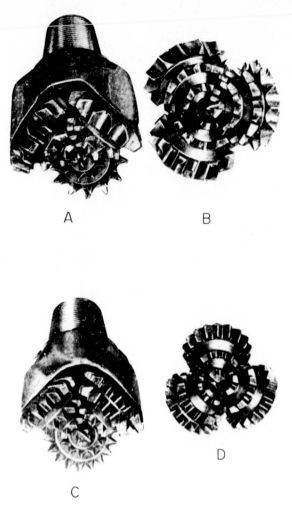

A   B

D

C

FIG. 6.1A. Rotary roller-bits for rock drilling. A and B, bits for soft rocks; C and D, bits for medium-hard rocks.

teeth spread the applied thrust over a larger area, and this is practicable because little thrust is needed to overcome the compressive strength of soft rocks. Large cuttings are produced, the bit is not subject to clogging, and the large teeth take a long time to wear down.

Bits for soft rock have the cones offset, and the cone axes do not intersect at a common focal point. When they rotate a drag-bit action is imparted to the teeth. This secondary shearing action aids penetration at low thrust, and it also helps to keep the hole straight.

Bits that are intended to drill in harder rocks have shorter, more closely spaced teeth, which permit of higher loading intensities. The cone offset is also reduced and may reach zero on bits designed for use in the hardest rocks. They may also have tungsten carbide inserts or buttons instead of teeth.

FIG. 6.1B. Rotary hard-rock roller-bit for rock drilling.

Performance Characteristics of Rotary Roller-bit Drills

In recent years roller-bit drills have largely replaced churn-drills in quarry blasting. Their application has made it possible to increase the diameter of the blast-holes from around 150–175 mm up to 300 mm or more, and this has had a marked effect on the economics of the operation, as indicated by productivity, powder factor, and overall cost (Figs. 6.2 A–D). Bauer quotes drill penetration rates averaging around 15 m per hour in hard taconite—three to four times that attainable in the same rock by using churn-drills [1–3].

In any form of rock drilling, good penetration performance depends on the ability of the system to clear away rock debris from the bit. This is done by pumping a fluid down the drill-rod so that the bit is flushed clean and the rock debris washed up the hole. When quarrying, the drill fluid is usually compressed air, for this keeps the hole dry, which helps blasting, and air does not freeze and thus block the hole, as would water in a cold climate. An air velocity of from 1200 to 1500 m/min is required to bring the rock debris up the hole.

In deep-hole drilling the flushing fluid is water, usually mixed with a proportion of bentonite to form a drilling mud. This provides a better means of transporting the rock cuttings than does air, and it also reduces the friction between the rotating drill-rods and the rock wall. The drilling mud also penetrates the pores and interstices of the wall, to produce a pressurized skin which helps to keep the hole intact in weak ground, pending the insertion of a casing to line the hole.

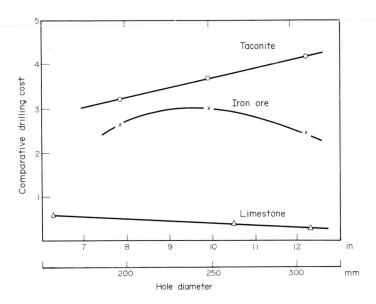

FIG. 6.2A. Comparative drilling costs versus hole diameter, in three rocks (Bauer [1]).

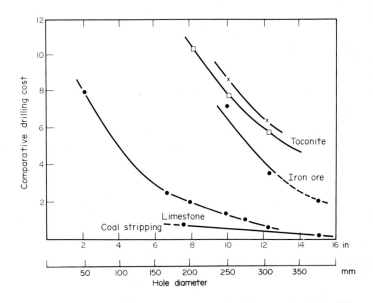

FIG. 6.2B. Comparative drilling costs per tonne, in different rocks (Bauer [1]).

By maintaining a pressure differential in the drilling mud, higher than the natural strata pressure at depth, movement of fluid and debris from the strata into the drill-hole is prevented during drilling. However, the skin is then a damaged zone which can collapse when drilling ceases if the casing is not strong enough to take the *in situ* strata pressures, which may increase due to creep of the rocks penetrated, for some considerable time after

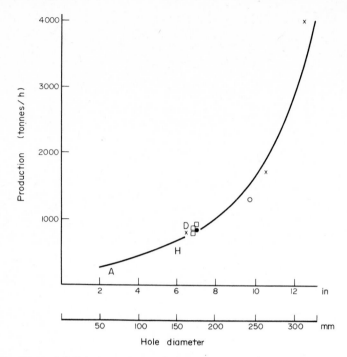

FIG. 6.2c. Tonnes drilled per hour in limestone for machines with different hole-diameter capacity (Bauer [2]).

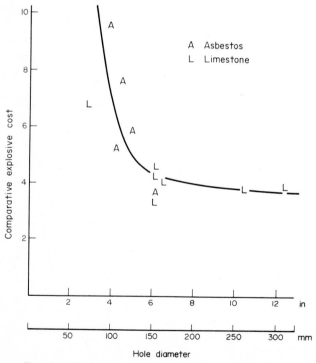

FIG. 6.2D. Explosives cost versus hole diameter (Bauer [2]).

Ej

in
spe
rela

whe
A
M
"per
betw
direc

Thi

derive
Hov
propor
can be
of the
and 250
116–130

Effect o

The pl
size, hol
velocity a
velocity l
margin, v
fluid prop
Chien p
velocity, a
[12].

The Rock-

Both bri
bits (Fig. 6.
the previou
terms of wea
fragments p
such as tootl
the surface b
Gnirk obs

the hole has been completed. The drill mud-strata pressure differential also has marked influence on the penetration performance of the roller-bit during drilling.

Murray and Cunningham report that increasing the confining the pressure of the mud column decreased the penetration rate by as much as 90 % in some cases [3].

Effects of Thrust

The general effect of thrust on the penetration rate of roller-bit drills is shown in Fig. 6.3. Gatlin [4] observed two distinct drilling regions:

(1) Drilling at bit loads below the compressive strength of the rock.
(2) Drilling at bit loads above this value.

Thrusts in regime (2) are applied wherever possible. The principle is illustrated in Fig. 6.4, which shows the increase in rate of penetration and the depth attained when sufficient thrust is applied to overcome the compressive strength of the rock.

For each rock type Gatlin approximated the relationship above the critical thrust as linear, and

$$R_p = a + bW,$$

where R_p is the instantaneous penetration rate, W is the thrust on the bit, a, b are the

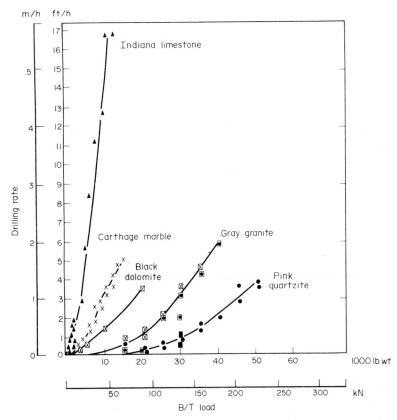

F𝐼G. 6.3. The effect of bit loading on penetration rate of roller-bits in various rocks (Gatlin [4]).

Fig. 6.5. Idealized roller-cone-bit (Cheatham [17]).

tooth forced into dry limestones and sandstones at confining pressures ranging from zero up to 52 MPa (7500 psi). His results are pictorialized in Fig. 6.6 for indexing distances of 6 mm and 12 mm, respectively, between adjacent craters.

At confining pressures below the brittle-to-ductile transition pressure, chips are formed and break free from the rock face between successive penetrations for an indexing distance which may be as great as 5 times the penetration depth. At confining pressures above the transition pressure the rock flows into the previous penetration crater when the ratio of the penetration depth to the indexing distance is of the order of 2.0–3.0. These ratios decrease somewhat for lower bit-tooth angles, and they increase for larger angles, both above and below the transition pressure. A combination of both brittle and ductile chip formation is observed at the transition pressure [14].

In a photoelastic model study of the stress distribution in a brittle material, resulting from the load applied by a sharp bit-tooth at a given distance from a simulated penetration crater, Garner shows a curved plane across which maximum tensile stress

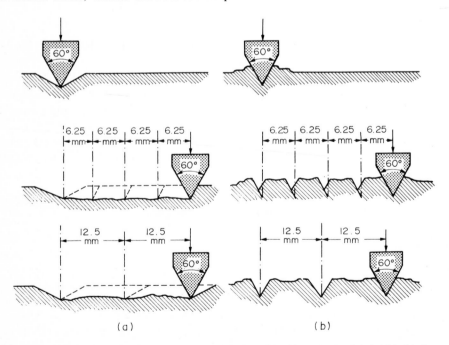

(a) (b)

Fig. 6.6. Features of chip formation at various indexing distances: (a) brittle; (b) ductile (Gnirk [14]).

interc

type,

Oth

practi

factors

remova

116

118

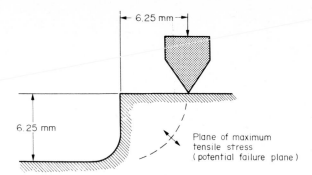

FIG. 6.7. The potential plane of failure (Garner [15]).

exists extending from near the rock–bit contact towards the bottom of the crater. This may be considered to be a potential failure plane (Fig. 6.7) [15].

Cheatham applies plasticity analysis for a solution of the problem of indexed loadings by a simulated dull bit-tooth on rock behaving in a ductile manner. Assuming that the dull bit-tooth can be effectively represented by a two-dimensional flat punch, an upper bound on the bit load that is required for the initiation of plastic flow in the rock can be obtained from the slip-line field shown in Fig. 6.8. The magnitude of load on the bit-tooth, required to overcome the power of dissipation on the extended linear discontinuity, is added to the load corresponding to the slip-line field without extension (Fig. 6.9). The hypothesis is advanced that slip will follow the path that requires the least power of dissipation or minimum work.

Thus all points on a given dissipation curve are equally probable slip paths to a free surface. Cheatham observed good agreement between a slip-line field, calculated by this method, and an observed failure pattern in damp sand loaded by a flat punch near a simulated penetration crater [16, 17].

Simplified bit-tooth penetration experiments by several investigators appear to indicate that the transition from brittle to ductile chip formation occurs in numerous limestones and sandstones at differential pressures that are sufficiently low to be of relevance in oil- and gas-well drilling and in hard-rock tunnelling. The force–displacement curves for sharp bit-tooth penetration, below the transition pressure for a particular rock, reflect crack and chip formation under the tooth, showing abrupt discontinuities and changes in slope. These curves are linear above the transition pressure and they are in agreement with penetration models proposed on the basis of plasticity theory.

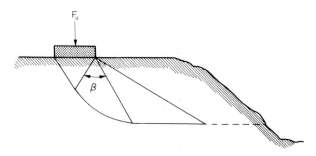

FIG. 6.8. Plastic slip-line field (Cheatham [17]).

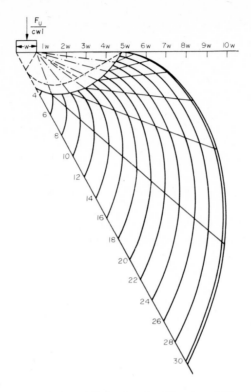

FIG. 6.9. Power dissipation plot (Cheatham [17]).

For multiple bit-tooth penetration above the transition pressure, interaction takes place with adjacent craters at finite depths of penetration. For relatively small indexing distances the rock may actually flow into an adjacent crater without loss of cohesion from the mass. In general, the optimum indexing distance between successive penetrations increases with an increase in the bit-tooth angle and with a decrease in the differential pressure [18].

Diamond Drilling

Rotary diamond-tipped drills are widely used for deep holes in hard rock. Hollow-walled, cylindrical bits are applied to extract core samples in geological site investigations and when prospecting for minerals and proving ore reserves. If solid cores are not required, a non-coring diamond bit will penetrate the hardest rock.

There are two main types of diamond-bit—impregnated bits and surface-set bits. An impregnated bit has a sintered, powder metal, matrix, with fragmented bortz or else whole diamonds of selected screen sizes uniformly distributed throughout the entire crown section. It is suitable for penetrating hard, homogeneous rock of fine and even grain, and should be rotated at a relatively high speed. As the matrix wears away, new sharp diamond points are exposed, but in some rocks it is necessary to sandblast the cutting surface periodically to produce new cutting edges. Once the impregnated portion has been worn away, the bit has no further value.

Surface-set bits are most commonly used in diamond-drilling operations in rock. They have stones individually set into the matrix. The sharper the edges and points which project from the matrix, the more rapid will be the penetration performance. However, as diamonds are brittle the very sharp edges rapdily wear away due to fracture and graphitization caused by localized overheating. For this reason excessive pressure is not applied to a new bit. As flat surfaces develop on each stone additional pressure can be applied in order to maintain a constant penetration speed. Most surface-set bits today have a hard-metal matrix.

Diamond Drag-bit

Diamonds may be added to the peripheral surface of a drag-bit for deep-hole drilling in hard rock. Besides providing the bit with some ability to scour the hole laterally, as an aid to controlling any deviation that may have occurred, the design is aimed to combine the cutting ability of the drag-bit with the resistance to lateral wear that is possessed by the diamonds. A high penetration rate, combined with the maintenance of hole diameter and a useful bit-life, are the objects to be attained [19].

Performance Characteristics of Diamond Drills

A diamond drill will penetrate any rock in time, but because of the high cost of bits and the need to minimize bit-wear and loss of diamonds, the operation requires experience, judgment, and skill if it is to be conducted economically. Needless to say, these qualities once having been gained by trial and error, are highly prized and jealously guarded. Very little fundamental research on the technique has been published.

Such research as has been devoted to the subject has been directed mainly towards methods of setting and securing diamonds at optimum cutting angles in the bit. However, Paone and Madson determined thrust–penetration characteristics at rotary speeds ranging from 500 to 1100 rev/min in three rock types, as illustrated in Fig. 6.10 [21].

Rock Penetration by Diamond Drill-bit

The literature on diamond drilling often includes such terms as "grinding", "wearing", "cutting", "breaking", "shearing", "scraping", "melting", and "chipping", but these terms are seldom defined. As with rotary roller-bit drills, it is probable that both brittle and plastic yield processes are operative.

Grodzinski described the cutting action of a single diamond as "breaking out chips of the material. Brittle materials break as small separate chips, and tough materials give a continuous chip" [22].

Derby described diamond drilling in the following terms: "When diamonds are forced into the formation and rotated, they either break the bond which holds the rock particles together, or they cause conchoidal fractures of the rock itself. The former action occurs when drilling in sandstones, siltstones, shales, etc., and the latter action when drilling in chert, flint, or quartz" [23].

Pfleider and Blake explain the cutting action in diamond drilling thus: "The thrust on the bit causes diamonds to penetrate the rock and minerals, by a compressive force. This force sets up internal stresses. In brittle minerals or rocks, these stresses are not relieved readily by strain, but rather by the formation of tension or shear fractures. The rotation of the bit produces both impact and shearing forces. The first stage of an elastic impact is

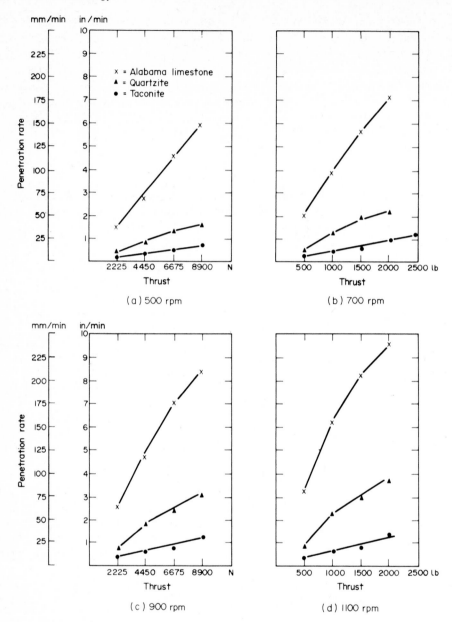

FIG. 6.10. Thrust–penetration curves for diamond-core drills (Paone and Madson [21]).

compression; the second stage an elastic restoring force. Both stages may be relieved by the development of tension and/or shear fractures. The shearing force, likewise, may cause these fractures."

The same authors recommend the use of small diamonds, with slight protrusion, for drilling hard, dense rocks, while larger diamonds, well exposed and with ample water passages, should be used in bits designed to drill soft rocks [24].

Paone and Madson observed rock failure under a diamond point to be due to:

(1) Surface failure resulting from the vertical or the normal force which causes plastic deformation at the contact surfaces and produces failure; this is the so-called crushing action.

(2) Surface failure due to the ploughing action of the diamonds which shears-off the rock surface by a tangential force.

The reaction of the rock to these forces includes the bearing resistance acting against the applied vertical force, and the frictional resistance between bit and rock, acting against the applied torque. Design of the impregnated bit does not permit the occurrence of the high stress concentrations that are found under the diamond points of a surface-set bit.

The small diamond chips in the impregnated bit do not protrude from the matrix as they do on a surface-set bit, and therefore, at any given time an indeterminable amount of matrix material will also be in contact with the rock surface. This matrix must be worn away in order to expose new diamonds. Because of the greater area of contact between bit and rock surface the crushing action is reduced, and the frictional resistance acting against the torque is most pronounced [21].

References and Bibliography: Chapter 6

1. BAUER, G. W., Trends in surface drilling and blasting, *Proc. CIM 70th AGM, Vancouver, April 1968.*
2. BAUER, G. W., Rotary bits in taconite and other hard rocks, *6th Int. Symp. on Drilling and Blasting, Univ. of Minnesota, 1958.*
3. MURRAY, A. J., and CUNNINGHAM, J., Effect of mud column pressure on drilling rate, *Trans. Am. Inst. Min. Engrs Petroleum Div.* **204**, 196 (1955).
4. GATLIN, C., *Petroleum Engineering*, Prentice-Hall, London, 1960.
5. MUIRHEAD, I. R., and GLOSSOP, L. G., Hard rock tunnelling machines, *Trans. Inst. Min. Metall.*, Section A, **77** (734) 1–21 (1968).
6. GATLIN, C., Rotary drilling rate, *Oil Gas J.* **55** (20) 193–8 (1957).
7. BIELSTEIN, W. I., and CANNON, G. E., Factors effecting the rate of penetration of rock bits, *API Drilling and Production Practices*, p. 61, 1950.
8. SPEER, J. W., Drilling time reduced 31 per cent, *Oil Gas J.*, p. 130 (11 Oct. 1954).
9. ECKEL, J. R., Effect of mud properties on drilling rate, *API Drilling and Production Practices*, p. 119, 1954.
10. WARDROUP, W. R., and CANNON, G. E., Some factors contributing to increased drilling rate, *Oil Gas J.* (30 Apr. 1956).
11. MAURER, W. C., The perfect cleaning theory of rotary drilling, *J. Petrol. Tech.*, pp. 1270–4 (1962).
12. CHIEN, S. F., Annular velocity for rotary drilling operations, *Int. J. Rock Mech. Min. Sci.* **9**, 403–15 (1972).
13. BAUER, A., and CALDER, P., Open-pit drilling—factors influencing drilling rates, *4th Canadian Symp. on Rock Mechs., CIM AGM, Ottawa, March 1967.*
14. GNIRK, P. F., An experimental study of indexed single bit tooth penetration into dry rock, *Proc. Ist Congr. Int. Soc. Rock Mech., Lisbon, 1966,* **2**, 121–9.
15. GARNER, N. E., The photoelastic determination of the stress distribution caused by a bit tooth on an indexed surface, MSc thesis, Univ. of Texas, 1961.
16. CHEATHAM, J. B., and PITTMAN, R. W., Plastic limit analysis applied to a simplified drilling problem, *Proc. 1st Congr. Int. Soc. Rock Mech., Lisbon, 1966,* **2**, 93–97.
17. CHEATHAM, J. B., Indexing analysis for plastic rock, *Trans. Am. Inst. Min. Engrs* **232**, 316–21 (1965).
18. HARRISON, J. W., The response of rocks to attack by cutting and abrasion, MEng thesis, Univ. of Sheffield, 1968.
19. PETERS, R. I., A new approach to bit design—the diamond drag bit, *Colo. Sch. Mines Q.* **58**, 153 (1963).
20. SASAKI, YAMATADO, SHIOHARA, and TOBE, Investigations of diamond core bit boring, *Indust. Diamond Rev,* **22** (259) (June 1962).
21. PAONE, J., and MADSON, D., *Drillability Studies*, US Bur. Min., Rept. Invest. No. 6776, 1966.
22. GRODZINSKI, P., *Diamond Tools*, Anton Smith & Co., London.
23. DERBY, C., Diamond core drilling methods and problems, *Petrol Engr* **18**, 98–99 (June 1947).
24. PFLEIDER, E. P., and BLAKE, R. L., Research on the cutting action of the diamond drill bit, *Min. Engr* 157 (Feb. 1953).
25. EVANS, I., Relative efficiency of picks and discs for cutting rock, *Proc. 3rd Congr. Int. Soc. Rock Mechs, Denver, Colo., 1974,* **2B**, 1395–9.

CHAPTER 7

Excavation Technology—III

Rock Ripping and Cutting

Rock-cutting Machines

Cutting machines have been used for many years in relatively soft rocks such as coal and rock-salt, and their use is being extended to harder rocks such as limestones and even to quartzite [1]. The mechanical construction of the conventional coal cutter is akin to that of the familiar chain saw, in principle, in that the rock is attacked by a number of picks, mounted on a chain and carried around a tracked jib.

The chain is caused to move at high velocity so that the jib can slice into the rock. The cutting action of each pick is essentially similar to that of the drag-bit, both wedge penetration and impact forces being operative.

Other machines employing drag-bit mechanical attack include shearers, rippers, and rock-ploughs, in which the size and mass of the penetrating tool is increased while the velocity of attack in the machines decreases in the order given (Fig. 7.1).

Chain Cutters

The performance characteristics of chain-pick machines have been extensively studied in connection with coal cutting. The characteristic nomenclature for such a pick is as shown in Fig. 7.2. Rake angles range from 3 to 55 degrees, and clearance angles from zero to 40 degrees.

When the picks are sharp the required cutting force decreases with increase of the rake angle. The force generated perpendicular to the line of attack also varies with the rake angle. It falls to zero at 20 degrees rake, and may become negative for larger angles, so that the pick tends to pull towards the rock. A minimum clearance angle of about 5 degrees is required to eliminate friction on the trailing edge.

While a rake angle of around 30 degrees would provide excellent cutting ability, this would leave too narrow a wedge angle to give acceptable resistance to wear. A wedge angle of around 70 degrees is therefore more commonly used.

A broad, flat-faced, chisel aspect is, in general, more efficient than narrow or pointed picks, in terms of the energy required to break unit volume of rock [2, 3].

The picks are so arranged on the chain that successive picks track on different lines of attack. The depth of cut should be about half the tracking interval between pick lines to provide maximum break. The distance between pick lines should be about 3–4 times the pick width, for chisel-shaped picks [32].

124

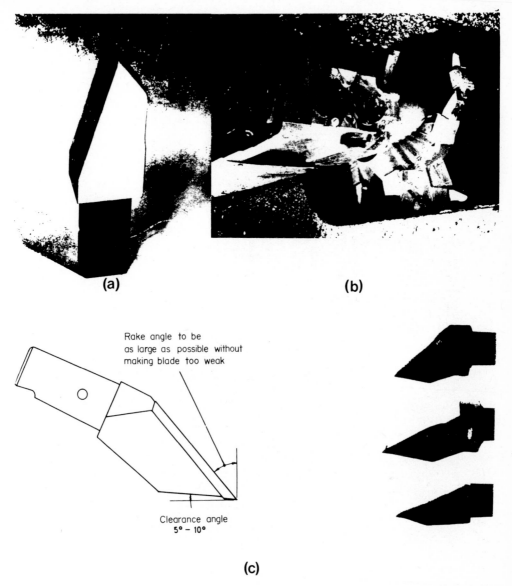

(a)

(b)

Rake angle to be
as large as possible without
making blade too weak

Clearance angle
5° – 10°

(c)

FIG. 7.1. (a) Chisel-shaped pick for rock-ripping machine; (b) coal-shearing machine fitted with rotating pick-drum; (c) rock plough blades.

The depth of penetration should be large enough to generate a torsion crack which can run to the rock surface and so split-off the largest possible fragment (Fig. 7.3). If the penetration depth is too short such a tension fracture is not generated, and energy is wasted in inefficient crushing and comminution. Ridges of rock build-up between adjacent pick lines, and as the picks track in the bottom of the deepening groove there is a progressive increase in the required cutting force. The appearance of the rock face then displays "coring", as shown in Fig. 7.4. The depth of cut should therefore not fall below one-third of the pick-line spacing (Fig. 7.5).

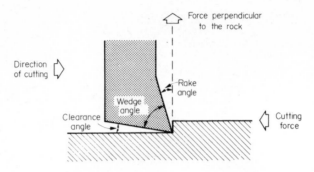

FIG. 7.2. Rock-pick angles and forces.

FIG. 7.3. Wedging action of a rock-pick (Evans [4]).

Disc Rock-cutters

Attempts to scale up the mechanics of impact and crushing to large-diameter tunnel-boring machines have met with some success, particularly in the weaker and medium-strength sedimentary rocks. But when applied to harder rocks, high-speed picks prove to be too fragile, and their cutting action too localized, to permit simultaneous excavation over a wide cross-sectional area at an economic cost in terms of power requirements and machine maintenance. And although roller-bits are more easily applied to tunnel boring, their penetration rate is slow, while bit-wear is rapid at the high values of thrust required.

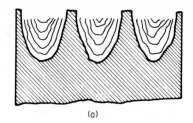

(a)

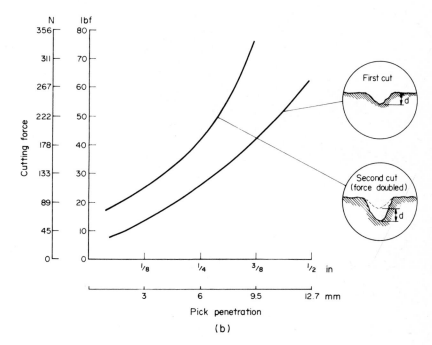

Pick penetration

(b)

FIG. 7.4. (a) Coring caused by successive shallow cuts by rock-picks on parallel cutting tracks; (b) effect of coring on the required cutting force (Evans [4]).

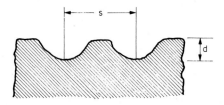

FIG. 7.5. Relationship between depth of cut and line spacing for maximum benefit (Evans [4]). Depth of cut, $d = s/2$ for best cutting (d should not be less than $s/3$).

In the search for alternatives, wheel or disc cutters have been introduced with success in rocks of 138 MPa (20 000 psi) compressive strength, and it appears that discs may replace roller-bits for this type of operation.

A typical disc cutter is shown in Fig. 7.6. When discs are used on the cutting head of a machine they are mounted so that the periphery of each disc is thrust against the rock face.

127

FIG. 7.6. Disc rock-cutter.

As the head is rotated the alloy steel disc revolves and its edge, which is armoured with tungsten carbide, cuts narrow grooves into the rock face (Fig. 7.7). The ridges of rock left between the grooves are then broken away by the wedge action of the disc periphery.

Discs have several advantages over picks. Since they tend to roll against the rock surface

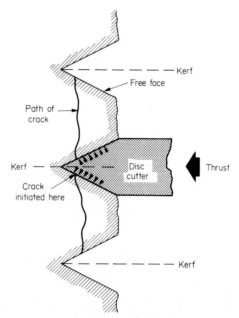

FIG. 7.7. Rock disc-cutting action.

the relative speed of disc and rock at the point of attack is quite small, whereas a pick tends to burst its way through the rock at a relatively high speed, causing intense local heating. After it has cut into the rock, each part of the disc perimeter rolls out of contact with the rock and is cooled by the ambient air stream. Moreover, due to its large surface area, a disc is better able to dissipate the heat generated by cutting. These factors give the disc a much greater wear resistance and prolong its useful life [2].

Rock-ploughs

Massive drag-bits, operated at slow speed, are employed as rock-ploughs in underground mining and as ripping tools with which to attack the bedrock in surface excavations and in earth-moving operations. Evans's theory of tensile fracture by wedge penetration is considered to be generally applicable to this process in weak to medium-strong rocks, but less applicable to stronger materials.

Referring to Fig. 7.8a and b, a tensile fracture is assumed to occur along a circular path *ab* to which the wedge movement is tangential at the point *a*. The resultant force *R* then acts at an angle $(\frac{1}{2}\pi - \varphi)$ with the surface of the wedge, where φ is the angle of friction between the wedge and the rock.

The resultant *T* of the tensile forces acting perpendicular to *ab* passes through point *O*, and a force *S* acts through point *b*, in the nature of a "hinge reaction". Using moments and the minimum work hypothesis, Evans determined that the force *F*, required to form a chip, is equal to

$$F = 2 S_t d \sin (\beta + \varphi')/1 - \sin (\beta + \varphi'),$$

where S_t is the tensile strength of the rock, d is the cutting depth, and β is half the wedge angle.

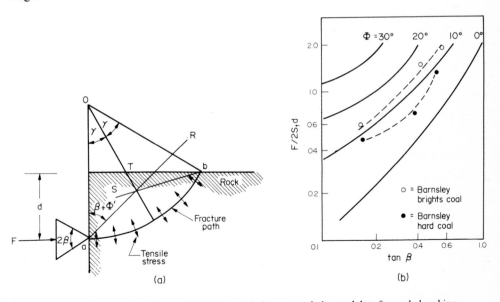

FIG. 7.8. (a) Coal-ploughing model; (b) theoretical curves and observed data for coal ploughing (Evans [3]).

Wedge Penetration in Rock Machining

It will now be apparent to the reader that all the mechanical methods of attacking rock embody the principle of penetration by indentation from a wedge or wedges. Sometimes this is accompanied by impact, sometimes by a continuous thrust, and sometimes by both impact and thrust together. However, there is some disagreement of opinion as to the manner in which the wedge acts in penetrating rocks as between hard and soft materials.

For coal, Evans and Murrel describe the process in terms of the model (Fig. 7.9). This assumes that the contact pressure on the wedge is equal to the unconfined compressive strength of the rock, and the model predicts that

$$F = 2\,bhS_0\,(f + \tan\beta),$$

where F is the force on the wedge, b is the wedge contact length, h is the depth of penetration, f is the coefficient of friction, wedge to rock, β is half the wedge angle, and S_0 is the unconfined compressive strength of the rock [5].

Theoretically, and experimentally, according to Evans and Murrel, the force required to push the wedge into the rock increases rapidly with increase in the coefficient of friction and increase of the wedge angle (Fig. 7.10).

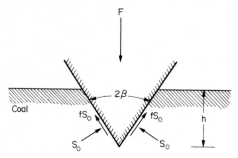

FIG. 7.9. Model for wedge penetration into coal (Evans and Murrel [5]).

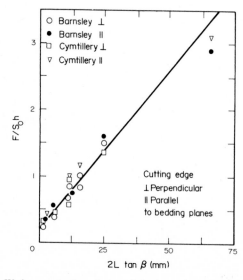

FIG. 7.10. Wedge-penetration characteristics in coal (Evans and Murrel [5]).

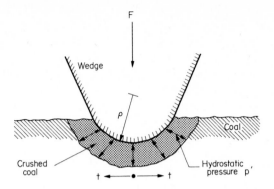

FIG. 7.11. Blunt wedge-penetration model (Dalziel and Davies [6]).

Dalziel and Davies [6] observed that the force required to produce fracture in coal increases as the radius of the wedge-tip raised by an exponent of approximately 0.5. They suggest that a layer of crushed coal beneath the wedge exerts a uniform pressure p' on the coal, producing a tensile stress concentration t at the tip of the wedge (Fig. 7.11). They show that this tensile stress should vary as, approximately,

$$t \propto p'/\sqrt{\rho}$$

where ρ is the radius of the tip.

The hydrostatic pressure should be proportional to the force on the wedge ($p' \propto F_f$) so if failure occurs when t exceeds the tensile strength of the rock, the fracture force should vary as

$$F_f \propto \sqrt{\rho}.$$

While the wedge-penetration theory for coal predicts tension failure it is often observed in harder rocks that the fractures appear to change from shear failure near the crushed zone to tensile failure at the rock surface. This is particularly so in deep-hole drilling and when boring into rock at high pressure, when craters are formed primarily by shear failure.

TABLE 7.1. *Transition pressures. Transition from brittle to ductile behaviour during single, sharp bit-tooth penetration under static load (Gnirk and Cheatham [7])*

Rock	Transition pressure (MPa)
Salt	1.7– 3.4
Indiana limestone	3.4– 5.2
Carthage marble (limestone)	5.2–10.3
Berea sandstone	13.8–17.2
Ohio sandstone	17.2
Virginia greenstone (schist)	17.2–20.7
Danby marble	27.6–34.5
Kasota dolomite	34.5–41.4
Hasmark dolomite	41.1

Gnirk and Cheatham, when investigating wedge penetration into a number of dry rocks at varying confining pressures found that, in general, the penetration mechanism is of a brittle character at low confining pressures and of a ductile nature at high confining pressures, with a transition from predominantly brittle to predominantly ductile behaviour at intermediate confining pressures. Transition pressures, observed for a number of rocks, are detailed in Table 7.1 [7].

The force–displacement characteristic is linear when the mode of penetration is ductile or plastic but is discontinuous and stepped when chips are broken off the face of the rock in a brittle manner (Fig. 7.12). A force–displacement curve, which exhibits only a single gradual discontinuity, with the same slope before and after the break, is representative of a transition between predominantly brittle and predominantly ductile behaviour (Fig. 7.13). For this situation the linear portions of the curves correspond to ductile behaviour and the single discontinuity to the formation of an essentially vertical crack or fracture near and below the wedge apex [8].

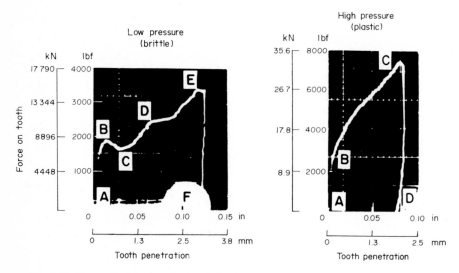

FIG. 7.12. Force–displacement characteristics for wedge penetration into rock (Maurer [8]).

Rock–Wedge Interaction in the Brittle Regime

For confining pressures below the brittle to ductile transition, Reichmuth suggests that the relationship between the half wedge angle β and the coefficient of friction μ at the wedge–rock interface is

$$\beta = \arc\tan\frac{\pi - 2\mu}{2 + \pi\mu}.$$

If β is numerically greater than the expression on the right-hand side of the equation, the stress distribution is effectively localized on a region below the penetrating wedge, thereby causing considerable crushing and compaction of the rock (Fig. 7.14). Conversely, if β is numerically less, then fractures develop close to the rock surface, and these result in the formation of chips, with comparatively little crushing or compaction [9].

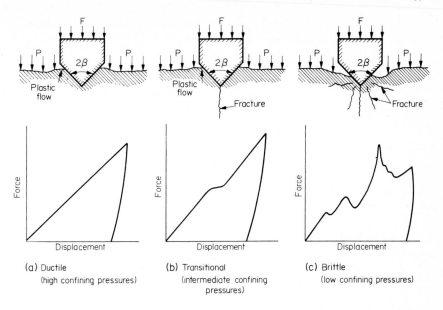

FIG. 7.13. Modes of rock failure and related force–displacement characteristics (Gnirk and Cheatham [7]).

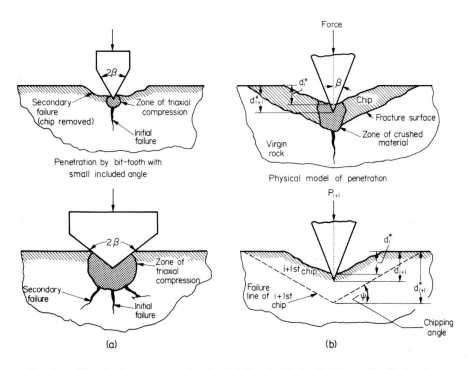

FIG. 7.14. Models of wedge-penetration into brittle rock; (a) after Reichmuth [9]; (b) after Paul and Sikarskie [11]).

Paul and Sikarskie postulate that failure of the rock is governed by the Mohr–Coulomb criterion, and they predict a cessation of brittle chip formation when

$$\beta > \tfrac{1}{2}\pi - \varphi,$$

where φ is the angle of internal friction of the rock. Since most rocks exhibit an angle of internal friction of approximately 30 degrees, this implies that the formation of chips will cease and only crushing will occur for total wedge angles around 75–90 degrees for several rock types.

These investigators further assume that brittle fracture will occur along planes extending from the wedge apex to the free rock surface and that stress equilibrium and the linear Mohr–Coulomb failure criterion are satisfied everywhere on these planes. This analysis has been extended by Benjumea and Sikarskie to explain the effects of bedding planes and non-isotropy on wedge penetration. Their experimental wedge-penetration tests on Indiana limestone showed a much greater difference in specific energy for wedges penetrating perpendicular and parallel to the bedding planes than that which is displayed by material properties in the corresponding directions [10, 11].

Elasticity theory has been applied to predict the force–displacement characteristics in wedge penetration. According to Maurer, the threshold force, i.e. the wedge force required to initiate rock failure, is proportional to the wedge width or the rock–wedge contact area for simulated borehole conditions. For granite and basalt at normal atmospheric conditions the pressure at the rock–wedge interface must be in the order of 3450–4140 MPa (500000–600000 psi) for brittle failure to occur.

Rock–Wedge Interaction in the Ductile Regime

Cheatham has applied plasticity theory to provide a model for sharp wedge penetration into rock at pressures above the transition pressure. In his analysis the rock is assumed to be isotropic and homogeneous, and to yield when the principal stresses satisfy Mohr's theory of failure. The effect of friction between the wedge and the rock was taken into account using the slip-line field of Fig. 7.15, and failure was assumed to occur when the Mohr circle became tangent to a parabolic Mohr failure envelope.

For a constant confining pressure and wedge angle, Cheatham's equations predict a linear increase in force on the wedge with increasing depth of penetration, i.e. a linear force–displacement characteristic.

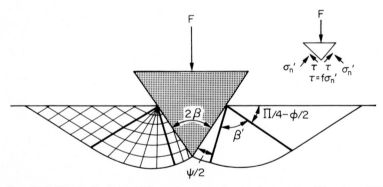

FIG. 7.15. Plastic slip-line field for rock penetration by sharp wedge (Cheatham [12])

Comparisons between calculated values of dimensionless force for a sharp wedge, on the basis of this model and experimental observations on several rocks, are in general agreement at confining pressures above the transition pressures. In general the comparisons are satisfactory for various limestones at confining pressures as low as 7 MPa (1000 psi). An example of the correspondence between calculated and experimental values of dimensionless force on a sharp wedge is shown in Fig. 7.16, for Carthage marble at a confining pressure of 70 MPa (10 000 psi). The calculated values are based on a combined linear–parabolic failure envelope for the rock and various coefficients of sliding friction at the rock–wedge interface.

Plasticity analysis has also been applied to the problem of penetration by a dull wedge into a material that obeys the linear Mohr–Coulomb criterion for yield, as shown by Cheatham. For this idealized situation the force resisting the downward penetration of the wedge is a linear combination of a constant force acting against the wedge flat, and a force on the sloping side which increases linearly with the depth of penetration (Fig. 7.17) [12–14].

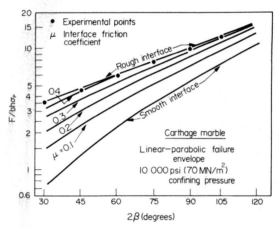

FIG. 7.16. Dimensionless force vs. bit-tooth angle for wedge penetration into marble (Gnirk and Cheatham [7]).

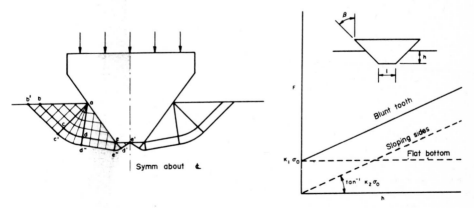

FIG. 7.17. Plastic slip-line field and force–displacement curve for a blunt bit tooth (Cheatham [12]).

Nishimatsu also examines the failure process in rock cutting in terms of a linear Mohr–Coulomb criterion, and assuming that shear strength governs the formation of a rock-chip. He derives equations for the normal and the tangential components of the resultant stress on an assumed line of failure, and shows that

$$F = \frac{2}{n+1} S_s t \frac{\cos k}{1 - \sin(k - a + \varphi)},$$

where F is the resultant force per unit length of tool, n is the stress distribution factor, S_s is the shear strength of the rock, t is the depth of cut, a is the rake angle of the cutting tool, k is the angle of internal friction of the rock, and φ is the angle of friction between bit and rock. The stress distribution factor is a constant associated with the state of stress in the rock-cutting process, determined primarily by the rake angle of the cutting tool, and it is deduced from the force–penetration characteristic [18].

Effect of Rate of Loading

The effects of loading rates have been studied by Maurer [8], Garner *et al.* [16], and Podio and Gray [19]. Since the craters underneath roller-bit teeth under load are formed in about 0.01–0.1 s, Maurer conducted "static" tests in the course of which craters were driven into Indiana limestone in 0.1–5.0 s. In other tests impact craters were driven in 0.001–0.005 s.

The force–displacement curves for the static and impact tests with differential pressure were similar. However, at a given pressure the threshold force was greater for impact loading, as shown in Fig. 7.18. The brittle–ductile transition pressure occurred at higher differential pressures in the static tests than in the impact tests. The force–displacement curves for impact tests with hydrostatic fluid pressure were nearly the same as those for impact under differential pressure. In the tests performed by Garner *et al.* an impact velocity of 2–3 m/s had little effect at atmospheric pressure, but at high differential pressures there was an appreciable increase in crater volume with high impact velocity.

Rock Cutting by Impact

Rock cutting by impact in the brittle regime below the transition pressure is of particular importance in coal and salt mining. Evans's wedge-penetration theory assuming tensile failure is usually applied, and this has been extended by Evans and Pomeroy to consider the breaking of coal by an impact blow. It is assumed that the force required to drive a symmetrical wedge into the rock is a function of the bearing area of the blade, and is given by

$$F \propto 2h \tan \theta,$$

where h is the depth of penetration and θ is half the blade angle.

If the resistance to penetration equals the compressive strength of the rock S_c, then

$$F = 2h \, S_c \, w \tan \theta,$$

where w is the width of the blade.

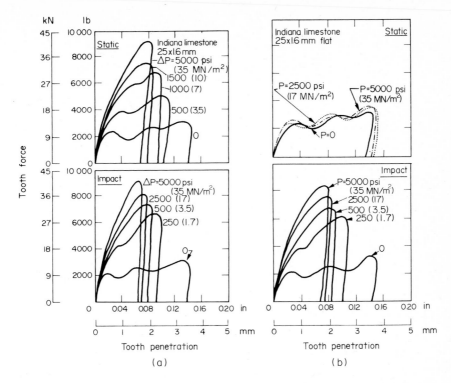

FIG. 7.18. Force–displacement characteristics for wedge penetration into limestone at various loading rates and pressures. (a) Differential pressure; (b) hydrostatic pressure (Maurer [8]).

If the energy required to drive the wedge through the penetration depth h is U, then

$$U = F\,h/2.$$

Hence

$$U = \frac{F^2 \cot \theta}{4\, S_c w}$$

and a general equation for the energy of the percussive blow becomes

$$U = \frac{td^2}{Aw} \cot \theta \left[\frac{\sin (\theta + \varphi)}{1 - \sin (\theta + \varphi)} \right]^2,$$

where t is the tensile strength of the rock, A is the ratio compressive strength–tensile strength, and φ is the angle of friction between rock and wedge [20].

Whittaker and Szwilski studied rock cutting by the action of a pick mounted on a radial swinging arm, impacting upon a number of rocks and also on to a rock–plaster compound. The general pattern of rock breakage was observed to be comprised of two zones in which the boundary of zone 1 was judged to be determined by tensile cracks developed according to Evans's failure theory. Zone 2 was seen to be subsequently removed by the ploughing action of the pick as it continued its swing, since the radius of the swinging arm was greater than the radius of curvature of the initial tension crack. It was suggested that the fracture process is dominated by the zone of triaxial compression that is generated close to the cutting point of the impacting pick, extending into a zone of secondary tensile failure. This

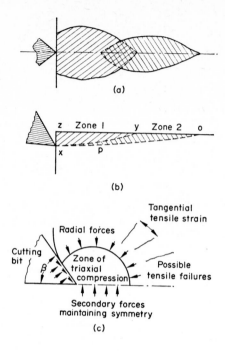

FIG. 7.19. Mode of rock breakage by impact action (Whittaker and Szwilski [21]).

follows the lines of Reichmuth's penetration model (Fig. 7.19) [21]. The mass of rock displaced per blow was directly proportional to the velocity of impact, and the energy absorbed in impact was seen to follow a general law

$$E_0 = KD^n,$$

where E_0 is the absorbed energy, D is the depth of cut, and K is a rock characteristic constant.

For all the natural rocks tested, $n = 1.35$.

Penetration of Rock by Three-dimensional Indentors

The rock-penetrating characteristics of three-dimensional indentors are discussed by Miller and Sikarskie [22]. The shapes of the indentors studied were cones, pyramids, and spheres, and these may be considered to correspond with the tungsten-carbide inserts on roller-bit cutters. The indentors were compared on the basis of the force necessary to penetrate a given distance by crushing, the stress required to cause chipping, and the amount of energy required to fracture a unit volume of rock. Experimental tests carried out on granite and limestone were compared with an extension of the Paul and Sikarskie wedge-penetration theory [11].

The load on the indentor was seen to be proportional to (penetration)^2 with the constant of proportionality independent of penetration. There appeared to be a universal critical stress necessary to initiate chipping. As would be expected, the required specific fracture energy was least with sharp cones and most with indentors of spherical profile.

The indentors exhibited non-linear chipping envelopes, with the indication that only crushing could occur after the indentor reaches a certain critical penetration. While the experimental results were insufficient to permit any firm conclusions, one would expect this critical penetration to be less for spherical indentors than for cones and pyramids, and to decrease with increase in radius of the sphere.

In this connection it is pertinent to bear in mind that there will be an optimum number of indentors that the cutting head can bear, and still break the rock. If the number of indentors is increased, ultimately a point will be reached at which the cutting effect is nil. At that stage the application of the cutting machine will actually be strengthening the rock, and not breaking it, because the application of thrust to the machine head adds confining pressure to the rock face.

Assessment of Rock Reaction to Mechanical Attack

The force–penetration characteristic identifies the nature of the rock's reaction to a particular mode of mechanical attack, but its use presupposes that the mode of attack has been selected. A perennial problem, which is fundamental to geotechnology, is to determine, before selecting the penetration process, what is the most suitable mode of attack for a particular rock. The choice is usually limited to a specific type of operation, such as blast-hole drilling, deep-hole drilling, tunnel boring, rock cutting, and ripping. Hence the need to classify rocks in relation to their "drillability", "cuttability", "blastability", or "ploughability". However, in each of these broad classes of operation there are several alternative processes from which to choose, while the excavating ability of a particular process depends on the nature of the process as well as on the physical and engineering properties of the rock.

For example, Paone and Bruce define drillability as "the real or projected rate of penetration in a given rock type with a given drilling system", and then proceed to apply this definition to a study of diamond drilling [23]. But, in a wider sense, for a particular size of excavation, or hole diameter, the drillability of a particular rock defines the type of system that is best suited to drill that rock. Alternatively, given the hole dimensions and a drilling system, it defines the type of rock that may be drilled.

A drillability index should therefore enable selection of the best drilling system, as between rotary, percussive, or percussive-rotary, for a specific rock, in combination with which the operating characteristics of the drill will determine the force–penetration characteristic and provide a projected range of penetration rates.

The relevant rock properties to be measured or assessed may be grouped:

(1) Engineering factors. Hardness, toughness, porosity, moisture content, density, etc.
(2) Geological factors. Rock fabric and structure, bedding, stratification, joints, cleavage, etc.
(3) The *in situ* state of stress. Overburden pressure, tectonic pressure, and formation fluid pressure (this is sometimes not important in shallow drill-holes but is critically important in deep-hole drilling).

Early discussions on the subject hinged largely on definitions of rock hardness and toughness, and it is of interest that Gardner, in 1927, suggested that hardness should be defined as the resistance of a rock to penetration by diamond drill, while toughness should be defined as its resistance to rupture by impact, as in blasting [24]. Since that time numerous investigators have reiterated this theme, which is represented in contemporary rock mechanics by the measurement of indentation hardness, such as by the Franklin point load test, and rock toughness as measured either by the ASTM rock-toughness test or by the Protodiakonov method.

The rock quality designation (RQD), as described by Miller and Deere, may be used to assess the engineering properties, including the strength, of the rock mass. The crushing strength of the rock material is indicated by the uniaxial compressive strength and the modulus ratio. The mass RQD is probably of more significance in relation to large-scale excavation, such as tunnel boring, but the uniaxial compressive strength is widely regarded as a prime index to describe rocks in connection with drilling, blasting, rock excavation, and rock mechanics in general.

An essentially practical line of approach is to assess the workability of rocks by setting up a prototype drilling or cutting or indentation process by means of which various rocks may be classified. This was the approach taken by Gyss and Davies, Shepherd, Alpan, Frannsen, Sievers, Head, and many others [24–34]. The approach finds its best application when the object to be served is to assess the applicability of a particular mode of attack to various rocks, e.g. the reaction of rocks to rotary drilling by the Goodrich method [33] or to diamond drilling by the methods of Sasaki and by Paone and Bruce [23]. However, this does not provide a general means of assessing rock workability in an absolute sense and its results are clouded by the physical conditions of the test. For example, Alpan measured uniaxial compressive strength, indentation hardness, toughness, and resistance to abrasion in relation to rate of penetration, and found that any one of these could represent drillability to the extent that a high value would indicate difficult penetration conditions, while a low value would represent easy penetration conditions.

Fish noted that the properties having a major influence on the drilling process were rock strength, which determines the drilling forces, rock abrasiveness, which determines the rate of bit-wear, and rock petrofabric structure, which determines the characteristics of the rock fragments. He also concluded that, from a practical viewpoint, a realistic assessment of rock drillability would need to show the projected penetration rate at a given thrust and an economic bit-life [35].

In the search for an absolute measure of rock workability an alternative approach is to apply energy concepts. As early as 1926, Hurley proposed a method of rock-hardness classification based on the energy required to cut a unit volume of rock [25], and about the same time Zaidman and Mazarov were conducting experimental laboratory and field tests, which provided the data for a measure of drillability based on the specific work done in various rocks [26]. Since that time specific energy has been used by Teale to establish the relative efficiency of various tools, machines, and cutting processes in a given rock, and, alternatively, to establish the relative resistances of various rocks to a given tool. It still remains to be shown, however, whether the performance of a given tool in various rocks can be quantitatively related to the performance of other tools in the same rocks.

Such a translation would be invaluable as a means of scaling-up the results of prototype laboratory tests to field applications of excavation processes. For example, Gaye, extending Teale's application of specific energy, proposes to scale up small-hole drilling tests, using as an index the rock number N_r, which is defined as compressive strength/specific energy (f_c/E_s).

Gaye found that in a given rock-cutting process the size of debris produced is a function of the specific energy and the rock number. The rock number is approximately constant for any given cutting process operating under efficient working conditions. (For example, in the case of rotary rock drilling, operating with a sharp bit between the extremes of too-low thrust, which causes the bit to "ride" the rock, and too-high thrust, which causes clogging at the bit.) Therefore, supposing a sample of rock taken for test from a field excavation situation be drilled in the laboratory by a small-scale test drill, and the rock number N_{r1} determined, then

$$\frac{N_{r1}}{N_{r2}} = \frac{f_c/E_{s1}}{f_c/E_{s2}} = \frac{E_{s2}}{E_{s1}},$$

where the suffix 1 refers to the laboratory test and suffix 2 refers to the field excavation process. Hence

$$E_{s2} = \frac{N_{r1}}{N_{r2}} E_{s1}.$$

The rock number thus becomes a dimensionless performance index by means of which various excavation processes may be compared [37].

Mellor [36] has shown that there are reasonable theoretical arguments that can be used to normalize specific energy with respect to uniaxial compressive strength, but advises caution in its use until more experience has been gained. Hughes, reporting on the results of many hundreds of tests at the NCB Mining Research Establishment, states that optimum ratios of f_c/E_s are remarkably constant, whatever the strength of the rock, for a given design of rotary machine. Typical values are quoted in Table 7.2.

In a review of rock-machining fundamentals, Hughes shows that if a machine removes a volume of rock V and creates debris of size D in a field site, the work done is $E_s V$ and

$$E_s V = \frac{GD}{V} \quad \text{or} \quad E_s = \frac{G}{D},$$

where G is a constant for the site where the machine is at work. G is proportional to the energy required at the site per unit new surface area.

In laboratory tests on the rock concerned, let the uniaxial compressive strength as measured be f_c, the specific energy e_s, and the size of debris d.

Then
$$e_s = g/d,$$

where g is a similar constant for the laboratory test conditions.

TABLE 7.2. *Typical rock numbers (Hughes [37])*

Process	f_c/E_s	Debris size (mm)
Laboratory small drill test	0.25	Dust
Rotary small blast-hole drill	1.0	8
Tunnel machine with rotary roller-bits	2.0	25
Tunnelling machine with disc cutters	3.0	50–75
Rotary impact heading machine	8.0	125
Selective impact heading machine	20.0	300
Uniaxial compression test on a "soft" testing machine (specimen shattered at failure)	700.0	

If the specific energy in a uniaxial compression test, assuming Young's modulus to be about 350 times the compressive strength, is 700, then

$$f_c = 700e_s = 700g/d.$$

The comparative efficiency of the rock excavation process is then indicated by

$$\frac{f_c}{E_s} = \frac{700gD}{Gd}.$$

D/d, being a relative indication of debris size produced by the field process and in the laboratory, is denoted by the rock number N_r.

g/G is a relative indication of the specific energy requirements for rock excavation, in the laboratory and on the site.

Replacing $700g/G$ by η, where η is the efficiency factor of the excavation process,

$$f_c/E_s = \eta N_r.$$

Hence the larger the average size of debris produced, the less work is required and the less need be the power of the machine. Specific energy reduces as the size of debris increases. The stronger the rock and the smaller is the size of debris that can be broken and the more power is required. Thus, in a strong rock not only is f_c large, but N_r is small, and the heavier and more powerful is the machine required to be. Thus, as rocks get stronger, so is it necessary to progress, first with rotary machines from cutter picks to discs, and then on to roller bits. Alternatively, in hard rocks, to go over to impact penetration.

The great advantage of selective impact machines is their flexibility. They can be operated to exploit geological discontinuities, to take advantage of any planes of weakness that may exist. The potential application of tensional impact forces offers the possibility of breaking down the rock with a minimum of effort.

References and Bibliography: Chapter 7

1. COOK, N. G. W., JOUGHIN, N. C., and WIEBOLS, G. A., Rock cutting and its potentialities as a new method of mining, *J. S. African Inst. Min. Metall.*, pp. 266–71 (Jan. 1969).
2. HARRISON, J. W., The response of rocks to attack by cutting and abrasion, MEng thesis, Univ. of Sheffield, 1968.
3. EVANS, I., *Theoretical Aspects of Coal Ploughing. Mechanical Properties on Non-metallic Brittle Materials*, Interscience, New York, 1958.
4. EVANS, I., The force required to cut coal with blunt wedges, *Int. J. Rock Mech. Min. Sci.* **2**, 1–12 (1965).
5. EVANS, I., and MURREL, S. A. F., Wedge penetration into coal, *Colliery Engng* **39**, (455) 11 (1962).
6. DALZIEL, J. A., and DAVIES, E., Initiation of cracks in coal by blunted wedges, *The Engineer, London* **217**, 217 (1964).
7. GNIRK, P. F., and CHEATHAM, J. B., An experimental study of single bit tooth penetration into rock at confining pressures 0–5000 psi, *Trans. Am. Inst. Min. Engrs* **234**, 11–130 (1965).
8. MAURER, W. C., The state of rock mechanics knowledge in drilling, *Proc. 8th Symp. on Rock Mech., Univ. of Minnesota*, September 1966.
9. REICHMUTH, D. R., Correlation of force–displacement data with physical properties of rock for percussive drilling systems, *Proc. 5th Symp. on Rock Mech., Univ. of Minnesota, 1963*.
10. BENJUMEA, R., and SIKARSKIE, D. L., A note on the penetration of a rigid wedge into a non-isotropic brittle material, *Int. J. Rock Mech. Min. Sci.* **6**, 343–52 (1969).
11. PAUL, B., and SIKARSKIE, D. L., A preliminary model for wedge penetration in brittle materials, *Trans. Am. Inst. Min. Engrs* **232**, 372–83 (1965).
12. CHEATHAM, J. B., An analytical study of rock penetration by a single bit tooth, *8th Int. Symp. on Drilling and Blasting, Univ. of Minnesota, 1958*, pp. 1A–25A.
13. CHEATHAM, J. B., Rock bit tooth friction analysis, *Trans. Am. Inst. Min. Engrs* **216**, II-37–II-332 (1963).

14. CHEATHAM, J. B., Indentation analysis for rock having a parabolic yield envelope, *Int. J. Rock Mech. Min. Sci.* **1**, 431–40 (1964).
15. GNIRK, P. F. and CHEATHAM, J. B., Indentation experiments on dry rocks under pressure, *Trans. Am. Inst. Min. Engrs* **228**, I-1031–I-1039 (1963).
16. GARNER, N. E., PODIO, A., and GATLIN, C., Experimental study of crater formation in limestone at elevated pressures, *Trans. Inst. Am. Min. Engrs* **228**, I-1356–I-1364 (1963).
17. TANDANAND, S., and HARTMAN, H. L., Stress distribution beneath a wedge shaped drill bit loaded statically, *Proc. Int. Symp. Min. Res., Univ. of Missouri, Rolla (1961)*, **2**, 799–800.
18. NISHIMATSU, Y., The mechanics of rock cutting, *Int. J. Rock Mech. Min. Sci.* **9**, 261–70 (1972).
19. PODIO, A., and GRAY, K. E., Single blow bit tooth impact tests on saturated rocks under confining pressure, *Trans. Am. Inst. Min. Engrs* **234**, 211–24 (1965).
20. EVANS, I., and POMEROY, C. D., *The Strength, Fracture, and Workability of Coal*, Pergamon Press, Oxford, 1966.
21. WHITTAKER, B. N., and SZWILSKI, A. B., Rock cutting by impact action, *Int. J. Rock Mech. Min. Sci.* **10**, 659–71 (1973).
22. MILLER, M. N., and SIKARSKIE, D. L., On the penetration of rocks by three-dimensional indentors, *Int. J. Rock Mech. Min. Sci.* **5**, 375–98 (1968).
23. PAONE, J., and BRUCE, W. E., *Drillability Studies; Diamond Drilling*, US Bur. Min., Rep. Invest. No. 6324, 1963.
24. GYSS, E. E., and DAVIES, H. E., The hardness and toughness of rocks, *Min. Metall.*, pp. 261–6 (June 1927).
25. HURLEY, G. T., Proposed ground classification for mining purposes, *Eng. Min. J.* **122**, 368–72 and 413–15 (Sept. 1926).
26. ZAIDMAN, Y., and MAZAROV, P., Classification of strata for the standardisation of cable drilling, *Gornij Zh.*, p. 17 (1948).
27. SHEPHERD, R., Physical properties and drillability of mine rocks, *Colliery Engng*, pp. 468–70 (1950); 121–6 (1951).
28. FRANNSEN, H., Experiments to determine the workability of coal, *Glückauf* **86**, 1129 (1950).
29. SIEVERS, H., The determination of the resistance to drilling of rocks, *Glückauf* **89**, 776 (1950).
30. PROTODIAKONOV, M. M., The determination of the strength of coal in mines, *Ugol'* **25**, 20 (1950).
31. HEAD, A. L., A drillability classification of geological formations, *World Oil*, pp. 125–38 (Oct. 1951).
32. EVANS, I., Line spacing of picks for effective cutting, *Int. J. Rock Mech. Min. Sci.* **9**, 355–61 (1972).
33. GOODRICH, R. H., High pressure rotary drilling machines, *Proc. 2nd Ann. Symp. on Mining Res., Univ. of Missouri (1956)*, p. 25.
34. ALPAN, H. S., Factors effecting the speed of penetration of bits in electric rotary drilling, *Trans. Inst. Min. Engrs* **109**, 1119 (1950).
35. FISH, B. G., *Studies in Rotary Drilling—The Basic Variables*, National Coal Board, MRE Rep. 2161, 1960.
36. MELLOR, M., Normalization of specific energy values, *Int. J. Rock. Mech. Min. Sci.* **9**, 661–3 (1972).
37. HUGHES, H. M., Some aspects of rock machining, *Int. J. Rock Mech. Min. Sci.* **9**, 205–11 (1972).
38. FARMER, I. W., *Engineering Properties of Rocks*, Spon, London, 1968.

CHAPTER 8

Excavation Technology—IV

Rapid Tunnelling

The Evolution of Tunnelling Techniques

The evolution of technology in underground excavation has been marked by periodic advances of a limited character in a long history of slow and difficult progress. Throughout recorded history, in the ancient and medieval worlds, and on into the early nineteenth century, tunnelling was largely a process of laboriously applying hand-tools to rock, sometimes assisted by fire-setting, in which the rock face was alternately heated by a wood fire then rapidly cooled by applying water.

Gunpowder began to be used for blasting in mines in the early seventeenth century, and mechanical drills were first introduced about the mid-nineteenth century.

By such means the average rate of advance in hard-rock tunnel driving was increased very little, and could be measured in terms of a few centimetres per week until the introduction of gunpowder for blasting increased the advance rate to from 1 to 4 m per week. Subsequently the introduction of mechanical drills brought an average rate of progress of about 7 m per week in the Hoosac Tunnel (1855–74), while improved drills, aided by nitroglycerine dynamite, lifted the average advance rate to around 24 m per week in railroad and irrigation tunnels about the turn of the twentieth century.

Since that time further progress, although aided by the use of improved materials and systems, has been restricted by the limitations imposed by space and accessibility to the tunnel face, and by the cyclic nature of the system generally applied. In this system the component operations of drilling, blasting, loading debris, and erecting supports, all follow one another separately, each in its turn taking up the whole of the working space over a proportion of the total work period.

Within this context, any improvements of the system as a whole depend on the efficiencies of each component operation. For example, by using AN–FO explosives, loaded into the holes pneumatically, in combination with millisecond-delay detonators, a saving of about 20 % blasting time resulted when driving the Canyon Tunnel through the granite of the Sierra Nevada. In conjunction with improvements in the air-drilling techniques, involving remotely controlled multidrill jumbos, and machines to pick up and load away the debris, an average advance rate of 146 m per week, over a diameter of $4\frac{1}{2}$ m, was maintained.

The alternative to cyclic systems in tunnelling is to apply continuous boring machines. Ideally, such a machine, mounted in a capsule to fit the excavated perimeter, would advance at a uniform rate into the space cut for itself, pouring its debris continuously into a waste disposal and transport system behind. Concurrently, at the rear end of the capsule the tunnel lining would be continuously formed and inserted.

Real systems that have so far been devised fall short of the ideal in several respects, and particularly in relation to the method of insertion of the supports that are required in weak ground, and in the limited capability of mechanical cutters to attack the hardest rocks over the whole cross-sectional area at an economic rate. Nevertheless, the application of rock-boring machinery in recent years has approximately doubled the rate of advance in tunnelling, from the range of 21–146 m per week, by drilling and blasting, to the range 46–244 m per week by tunnel-boring machines, driving tunnels ranging from 5 to 10 m in diameter [1].

Soft-ground Tunnelling Machines

Techniques devised for tunnel boring in soils and soft rock more closely approach the ideal and have evolved from Brunel's original device. Such systems are comprised of a cylindrical shell to contain the machinery. At the head of the cylinder is mounted the clay-cutting shield or, when harder materials are to be penetrated, a rotary cutting or reciprocating head. As the shield advances it moves into the excavated perimeter and a temporary lining, which usually consists of timber-lagged steel rings, is extended at the tail.

It is usual for the tunnel excavation to be completed over a considerable length before the permanent lining, of concrete or steel, is inserted, although some newer installations have experimented with extruded monolithic concrete placed concurrently with the primary advance. Another alternative procedure, if the walls are suitable, is to insert a lining of steel wire mesh, pinned to the perimeter by rock-bolts extending radially into the walls, sometimes covered also by shotcrete. This can also be done concurrently with the primary excavation.

In general, soft ground is easy to excavate but difficult to support at the head of the tunnel. The main purpose of the cylindrical shell is therefore to give support to the face equipment and to protect the workers. It is usually forced forwards by hydraulic jacks thrusting against the lining, or against an expanding push-ring at the back of the shield.

Hard-rock Tunnel Borers

Soft-ground tunnel borers with rotary cutting heads usually employ drag-bits, which can cope successfully with the weaker rocks such as chalk and the argillaceous sediments, having compressive strength up to around 21 MPa (3000 psi). The stronger the rock the greater is the thrust that must be applied to penetrate it. The basic principles of rock penetration, as already described in connection with rock drilling and hole boring, are also applicable to tunnel borers.

With increase of rock strength and required thrust, the necessary cutting-tool changes from drag-bits to toothed rollers for the harder sediments, up to about 40 MPa (6000 psi) compressive strength, then to disc cutters, which can deal with the medium-hard rocks such as sandstones, limestones, and metamorphics up to a limit ranging from 80 to 138 MPa (12 000–20 000 psi) compressive strength. Hard and very hard crystalline rocks, of strength more than around 140 MPa (21 000 psi), may require cutters fitted with three-dimensional indentors termed "button rollers" [2, 3].

The increasing scope and range to which mechanical tunnel borers are being applied is shown by the data listed in Table 8.1. It will be seen that within the space of 9 years the type of rock that was being bored increased in strength from 7 to 345 MPa (1000–50 000 psi), and while the mechanism applied to the weaker rocks required about 600 000 N (135 000 lb) thrust and 75 000 N m (55 000 ft-lb) torque, the hardest rocks were being attacked by machines delivering up to 2400 kNm (1.75 million ft-lb) torque, at up to 7 MN (1.5 million lb) thrust.

The manner in which the machines are anchored in order to generate such thrusts is of considerable importance. In soft rocks sufficient reaction may be generated by the weight of the machine itself, standing on crawlers or braced by struts against the tunnel wall, as in Beaumont's rock-borer, which successfully drilled an exploratory tunnel under the English Channel, 2 m in diameter and nearly 1000 m long, at a rate of 15 m per day, nearly 100 years ago.

The high levels of thrust that are required in the harder rocks are generated by pressure pads operated by hydraulic rams. Alternatively, the boring head may include a pilot bit by means of which an advance anchor can be inserted to facilitate progression on a push–pull principle.

Experiments with Disc Cutters

While the fundamentals of wedge penetration into rock can be expected to apply to all mechanical rock-cutting devices, there is, as yet, considerable doubt as to the extent to which the results of laboratory experiments can legitimately be extrapolated to a full-scale tunnelling situation. Nevertheless, some manufacturers and institutions have sponsored research on the performance of rock-cutting tools as a basis of machine design. Experiments with disc cutters are described by Teale, and also by Bruce and Morrell [3, 4].

For discs having cutting edge wedge angles of 60 degrees and 90 degrees, Bruce and Morrell report a linear force–penetration characteristic, and for the 90-degree cutting edge a relationship between the applied energy and the crater volume produced:

$$V = 0.0437 + E_n \left[\frac{536.7}{\gamma} + \frac{1.2185 \times 10^{-4}}{\rho} - \frac{1.001 \times 10^{-2}}{\sigma_t} \right],$$

where V is the crater volume, E_n is the total input energy, γ is Young's modulus of the rock (static value), ρ is the rock density, and σ_t is the rock tensile strength.

Teale's experiments with disc cutters were motivated by the need to design a machine capable of tunnel boring in carboniferous sandstones and grits. Consideration of the action of toothed-wheel cutters, according to the principle illustrated in Fig. 8.1, showed that a limit in the depth of penetration occurred when

$$p = r (\cos \alpha - \cos 3\alpha).$$

Up to this limit, penetration was directly proportional to thrust, the exact relationship being determined by the geometry of the roller and the properties of the rock. For a given depth of penetration the volume of rock broken by a toothed roller appeared to be governed more by geometrical considerations than by any physical properties of the rock.

TABLE 8.1. *Mechanical tunnel borers, 1960–9 (Bruce and Morrell [3])*

Date	Tunnel diameter		Rock compressive strength		Penetration rate		Rotation speed rev/min	Thrust		Torque	
	ft in	m	psi	MPa	ft/h	m/h		lb	kN	ft-lb	kN m
1960	9–0	2.74	1000–2400	6.9–16.5	2–3	0.6–0.9	8–10	100 000–170 000	444.82–756.19	55 400	75.11
1961	4–6	1.37	17 000	117.2	5–20	1.5–6.1	16.5	226 000	1005.29	—	—
1963	7–0	2.13	8000–14 000	55.2–96.5	3	0.9	10	393 000	1748.14	85 000	115.24
1968	9–0	2.74	17 000–50 000	117.2–344.75	4.5	1.4	15.5	559 250	2487.66	132 405	179.51
1966	7–0	2.13	12 000–31 000	82.7–213.7	5.5	1.7	15	600 000	2668.92	120 000	162.70
1967	10–0	3.05	10 000–35 000	69–241.3	1–4	0.3–1.2	9.2	866 000	3852.14	235 000	318.61
1967	13–0	3.96	10 000	69	5	1.5	9.5	866 000	3852.14	275 000	372.86
1969	13–0	3.96	—	—	—	—	8–25	1 000 000	4448.22	347 000	470.47
1969	12–6	3.81	12 000–20 000	82.74–137.9	—	—	9	1 500 000	6672.30	450 000	610.11
1968	18–0	5.49	12 000–31 000	82.74–213.7	6–7	1.8–2.1	4.5	1 580 000	7028.16	1 720 000	2331.98

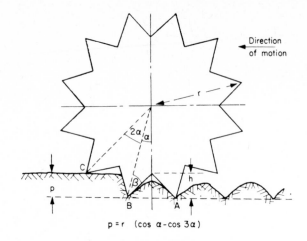

$$p = r \ (\cos \alpha - \cos 3\alpha)$$

FIG. 8.1. Penetration by a toothed roller cutter in relation to its radius and number of teeth (Teale [4]).

For a given wheel, the volume of rock broken in the first run over a new surface was approximately proportional to the square of the depth of penetration.

Teale concluded that bevelled or disc/wheel cutters would prove to be at least as effective as toothed rollers, especially in the more brittle rocks.

The general evolution of hard-rock tunnel boring has, however, proceeded mainly upon empirical lines. Muirhead and Glossop [2] listed in 1968 details of sixty rock tunnel-boring operations installed after 1954, from which it was apparent that the experience gained in early installations was applied to extend the limits of the technique in later development. Today it is fair to say that tunnels up to 10 m in diameter can be bored through the hardest rocks. Whether it will always be economically feasible to bore rock tunnels in this way will depend upon many factors, not all of which are concerned with rock mechanics. Nevertheless, it can confidently be expected that rock tunnels up to 10 m in diameter can be bored by machine at approximately double the rate of advance that can be achieved by cyclic excavation systems under similar geological conditions [5].

Ancillary Problems in Hard-rock Tunnel Boring

The energy applied in rock-tunnel-boring machines ranges from around 149 kW to 1.12 MW, with an increasing proportion of the later machines approaching the higher figure, as harder and more difficult conditions are faced. Only a small fraction of this energy is used in rock fracture and displacement. The greater part of it is used in overcoming friction and generating heat.

The rock-breaking process becomes progressively less efficient as the penetration process changes from tension splitting to hard-rock grinding. The size of the fragments displaced in brittle fracture becomes progressively smaller in the stronger rocks. More dust is generated, which forms a health hazard to personnel and requires the application of dust-suppression techniques, involving water sprays.

The high levels of thrust that are required to attack strong rocks may take the pressure at the bit–rock contacts above the transition pressure, so that in the higher-thrust

applying a compu
given by

where *AR* is the pr
compressive stren;
groundwater inflo

New Methods of]

Penetration by Vi

The frequency o
Vibrations at much
penetrate rocks. U]
years to shaping and

Ultrasonic vibrai
magnetic field to a
principles of constri
along a conical prot
to penetrate the mat
aided by carrying a

Most of the repoi
Union. In general, tl
500-W machine, op
penetration rate of ;

One of the major
cutting tool necessa;
practically useful si;
frequency into the au
for the operators in u
the-hole drills used f(
possibilities for vibra
100–1000 Hz on to tl

Magnet

Cham
coolir

machines an increasing proportion of the applied energy is dissipated in plastic deformation. Atmospheric heat, humidity, poor visibility, noise, and dust, all contribute towards an environmental problem, involving the health, safety, and functional efficiency of working personnel. With the high-thrust machines in strong rock this may be so extreme as to necessitate placing the operatives in protective suits, carrying their own supply of cool, clean, air.

Automation of the system permits of controlling the operations from a vantage point remote from the tunnel-head environment, but poses other problems of its own. Automated machines can only perform what they are programmed to do, and in the case of rock tunnelling a successful programme can only be formulated if all the conditions under which the machine must operate are known in advance.

A pilot exploratory tunnel driven in advance of the boring machine is sometimes constructed in the case of large-diameter tunnels in civil engineering, but as a general mode of procedure it is not feasible in underground mining. Much attention is therefore being directed to the improvement of techniques of exploration geology and geophysics for use as an essential preliminary to rapid excavation techniques underground.

Tunnelling machines have a very limited capability of adaptation to change in their geological surroundings. Rotary cutting heads can be powered by sensors responsive to overall rock resistance, but their response is limited to variation of rotary speed and thrust. They have great difficulty in dealing with geological discontinuities and changes in the lithology and petrology of the materials they encounter. So much so that they are unable to deal satisfactorily with rocks containing hard nodules or boulders in a matrix of weaker material.

The rotary boring principle must then be abandoned in favour of some other technique, such as one that permits of selective impact. However, those techniques cannot be automated to the extent that the machines can self-adapt when attacking the rock face. Alternative methods of rock penetration, other than the application of rotary and reciprocating mechanical tools, must then be sought.

Estimating the Projected Advance Rate of Mechanical Tunnel Borers

The maintenance of a sustained advance rate, at the highest possible economic figure, is an essential aim in tunnel boring, so as to reduce the overall cost of the operation. The effect of increased advance rate on the incremental operational costs is described by Robbins as a result of tunnel-boring experience under a variety of circumstances [5, 7].

The rate of increase of cutter wear, with increased rate of advance, has a major influence on determining what is the maximum advance rate that can be economically maintained in a given rock formation. This factor becomes increasingly critical in hard, abrasive rock [5].

The effect of rock strength on the projected advance rate may be summarized. Optimum conditions for mechanical tunnel boring exist in the medium-strength rocks of a uniform character. Conditions less than the optimum are likely to exist in hard rocks due to the cost of bits, while, on the other hand, tunnel boring in weak rock is likely to be more costly due to problems of strata control and support [6].

Assuming massive rock of uniform engineering properties, no complications engendered by pockets of excessively hard or weak ground, and little water inflow, Williamson

The use of electroacoustic devices to generate these vibrations offers the distinct advantage of eliminating sliding friction or roller-bearing surfaces in the drill mechanism, for these must be sealed off from the abrasive drilling fluid. On the other hand, they require some means of generating and controlling up to about 200 kW of electrical power and transmitting this down the drill rods [11].

Vibration devices are also applied to penetrate surface soils when pile driving. The drilling system then consists of an electrically driven vibrator, mounted on a tractor, and used to transmit vibrations of 20–50 Hz frequency to the pile. Penetration is then effected by the pile sinking under its own weight into the soil, which is liquefied by the vibration. Penetration rates of up to 60 m/h are reported in sandy topsoil and clay at a depth of 30 m [12].

Hydraulic Jet Cutting

Hydraulic jets, or monitors, have long been used for breaking-down and washing away surface alluvial deposits, sands and gravels, and to extract clay from quarries in decomposed igneous rocks. During recent years the technique has also been applied underground, to excavate weak sedimentary rocks, notably coal and sandstone, and to transport the broken material in hydraulic pipelines [13, 14].

Jets at pressures of up to 30 MPa, with a flow rate of 180 l/m, giving a nozzle velocity of 180 m/s, cut into coal to a depth of 0.6 m. If the energy in the jet is increased, either by increasing the pressure or the quantity of water applied, the rate of rock degradation is increased, but the extent to which this is possible is limited if hand-held monitors are used. The maximum controllable reaction force in such a device is about 180 l/min at 30 MPa pressure. If pressures higher than this are to be used, then the monitor must be mechanically controlled.

Higher pressures are required to cut harder rocks, and in recent years several investigators have explored the possibilities and limitations of hydraulic jet cutting in hard rock. The method has obvious attractions as a possible substitute for, or supplement to, mechanical rock boring in rapid tunnel excavation. As has already been noted, mechanical borers have great difficulty in dealing with a rock face that includes materials of different hardness, such as nodules and boulders, in a softer matrix. A hydraulic jet could more easily detach such materials by cutting into the softer matrix around them. Also, the presence of pyritic nodules in the strata may cause a dangerous gas-ignition hazard when tunnel boring, due to frictional sparking at the bit-tips. The provision of water-jets is required to suppress both dust and frictional sparking, and it is only logical to increase the hydraulic pressure applied so that the jets may also do useful work in excavation.

In general terms two methods of approach are being pursued. The first employs intermittently injected slugs of water at very high velocity, while the second applies a continuous jet at the highest pressure feasible in hydraulic pump design (which is lower than the impulse jet). The impulse jet approach is typified by the experiments of Farmer and Attewell, in which single-shot jets, in the velocity range 5–13 m/s, were projected on to a number of low-strength sediments and limestones (Fig. 8.5). The method is analogous to drilling, and a study of the craters produced provided empirical relationships

$$S = k d_c \left(\frac{v_0}{C}\right)^n \quad \text{and} \quad \frac{S}{t} = k'Q,$$

FIG. 8.3

estimates the m
strength of the r
maximum pene
maximum penet
Wheby and (

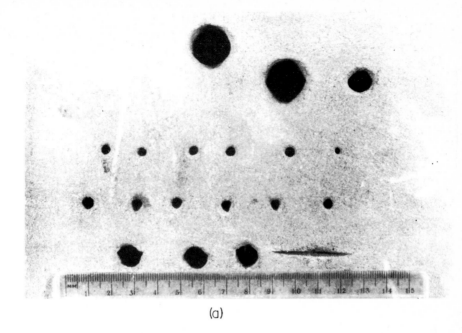

(a)

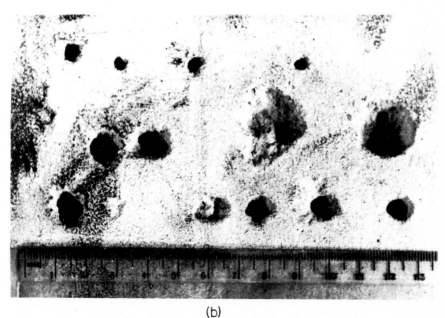

(b)

FIG. 8.5. Hydraulic jet impact craters; (a) in limestone; (b) in sandstone (Farmer and Attewell [15]).

where k and k' are constants, d_c is the crater diameter, C is the longitudinal wave velocity for the rock, $n = 2/3$ (approx.), S is the penetration distance, v_0 is the impact velocity, and Q is the rate of flow.

Above a critical velocity (below which there was no penetration), the penetration was proportional to impact velocity, and the rate of penetration was proportional to the rate of water flow. At impact velocities between 240 and 340 m/s there occurred a transition, which was considered to be due to increasing turbulence in the impact crater, with increase in the stand-off distance as the hole deepened (Table 8.2) [15]. It was therefore apparent that the limitations of space will determine how much energy can be applied by a water-jet when applied as a drilling mechanism if the jet has to follow the deepening hole (Fig. 8.6). A more promising line of approach would appear to be the incorporation of water-jets to supplement the drilling mechanism of rotary drag-bits, and Maurer reports that field tests on prototype drills, which embody these principles, drilled 2–3 times faster than conventional drill-bits in the same rock. The jets applied in these drills are continuous and not impulse jets. Harris and Mellor [17] note that there is, as yet, no convincing evidence to show that the specific energy required to penetrate rock by projectiles decreases significantly with increasing impact velocity. However, Epshteyn *et al.* have shown that there is every possibility that high-pressure, short-impulse water-jets may be useful to break hard rock by surface impact rather than by continuous flow penetration [16].

TABLE 8.2. *Average penetration rate of hydraulic set into three rocks: 3-mm diameter nozzle*
(Farmer and Attewell [15])

Rock	Compressive strength		$v_o = 500$ m/s (25 000 psi: 172 MPa pressure)	$v_o = 400$ m/s (20 000 psi: 138 Mpa pressure)
	psi	MPa		
Sandstone	4000	27.58	2.60 m/s	1.87 m/s
Gritstone	8400	57.92	2.09 m/s	1.41 m/s
Limestone	7900	54.47	2.11 m/s	1.43 m/s

At the present time feasibility studies and field trials are being conducted of high-pressure jet drills in which explosive propellants are used to produce the jet impact. Capsules containing the propellants are pumped down the drill-stem to the hole bottom and ignited in a combustion chamber located above the rotary bit. The high-pressure

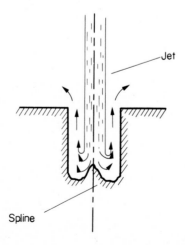

FIG. 8.6. Idealized symmetrical flow in a hydraulic jet crater (Farmer [9]).

exhaust products then propel the drilling mud through nozzles in the bit at pressures of from 410 to 480 MPa (60 000–70 000 psi). A check valve at the top of the combustion chamber prevents the jet pressure from being transmitted up the inside of the drill-stem.

Other research on water-jet rock cutting has employed continuous jets traversing over the surface of the rock to cut a slot. The practical results, reported by Harris and Mellor, bear a strong similarity to the grooves cut mechanically by the revolving disc cutters in Teale's experiments on mechanical cutting.

Contemporary high-pressure pumps in commercial use can produce continuous flows at delivery pressures of 480 MPa, and experimental units delivering at 690 MPa exist. There is, therefore, every indication that hydraulic jets can be used to bore tunnels by cutting parallel slots, using a secondary wedging system to break out the interlying ribs, as is now done with rotary disc cutters [17].

Crow describes the mechanism of steady slot cutting by a high-pressure water-jet as shown in Fig. 8.7.

The rock feeds at speed V under the jet nozzle impinging on to the rock at the angle θ. The jet flow turns until $\theta = 0$ at which point the slot has reached its terminal depth h. The depth increases as the feed rate exceeds an intrinsic speed C. This intrinsic speed embodies the interaction of turbulence and cavitation in the fluid, and brittle fracture, permeability, and porosity in the rock, which are all operative factors in the cutting process [18, 19]. Crow's experiments showed

$$C = \frac{\kappa \tau_0}{\eta f U_r g},$$

where η is the viscosity of the jet fluid, κ is the permeability of the rock, f its porosity, τ_0 its shear strength, U_r its coefficient of internal friction, and g is a typical grain diameter. Harris and Mellor used jet-nozzle pressures ranging from 7 to 400 MPa (1000–60 000 psi) and rock traverse speeds of up to 1.7 m/s. The target materials were sandstone, limestone, and granite. Their results suggest that the efficiency of multipass operations will increase with increase of traverse speed, so that optimum conditions for deep cutting will lie in the direction of high-speed multiple passes.

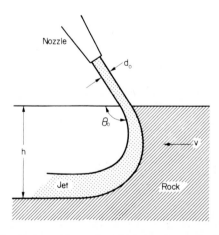

FIG. 8.7. Steady slot cutting under a high-speed water-jet (Crow [18]).

Applied Geotechnology

If this is so then the criterion of specific energy may not be appropriate, since large volume removal is not required of the jet. Volume removal would come by virtue of the secondary wedging mechanism which, in a practical machine, would be combined with the jet. If it is required to effect more efficient volume removal by the jet itself, then a system which gives low penetration, utilizing high traverse speeds or relatively low-jet pressures, might be preferable, because surface spalling on each side of the cut was observed under conditions of low penetration for all the rocks tested [17–19].

Cavitation Erosion

Cavitation is a phenomenon which is associated with the formation and collapse of bubbles in a liquid. The energy released by the collapse of a single bubble is small, but the spherical convergence of the collapsing bubble generates an energy which is capable of eroding very hard materials. This cavitation erosion is caused by the shock wave which is generated as the bubble collapses.

The possibility of using the phenomenon as a technique for deep-well drilling has been explored by several investigators in recent years, notably by Angona [20]. Cavitation is potentially a bit-less system, involving no physical contact between drill and rock, and hence no wear on the drill.

At a drilling depth of around 1000 m the hydrostatic pressure is about 100 atm, at which pressure Angona's experiments indicated an erosion rate of 1.2 kg for a 5-s exposure. If this were continuously maintained it would be equivalent to drilling a 250-mm diameter hole at a penetration rate of 8 m/h, which is comparable with the performance of deep-well rotary drills.

In contrast with rotary drilling, in which the attainable penetration rate decreases with depth, in cavitation erosion the penetration rate would increase with increase of hydrostatic pressure. Extrapolation of Angona's results indicates an erosion potential penetration rate of around 274 m/h at a depth of 3000 m. However, an acoustic pressure around 340 atm and a down-hole transducer power output around 75 kW would be required at this depth, and this is outside the range of acoustic transducers that are available commercially at the present time.

Ground Penetration by Projectile Impact

The use of projectiles to penetrate earth materials is not solely a matter of military interest, for it also has applications in civil engineering and in industry, e.g. when used directly for the rapid excavation of cuts and craters in frozen ground, and in the form of pellet drills, to impact and to break rock [29].

The impact velocities that are applied by such projectiles provide an intermediate range between those generated by hydraulic jets, on one side, and mechanical percussion, on the other. Maurer and Rinehart [30] describe the processes of impact crater formation in rock, and their study has been extended by Singh [31]. Using a light gas-gun to fire solid and water-filled capsules, Singh obtained impact velocities ranging from 2.2 to 6.9 km/s.

Observations of crater volume and of crater depth, for various values of impact velocity, indicated an optimum impact velocity, which for aluminium projectiles on to basalt was around 5 km/s (Fig. 8.8).

156

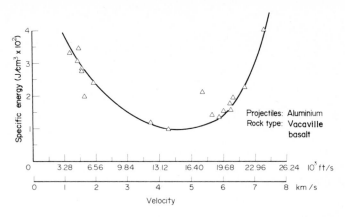

FIG. 8.8. Projectile impact on rock. Effect of impact velocity on specific energy (Singh [31]).

More recently, Newsom *et al.* [29] have reported on the Terra-drill, which applies projectiles to impact upon rock ahead of a roller-bit. The object is to weaken the rock and increase the rate of penetration of the roller-bit. The hardened-steel projectiles are 64 mm in diameter and are fired from guns at impact velocities up to 1130 m/s. A prototype drill fires two projectiles per minute from a magazine which carries 4800 rounds, corresponding to 40 h of operation before the system must be withdrawn for re-load. The advance rate in limestone was reported to be doubled after the projectile attack as compared with a rotary rock-bit without projectile assistance.

Penetration of Rock by Explosive Shaped-charges

The term "shaped-charge" is applied to cylindrical charges of high explosive with a cavity formed at the end opposite to the point of initiation. The effect of the cavity is to produce an intensified pressure which is projected as a jet in the direction of initiation. If the cavity is lined by a metal, then this is fragmented by the detonation to produce a high-velocity jet (Fig. 8.9).

Shaped-charges have been applied to industrial use, e.g. to penetrate the casing of oil- and gas-wells, to tap furnaces, and to cut metals. Their application to geotechnology has also been explored as a method of rock fracture and penetration of frozen ground [32, 33].

Rollins *et al.* have described the application of the technique to granite, using various metal liners, and composition C4 as the explosive [34]. For all the materials an optimum stand-off distance, giving maximum penetration, was observed. The jet-tip velocities in granite ranged from 1 to 10 mm/μs, with the velocity decreasing to a value at which a maximum penetration rate was observed.

Typical hole profiles, obtained by cylindrical charges, with a conical cavity, charge length, two cone diameters, and charge/cone diameter ratio 1.04, the explosive loaded to a density of 1.6 g/cm^3, with liners as detailed in Table 8.3, are shown in Fig. 8.10.

The penetration of shaped-charges into frozen ground was investigated by Benet in two series of tests. In the first series, charges weighing from 80 g to 18 kg were fired into permafrost, rock, and ice. The penetration achieved at optimum stand-off was

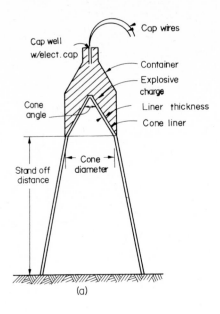

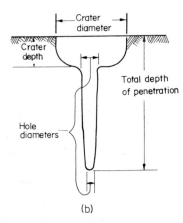

Fig. 8.9. (a) Explosive shaped-charge in firing position; (b) cross-section of typical hole produced (Benet [32]).

TABLE 8.3.   *Explosive shaped-charge: metal liners (Rollins et al. [34])*

Metal	Weight (g)	Liner thickness (cm)
Aluminium 2011 (T-3)	32.5 ±0.25	0.3500 ±0.002
Aluminium 7075 (T-6)	32.0 ±0.25	0.3480 ±0.002
Yellow brass	36.1 ±0.25	0.1150 ±0.002
Maraging steel	34.7 ±0.25	0.1161 ±0.002
Monel	34.2 ±0.25	0.1065 ±0.002
Copper (42°)	47.9 ±1.00	0.1050 ±0.002

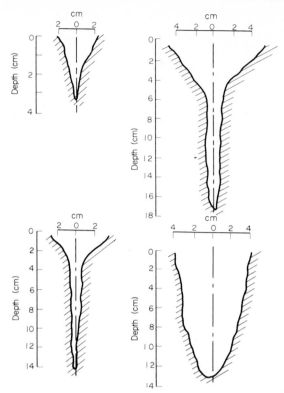

F IG. 8.10. Typical hole profiles in granite (Rollins *et al.* [34]).

approximately proportional to the cone diameter, with penetration about 11 times the cone diameter. The average crater diameter was about 0.61 times the cone diameter for 60-degree cones and about 0.32 times the cone diameter for 45-degree cones. For a given cone angle the diameter of the crater produced was approximately proportional to the cube root of the charge weight for geometrically similar charges [33].

In Benet's second test series, standard charges of RDX explosive, weighing 1.0 kg, were used in cones 100 mm in diameter, but the material, the thickness, and the angle of the cone liner were all varied. He concluded that machined aluminium cone liners with a 70-degree cone angle produced the best results. Benet also tested the effect of changing the charge weight while maintaining the cone diameter, the cone angle, and the stand-off distance constant. The results suggested that penetration of frozen rock or ice by shaped-charges tends to a limit as the charge weight increases. It appears that a 1-kg charge, which represents a shaped-charge of conventional proportions, may be close to the optimum size. Greater charges effected a disproportionately small increase in penetration, and increasing the charge weight beyond 1 kg produced no significant change in the hole diameter.

Explosion Drilling

The use of explosive-propelled high-pressure water-jets has already been mentioned, but at an earlier date, in 1960, Ostrovski described several techniques by means of which small explosive charges could be fed down a hollow drill-rod to emerge and explode at the drill-bit. One method was to use capsules containing two liquid ingredients which formed an explosive mixture when a separating diaphragm was broken. The mixture was detonated by a percussion pin as the capsules emerged from the drill-bit.

In another system the explosive ingredients were fed separately down the drill-pipe and then mixed automatically in a detonating chamber at the drill-head. Ostrovski quotes achieved drilling rates ranging from 1 to 10 m/h in various limestones and dolomites with hole diameters ranging from 50 to 80 mm [28].

The AAI Corporation, in the United States, are developing an explosive drill in which capsules, each containing a cluster of seven shaped-charges, are pumped down the drill-stem and detonated upon impact with the rock. The capsules are fed at the rate of 30 shots per minute from a magazine holding twenty-three capsules. Holes ranging in size from 30 cm to 61 cm are drilled in granite and limestone at advance rates of 27–36 m/h—much faster than can be obtained with conventional drills in the same rocks [44, 45].

Rock Fracture by Electrical Means

Mechanical and hydraulic jet techniques of rock fracture include several intermediate processes, each of which involves the transfer of energy. Ultimately, the proportion of the total energy which actually goes into rock fracture is quite small, so that the overall efficiency is low. This efficiency may be increased by utilizing electrical energy applied directly onto or into the rock. Various methods are available, some of which are in various stages of commercial exploitation, while others are still experimental.

The primary objective in the use of electrical energy for rock breaking is to induce direct tensile fractures and not to crush the rock by compressive forces. This can be accomplished by energy conversion systems applied either internally or externally to the rock. The conversion systems include electrothermal, electromagnetic, and electromechanical phenomena.

Electrothermal Techniques

Two methods of approach may be used: (1) the rock to be broken is placed within a high-frequency alternating electric or electromagnetic field; (2) the rock is punctured by applying conductors directly to it. In either case, certain physical properties, such as temperature, specific heat, thermal conductivity, electrical conductivity, and modulus of elasticity, come into play to generate physicochemical processes. These include differential thermal expansion and the chemical decomposition of rock-mineral constituents, the generation of break-down products, the expansion of gas and moisture, and mineral phase transformations. It is as a result of these processes that thermal stresses are generated within the rock material.

The thermal effect produced in the rock is determined by the electrical and mechanical properties of the mineral constituents. High silica minerals have a high electrical resistance and a low inductive capacity. Hence they are little affected by a surrounding magnetic field. Minerals such as the metallic ores magnetite and haematite, pyrite, galena, the ores of copper, nickel, and titanium, become electrically conductive at some critical voltage level. Rocks which contain these minerals in sufficient proportions can be heated internally by an inductive method in which high-frequency magnetic fields are induced through looped induction coils placed on the rock surface.

Rocks that contain fewer of the conductive minerals may be heated in a high-frequency electric field imposed between two electrodes applied externally to the rock. The material in the region between the electrodes is heated dialectrically, as a result of inter-molecular friction, under the influence of the rotating electromagnetic field.

Another principle is based on the dielectrical break-down of resistance between two electrodes. A high-frequency current is applied to puncture a non-conductive or a semi-conductive rock, and effect a break-down of its electrical resistance. A path of molten electrolyte is formed between the electrodes. Thereafter, heating of the molten channel can be accomplished by a low-frequency power source (Fig. 8.11) [43].

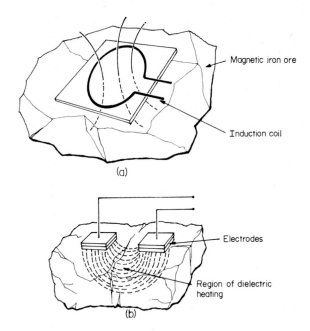

FIG. 8.11. (a) Induction heating of iron ore; (b) dielectrical heating of rock (Sarapu [43]).

Electroshock Impact

Electrical energy may be applied to generate shock waves for impact on to rock. Two methods are possible: (1) Electrohydraulic impact, and (2) direct condenser discharge impact. In electrohydraulic fracturing a high-energy spark is caused to discharge under water. The wave energy, being thus confined, generates a shock pulse which may be used as an agent for rock fracture, either to crush boulders immersed in the water or applied at the tip of a rotating drill (Fig. 8.12).

Microwave Drills

Another way of heating rocks electrically is by use of microwaves. Radiant wave energy at frequencies of 1000–3000 MHz is produced and directed along a tubular waveguide to

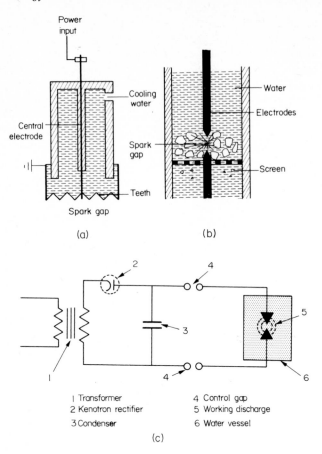

FIG. 8.12. (a) Electrohydraulic drill; (b) electrohydraulic crusher; (c) electrohydraulic rock breaking (Sarapu [43]).

the rock. The object is to produce in-depth heating within the rock, and so produce thermal fracture. Lauriello and Fritsch [22] describe the dielectrical heating effects in rock due to microwave radiation. The parameters that determine the rock-fracturing ability of a microwave system are the average power output, the beam concentration, and the attenuation of energy within the rock mass. Both collimation and attenuation are frequency dependent. With decrease in frequency the beam loses concentration and the attenuation decreases, which is not conducive to the generation of steep thermal gradients within the rock. As a result of this the highest microwave power outputs that are economically feasible at low frequencies, at the present time, are not high enough to induce rock fracture.

At higher frequencies the beam directivity and the rate of energy attenuation both increase. However, in spite of optimum system and rock attenuation conditions, the efficiency of microwave heating as a rock fracture medium is poor. Nevertheless, specific energies observed in experimental arrangements are orders of magnitude greater than those obtained in techniques of mechanical excavation.

Chemical Rock Weakeners

The effects of climate upon hard rocks that are exposed to the agencies of weathering are apparent in physical geology. The weakening effects of chemicals other than water on rocks have been a subject of speculation and study over the years. Rocks may be weakened mechanically due to the generation of internal pore pressures when certain clay minerals absorb water and swell as a result. Saturated rocks may thus be deformed. Also, various circulating fluids may affect molecular cohesion in the rock material.

Liquids with a greater molecular affinity for the solid will promote greater weakening and act as softeners. Adsorption softening then results, and the potency of the softeners increases when the rocks are subject to load [35]. Rehbinder *et al.* investigated a number of rock hardness reducers, listed in Table 8.4. More recently Moavenzadeh *et al.* [37] have reported on experiments concerning thermal weakening and chemical weakeners on a variety of rocks. Fifteen chemical reagents, listed in Table 8.5, were applied in water solutions and the strength of the rocks was observed in a notched beam test.

TABLE 8.4. *Recommended hardness reducers for rocks (Rehbinder* et al. *[35])*

Rock	Hardness reducer	Optimum concentration[a] (%)
1. Quartzites and rocks with a high content of quartz:		
(a) Quartzites, quartzitic sandstones with silicate cement and igneous rocks with a high content of quartz	1. $AlCl_3$ 2. NaCl 3. Naphthenic oil + Na_2CO_3	0.02–0.1 0.1–0.5 0.25–0.5 0.25
(b) Quartzitic sandstones with clay cement	1. NaCl	0.25
(c) Quartzitic sandstones with clay–lime cement	1. $NaCl + Na_2CO_3$	0.25 0.25
(d) Quartzitic sandstones with lime cement	1. Na_2CO_3 2. $Ca(OH)_2$	0.25 0.05–0.07
2 and 3. Silicated igneous and metamorphic rocks with an insignificant content of quartz, or with none at all (syenites, diorites, diabases, nephelines, and chlorite rocks, etc.)	1. Soap (oil soap) + Na_2CO_3 2. NaCl 3. $AlCl_3$	0.25 0.25 0.1–0.25 0.1
4. Carbonate rocks		
(a) Limestones, dolomites	1. Na_2CO_3 2. NaOH 3. $Ca(OH)_2$	0.25 0.05 0.05–0.07
(b) Silicated carbonate rocks	Same as (a) + NaCl	0.25
5. Sulphate rocks (anhydrates, gypsums)	1. NaOH 2. $Ca(OH)_2$	0.5–0.1 0.06
6. Ore rocks		
(a) Martite, haematite, magnesite, pyrite	1. $AlCl_3$ 2. $FeCl_3$ 3. NaCl	0.1 0.1 0.1–0.25
(b) Siderite	1. Na_2CO_3 2. $Ca(OH)_2$	0.25 0.05–0.07
7. Clay rocks		
(a) Shales and argillites	1. NaCl	0.25–0.5
(b) Silicated clay rocks	1. NaCl	0.25
(c) Clay rocks with carbonates	1. $NaCl + Na_2CO_3$	0.25 0.1–0.5

(a) In aqueous solutions.

TABLE 8.5. *Chemical rock weakeners* (*Moavenzadeh* et al. [37])

Reagent	% concentration in H_2O	Description	Possible effect
Distilled water	—	H_2O	Dissolution
Acetone	Pure acetone	Dimethyl ketone	Dissolution
Sodium hydroxide	0.05	NaOH	Dissolution
Aluminium chloride	0.1	$AlCl_3 . 6 H_2O$	Chemical attack or dissolution
Armeen 8D octaylemin	1.0	Quaternary amine	Surface energy reduction and possible chemical attack
Alipal CO-436	1.0	Quaternary amine	,,
Arquad 2C-75	1.0	Quaternary amine	,,
FC 170	1.0	Contains a relatively fluorinated and solubilizing group	,,
FX 172	1.0	,,	,,
Zonyl S-13	0.1	Fluorochemical surfactant fluoroalkyl phosphate free acid	,,
Zonyl A	0.1	Fluorochemical surfactant non-ionic	,,
Z 6020 silane	0.1	An amino functional compound (coupling agent)	,,
A 1110 silane	0.1	,,	,,
A 1120 silane	0.1	,,	,,
A 187 silane	0.1	Epoxy functional silane (coupling agent)	,,

The process of weakening was suggested as "stress activated corrosion", similar to that which is known to occur in brittle metals, and it was observed that a relatively small change in temperature and the application of ultrasonic energy to the chemical bath during mechanical testing considerably lowered the work necessary to fracture the rocks [36].

The possibility of combining the principles of mechanical penetration, to generate tensile stresses, with the use of rock softeners and thermal weakening, is therefore worthy of detailed and systematic exploration.

Direct Thermal Techniques

Rock spalling, melting, and vaporization

The application of heat to some rocks sets up thermal stresses that result in fracture by surface spalling. Norton has defined the "relative spallability" of refractories in accordance with which the relative thermal spallability of some rocks used as refractories is as shown in Table 8.6 [21]. Lauriello and Fritsch suggest that spalling is largely determined by the characteristics of energy attenuation in the material. They observed that when applying microwave heating over a limited energy range, in-depth fractures did not occur but spalling sometimes did take place. They noted that rocks which displayed high attenuation, resulting from entrapped moisture, were most susceptible to surface fracture or spalling.

TABLE 8.6. *Relative spallability of four refractories (Norton [21])*

Quartzite	10.5
Granite	9.5
Dolomite	7.8
Limestone	3.8

Other rocks, such as basalt, do not spall but melt under the applied heat source. Lauriello and Fritsch plot the various parameters, including the effects of spot radius, heat flux, and exposure time, that are possible before spalling occurs in granite, or melting in basalt, in Figs. 8.13A, B [22].

For a rock such as granite, spalling commences at a relatively low surface temperature (around 1100°F) and this limits the extent of in-depth weakening under thermal attack. On the other hand, basalt is susceptible to extensive weakening because spalling rarely occurs, so that surface temperatures up to 2300°F can be attained prior to melting.

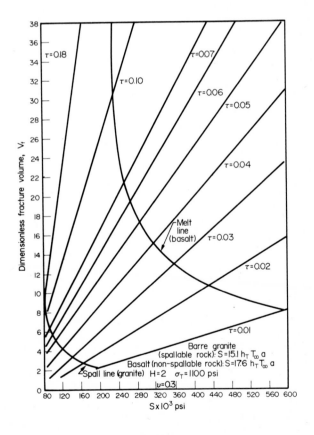

FIG. 8.13A. Thermal break-down of rocks: fracture volume–shear strength parameters.

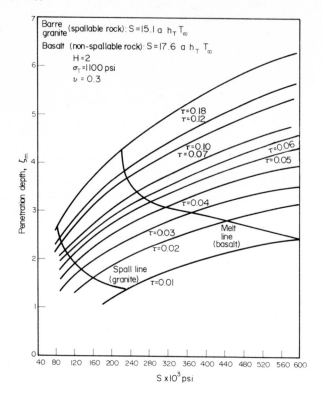

FIG. 8.13B. Thermal break-down of rocks: penetration depth–shear strength parameters (Lauriello and Fritsch [22]).

Rock Penetration by Flame Jets

Two major types of flame-jet gun are used for rock penetration—the postburning flame torch and the jet torch.

In the postburning torch, fuel and oxidizing agent are pre-mixed and burned in the atmosphere upon exit from the supply pipe. An example is the oxyacetylene torch which emits a flame of 6000° F with a high heat transfer coefficient. However, the fuel cost for this type of torch is high, and the consequent rock excavation cost, for a comparable volume of rock, is from three to six times the cost of mechanical excavation in hard rocks.

The jet torch, which uses cheaper fuels, has proved to be a more practical tool, and it is now widely used as a means of drilling rocks, particularly the hard taconite iron ores. In this device the fuel, which is a mixture of hydrocarbon oil and oxygen, or sometimes propane and air, is ignited in a combustion chamber. The shock waves generated by this explosive combustion emerge from the nozzle at supersonic velocities, and this contributes to the fracture process (Fig. 8.14). Typical performance figures are listed in Table 8.7. Anderson and Stresino measured the heat flux produced by oxygen – oil and also by methane – air torches, and a review of jet-piercing rock torches, published by the Canadian Department of Mines and Technical Surveys, quotes specific energies for excavation by various flame jets on dolomite as listed in Table 8.8 [23].

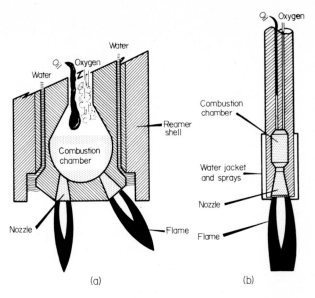

FIG. 8.14. Flame-jet torches: (a) twin-nozzle drill-head; (b) single-nozzle monitor.

TABLE 8.7. *Rock penetration by flame-jet drill (Farmer [9])*

Rock	Piercing speed (ft/h)	Spallability
Granite	25–30	Good—some melting
Quartzite	45–60	Good
Dolomite	35–40	Powdery
Conglomerate	20	
Sandstone	20	

TABLE 8.8. *Excavation by flame-jet torches (Canadian Department of Mines and Technical Surveys)*

Type of burner	Specific energy (joules/cc)
Propane–air	6280
Acetylene–air	5020
Propane–oxygen	4890
Kerosene–oxygen	5000

Plasma Jets

Plasma jets are produced by passing an electric arc discharge across electrodes in a gas-stream moving at high velocity. Temperatures in the range 10 000–20 000°C are generated, capable of fusing any rock. The high temperatures also fuse the electrodes, and to reduce the rate of electrode consumption inert gases such as argon and helium may be employed. However, nitrogen–hydrogen mixtures are also applied because of their easier availability and high heat transfer characteristics.

From 20 to 40 % of the input electrical power is consumed in cooling the electrodes by water. Thorpe describes typical plasma-arc drill operating characteristics, while some account of experimental applications to rock penetration are given by Farmer and also by Bouche [24, 25].

Electron Beams

Electron beams, producing power concentrations from 10^6 to 10^9 W/cm^2, which is capable of melting any rock, are widely used in precision engineering. The beam is produced by applying electrical potential to accelerate electrons from a tungsten carbide cathode, towards the anode, using grid bias and electrolenses to focus the beam on to the workpiece.

Electron beams of up to 60-kW power are available, and larger powered instruments are within the range of current technology. However, since the operator must be protected against X-ray emissions, usually by lead shielding, and the radiation hazard increases with increase of the power applied, beams of 30-kW power probably represent about the limit of practicability for rock penetration.

Schumacher applied a 9-kW electron beam to a variety of rocks to record the specific energies listed in Table 8.9 [25, 26].

Lasers

The energy concentration in a coherent light beam, known as a laser, can be focused to produce power densities in excess of 10^{12} W/cm^2. Early experiments on the use of this concentration of power met with limited success when the lasers were used to pierce holes in rock due to the rapid fall-off in efficiency with increase in stand-off distance. More success has attended the application of lasers for slot cutting and for weakening the rock face.

It has been amply demonstrated that the strength of hard rock is significantly reduced when heat is applied (Fig. 8.15). Williamson *et al.* recorded 90 % loss of original strength of small rock beams that had been subjected to laser heating. In experimental arrangements lasers have been applied to pierce holes 10 cm deep in trap rock. This was done in 5 s using

TABLE 8.9. *Rock penetration by electron beam (Schumacher [26])*

Rock type	Specific energy (joules/cc)
Sandstone	231
Limestone	230
Coarse granite	117
Fine granite	315
Granodiorite	240
Quartzite	244
Taconite	204

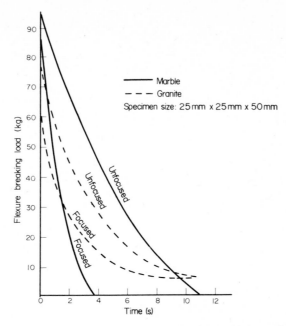

FIG. 8.15. Effect of laser heating on rock strength (Singh after Moavenzadeh *et al.* [27]).

a 6-kW laser. In other experiments, slots 1 cm deep were cut in basalt, granite, and quartzite by a 4-kW laser traversed at a little over 2 cm/s [27].

At the present time it appears that the inclusion of lasers as a supplement with mechanical cutters on tunnel-boring machines is technically feasible, but is not as yet an economic procedure due to the cost of the laser and the associated equipment that is required to deliver the laser energy to the rock face.

Rock Melting

From a theoretical standpoint, rock melting offers great potential as an earth-drilling and tunnelling technique since, not only does it offer an effective means of rock penetration, but the liquid melt, when chilled, forms a glass lining to the excavation and this helps to support the wall. Any excess of debris produced during excavation may also be in the form of glass frozen into pellets or fibres in a condition suitable for transport and disposal.

The method is, in principle, applicable not only to hard rock, but also to unconsolidated ground where, as a result of melt penetration, a wall of glass-cemented fragments can be produced with considerable structural strength and stability.

The Subterrene

Investigators at the Los Alamos Laboratory of the University of California have been experimenting with the technique for many years. A stage has been reached at which prototype electrical melting–consolidating, and extrusion, subterrene penetrators have been designed and applied to bore holes 50–66 mm in diameter. Field and laboratory tests with this equipment are providing the design data that will be essential to scale up the operations to practical and economic proportions.

In penetration by rock melting the physical strength and hardness of the rock are of no account. Therefore the obstacles which variable rock characteristics present to mechanical excavators do not exist. The penetrator is constructed of refractory metals, alloys, and ceramics, whose melting points are considerably higher than those of rocks, and the applied temperatures can be controlled to give an adequate and practicable service life to the penetrator components.

Various design procedures are possible, and these are classified according to the method by which the glass melt and the rock debris are disposed of. They are: consolidation, extruding, coring, and lithofracture.

Complete consolidation is possible in porous materials in which the whole of the material penetrated forms a glass wall during its transformation from porous rock into melt, and no material is discharged from the hole. In denser materials the principle of extrusion penetration is applied in which, not only is a glass wall formed, but debris must also be removed. The coring penetrator melts only an annulus of rock, and leaves an unmelted core, cased in glass. This core must be removed mechanically. The principle of lithofracture is to inject the melt into fissures and cracks in the surrounding rock wall (Fig. 8.16).

The prototype consolidation–melt penetrator used in the field tests included a 10-kW electrical power source, a 15-h.p. air compressor for the coolant, and a 2000–3000-lb hydraulic piston to apply thrust to and to retrieve the penetrator stem. No torque is necessary, so that no rotating machinery is required. This equipment bored 50-mm diameter holes horizontally and vertically in volcanic tuff to a depth of 25 m. The penetrator life was in excess of 100 h (Fig. 8.17).

Hanold analyses the power requirements and the operating characteristics of a kerfing penetrator of full-scale proportions in various rocks. The estimated power source requirements are around 10 MW for a soft-ground kerf-melting tunneller, and around 25 MW for a kerf-melting machine, supplemented by thermal weakening and stress fracturing, in hard rock. Alternative chemical, electrical, and nuclear heat sources are considered, and it is anticipated that economic and technical considerations may

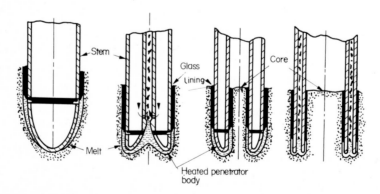

FIG. 8.16. Rock-melting penetrator principles: (a) consolidation melting; (b) extrusion melting; (c) consolidation coring; (d) coring and extrusion (Neudecker *et al.* [39]).

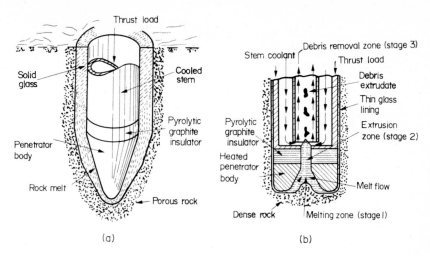

FIG. 8.17. (a) Consolidation melting penetrator; (b) extrusion melting penetrator (Gido [41]).

ultimately favour the use of compact nuclear power sources. In this connection the expertise developed during studies of compact nuclear power sources for space vehicles, submarines, and marine craft may have an important bearing on the future of underground tunnelling technology. The development of compact nuclear power sources of from 10 to 25 MW, shielded and suitable for large rock tunnellers, is well within the range of current technology, as also are the automated control systems with which to operate them [38–42].

Comparison of "Exotic" Rock-penetration Systems

When comparing the many novel rock-penetration systems that have been investigated, a sound basis of comparison is difficult to obtain. The concept of specific energy is probably the most widely used method. In recent years this has been supplemented by the dimensionless "rock number". However, there is no common agreement as to how the specific energy should be determined, and different procedures may produce widely different numerical values.

Also, the concept of specific energy is not always relevant. For example, from a specific energy point of view, techniques such as rock melting and water-jet cutting are inefficient because they utilize all their energy in completely destroying a relatively small volume of rock. But if this principle is applied to cut a narrow kerf, so as to form a surface of relief to which the whole rock surface of the tunnel can be broken into large pieces, say by wedging or by thermal cracking, then the system becomes potentially highly efficient.

Again, some of the techniques, such as melt penetration and flame jets, may have to be discounted for underground use on account of the unacceptable environmental conditions produced by radiation hazards, noxious gases, and heat. Environmental problems do not occur to the same extent in open workings, so that the rock-fusion and thermal-lance techniques may be more applicable to well boring and trench cutting at the ground surface. For underground work, substantial assistance may be gained from thermal weakening, using commercially available electric heaters.

We should not overlook the fact that, in the end, it may not be questions of engineering or of economic feasibility that decide whether or not a particular tunnelling project may proceed. The overriding criterion is more likely to be one set by the community at large and regulated by legal process. In order to meet such a criterion the project will have to be conducted in such a manner as to meet environmental standards framed to ensure the safety and health of working personnel and to protect the community at large.

From this point of view the underground environmental problems generated by rock fusion, radiation shielding, and nuclear power systems differ only in degree from those that are already approaching the intolerable in high-thrust mechanical tunnel boring.

We may therefore speculate that the future of rapid excavation technology probably lies in two directions—one towards fully automatic systems operated by remote control, typified by fusion melt boring; the other would make use of mechanical principles including hydraulic jet cutting, possibly incorporating chemical and thermal rock-weakening techniques, and manually operated at medium and short range.

However, excavation accounts for only approximately one-third of the cost of a tunnelling project. Materials handling accounts for a little less than one-third of the total cost when tunnelling through rock and about one-quarter of the total when excavating in soft ground. The remainder, and the bulk, of the total cost is attributable to the supports and linings.

References and Bibliography: Chapter 8

1. NASIATKA, T. M., *Tunnelling Technology—Its Past and Present*, US Bur. Min., Inf. Circ. No. 8365, 1968.
2. MUIRHEAD, I. R., and GLOSSOP, L. G., Hard rock tunnelling machines, *Bull. Instn Min. Metall.*, No. 734, pp. A1–A21 (1969).
3. BRUCE, W. E., and MORRELL, R. J., Principles of rock cutting applied to mechanical boring machines, *Proc. 2nd Symp. on Rapid Excavation, Sacramento State College, October 1969*, paper 3, pp. 1–43.
4. TEALE, R., The mechanical excavation of rock—experiments with roller cutters, *Int. J. Rock Mech. Min. Sci.* **1**, 63–78 (1964).
5. ROBBINS, R. J., Economic factors in tunnel boring, *Proc. S. African Inst. Tunnelling Conf., Johannesburg, 1970*.
6. ANON., *Feasibility of Flame Jet Tunnelling*, Report Fenix and Scissons Inc. Conventional Tunnelling Methods, May 1969.
7. WILLIAMSON, T. N., *Tunnelling Machines of Today and Tomorrow*, Highway Research Record, Rep. No. 339, Rapid Excavation, 1970.
8. WHEBY, F. T., and CIKANEK, S., *A Computer Program for Estimating Costs of Hard Rock Tunnelling (COHART)*, Harza Engineering Co. Inc., Chicago, May 1970.
9. FARMER, I. W., New methods of fracturing rock, *Mines Min. Engng*, pp. 177–84 (Jan. 1965).
10. MAURER, W. C., *Novel Drilling Techniques*, Pergamon Press, Oxford, 1968.
11. SIMON, R., Drilling by vibration, *Trans. Am. Inst. Min. Engrs*, paper No. 58 PET–21, July 1958.
12. GUMENSKI, B. M., and KOMAROV, M. S., *Soil Drilling by Vibration*, Consultants Bureau Translation, New York, 1961.
13. NASIATKA, T. M., and BADDA, F., *Hydraulic Coal Mining Research*, US Bur. Min., Rep. Invest. No. 6276, 1963.
14. MCMILLAN, E. R., Hydraulic jet mining, *Min. Engr* **14**, (6) 41–43 (1962).
15. FARMER, I. W., and ATTEWELL, P. B., Rock penetration by high velocity water jet, *Int. J. Rock Mech. Min. Sci.* **2**, 135–53 (1964).
16. EPSHTEYN, Y. F., ARSH, E. I., and VITORT, C. K., *New Methods of Crushing Rocks*, US Dept. of Commerce Translation, Washington DC, 1962.
17. HARRIS, H. D., and MELLOR, M., Cutting rock with water jets, *Int. J. Rock Mech. Min. Sci.* **11**, 343–56 (1974); and HARRIS, H. D., and BRIERLEY, W. H., A rotating water jet device, ibid., 359–65 (1974).
18. CROW, S. C., A theory of hydraulic rock cutting, *Int. J. Rock Mech. Min. Sci.* **10**, 567–84 (1973).
19. CROW, S. C., The effect of porosity on hydraulic rock cutting, *Int. J. Rock Mech. Min. Sci.* **11**, 103–5 (1974).
20. ANGONA, F. A., Cavitation—a novel drilling concept, *Int. J. Rock Mech. Min. Sci.* **11**, 115–19 (1974).

21. NORTON, F. H., *Refractories*, McGraw-Hill, New York, 1931.
22. LAURIELLO, P. J., and FRITSCH, C. A., Design and economic restraints of thermal rock weakening techniques, *Int. J. Rock Mech. Min. Sci.* **11**, 31–39 (1974).
23. ANDERSON, J. E., and STRESINO, E. F., Heat transfer from flames impinging on flat and cylindrical surfaces, *J. Heat Transfer* (Feb. 1963).
24. THORPE, M. L., The plasma jet and its uses, *Res. Dev.* **11**, (1) 5–15 (1960).
25. BOUCHE, R. E., Drilling rocks with plasma jets, MS thesis, Colorado School of Mines, 1964.
26. SCHUMACHER, B. W., *Electron Beams—A New Tool for Cutting and Breaking Rock*, Westinghouse Res. Lab., paper 68–102–EWELD–P2 (Nov. 1968); also *Yale Scientific Magazine*, Mar. 1969.
27. MOAVENZADEH, F., WILLIAMSON, R. B., and McGARRY, F. J., *Laser Assisted Rock Fracture*, MIT Research Rept., Dept. of Civil Engineering, PB 174245, Jan. 1967; and JUREWICZ, B. R., CARSTENS, J. P., and BANAS, C. M., Rock kerfing with high powered lasers, *Proc. 3rd Congr. Int. Soc. Rock Mech., Denver, Colo., 1974*, **2B**, 1434–40.
28. OSTROVSKI, N. P., *Deephole Drilling with Explosives*, Consultants Bureau, New York Translation, 1960; and BLACK, W. L., *Development of Equipment for Explosive Drilling*, AAI Corp. Cockeysville MD, Engineering Rep. ER-8724, DOD Contract DAAG53-76-C-0023, June 1976.
29. ECKEL, J. E., DEILY, F. H., and LEDGERWOOD, L. D., Development and testing of jet pump pellet impact drill bits, *Trans. Am. Inst. Min. Engrs* **207**, 1–9 (1956); and NEWSOM, M. M., ALVIS, R. L., and DARDICK, D., *The Terradrill Program*, Sandia Labs. Rep. No. SAND 76-0228, June 1976.
30. MAURER, W. C., and RINEHART, J. S., Impact crater formation in rock, *J. Appl. Phys.* **31**, 1247–9 (July 1960).
31. SINGH, M. M., Novel methods of rock breakage, *Proc. 2nd Symp. on Rapid Excavation, Sacramento State College, October 1969*, pp. 4.1–4.27.
32. BENET, R., *Penetration of Shaped Charges into Frozen Ground*, US Army CRREL Tech. Rep. 130, 1963.
33. VAN THIEL, M., WILKINS, M., and MITCHELL, A., Shaped charge sequencing, *Int. J. Rock Mech. Min. Sci.* **12**, 283–8 (1975).
34. ROLLINS, R. R., CLARK, G. B., and KALIA, H. N., Penetration in granite by jets from shaped charge liners of six materials, *Int. J. Rock Mech. Min. Sci.* **10**, 326–30 (1973).
35. REHBINDER, P. A., SCHREINER, L. A., and ZHIGACH, K. F., *Hardness Reducers in Drilling*, CSIR Translation, Melbourne, Australia, 1948.
36. HEINS, R. W., and STREET, H., Hardness reduction by wetting, *Trans. Am. Inst. Min. Engrs* **229**, 223–4 (June 1964).
37. MOAVENZADEH, F., WILLIAMSON, R. B., and WISSA, A. E. Z., *Rock Fracture Research*, MIT Dept. of Civil Engng. Res. Rep. R66–56, 1966.
38. KRUPKA, M. C., *Phenomena Associated with the Process of Rock Melting*, Rep. LA5208-MS-UC38, US Dept. of Commerce, NTIS, Feb. 1973.
39. NEUDECKER, J. W., GEIGER, A. J., and ARMSTRONG, P. E., *Design and Development of a Prototype Universal Extruding Subterrene Penetrator*, Rep. LA-5025-MS UC38, NTIS, Mar. 1973.
40. SIMS, D. L., *Identification of Potential Applications for Rock Melting Subterrenes*, Rep. LA-5026-MS UC38, NTIS, Feb. 1973.
41. GIDO, R. G., *Description of Field Tests for Rock Melting Penetration*, Rep. LA-5213-MS UC38, NTIS, Feb. 1973.
42. HANOLD, R. J., *Large Subterrene Rock Melting Tunnel Systems*, Rep. LA-5210-MS UC38, NTIS, Feb. 1973 and Rep. No. LA-5979-SR, 1977.
43. SARAPU, E., *Rock Breaking by Electrical Energy*, Electrofrac Corporation, Kansas City, Missouri, 1976.
44. FOGELSON, D. E., Advanced fragmentation techniques, *Proc. 3rd Congr. Int. Soc. Rock Mech., Denver, Colo., 1974*, **1B**, 1645–63.
45. MAURER, W. C., Advanced excavation methods. *Underground Space.* **2**, 99–112 (1977).

CHAPTER 9

Movement and Control of Groundwater—I

Environmental Aspects

AMONG the most urgent of the world problems of our time is how best to supply the raw materials and energy that are required as the basis of an industrialized social economy. As the emergent nations look more and more towards industrialization as a means of raising their living standards so the demand for materials increases, and it accelerates as the general standard of living improves. Of all natural resources none is more important than water, for, beyond the basic requirement to support life it appears that the per capita demand for domestic water increases with increasing sophistication of the social order. Also, greater industrial utilization of water ensues as a result of the evolution of new technologies, particularly in such fields as chemical engineering, hydrometallurgy, and heat-transfer processes.

Not only that, but among the questionable benefits of industrialization is the possibility of supporting urban populations of much greater density than was formerly the case in a rural agrarian society. When added to the benefits of improved preventive and social medicine, the consequences are seen in an explosion of population which the world's agriculture is strained to feed. In order to approach this task, agriculture itself has perforce to be organized in such a manner and on such a scale as itself to represent an industrialized agriculture, and this must extend beyond the lush farm regions of the world to incorporate more and more of the marginal farmlands and arid regions.

If we define an arid region as one where systematic agriculture is not possible without irrigation, then we should note that the arid regions of the world include vast areas which receive heavy rainfall. They include, for example, such territories as the subcontinent of India, where the even tenor of life is punctuated by periodic crises which alternate between disaster by flood and famine by drought, and tropical regions like Northern Queensland, Australia, which receives 20–25 in. of rainfall each year. However, this occurs in the summer, when evaporation rates are high, so that continuous irrigation is necessary for crop growth [1, 2].

But the need for planned management of water resources is not limited to arid regions. It exists in densely populated areas such as Britain, the Netherlands, and Western Germany, where rainfall is generally ample for systematic agriculture. Here the needs of urban populations and industry are satisfied at the expense of drowning rich agricultural land underneath surface reservoirs, while the river waters of the Rhine and the Thames must be used, re-cycled, and re-used several times between their source and the sea.

Water Management

The need for planned and controlled exploitation of natural water resources is therefore becoming generally recognized today, although it is far from simple to put into effect. There are those who maintain that water is unmanageable, since, like the air we breathe, it must be available to all who have access to it. While surface waters may be impounded and privately enclosed, there are no similar restrictions upon whom rain or other precipitation may fall, while the unseen store of underground water is constantly replenished by nature and can be reached by almost anyone who has the means either to dig a well or drill a borehole, or can pay someone else to do this for him.

Where control is exercised, it is usually through the medium of a publicly-owned or state-owned authority, or a private company financed by shareholders, with specific and limited powers and terms of reference. These do not always provide for control over water pumped from wells privately owned or operated by individuals in the area concerned. But systematic water management should embrace even wider powers than this. In urban and industrial areas the ground is frequently used as a depository for fluid wastes, as a heat-sink, and as a filter for sewage and chemical effluents. In rural and mountain areas such activities as deforestation and replanting of forests have a profound effect on groundwater hydrology, surface runoff, and soil erosion. Mining, surface drainage and irrigation, and the encroachment of urban development upon rural areas all affect the amount and quality of water that finds its way into the ground. The ground is thus a main source of potential water supply for the multifarious uses of modern society. The aim of water management should be to control this reservoir, and all its various modes of input and output, to the optimum benefit of society in terms of water quantity, quality, cost, and reliability of supply. We should aim to increase the proportion of atmospheric precipitation that is retained as groundwater. About 70% of the water that falls as precipitation on the United States is returned to the atmosphere through evaporation and transpiration. Of the remainder, only one-third, or approximately 10% of the total, is utilized directly by man [3].

The Hydrological Cycle

The general features of the natural hydrological cycle will be well known to all students of physical geology and geography. The underground portion of the cycle centres on the aquifer, a porous rock layer in which water can accumulate, with pores and other interstices through which the water may move under the influence of gravity. Water can be retained, to saturate the aquifer below the water table, if an impervious boundary exists below and to the sides of a catchment area. It is sometimes possible for the stored water to flow laterally into a zone where it is also bounded above by an impervious layer, in which case it is retained under the hydraulic pressure generated by the free water table. This water originates from the atmosphere either as direct precipitation upon the ground surface or indirectly as seepage from surface lakes and rivers. It percolates under the influence of gravity through permeable soil and unsaturated rock layers to join the saturated reservoir of underground water below the water table. Lateral flow in the saturated zone is promoted as a result of point-by-point differences in hydraulic potential, measured by piezometric pressure, so that drainage proceeds towards natural outlets which are visible as springs and surface seepage, ultimately reaching the ocean via streams and rivers. At all points where the water is accessible to vegetation and to the atmosphere, evaporation and transpiration to the atmosphere complete the cycle.

The groundwater flow system thus comprises three portions: (1) the recharge zone, (2) the lateral flow zone, and (3) the discharge zone. Techniques of water management and control may be applied specifically to one or more of these three zones to modify the natural relationships between them in accordance with some desired aim.

Determination of Groundwater Flow Characteristics

The first step in any control procedure is to determine the facts of the situation. In the lateral flow zone the movement of water in the aquifer is governed by two physical laws:

(a) Darcy's law of fluid flow through porous media, which states that the rate of flow through a porous and permeable medium is proportional to the gradient of the hydraulic potential or "hydraulic gradient".
(b) The fundamental principle of conservation of matter which states simply that water in an aquifer is neither created nor destroyed.

When considered as an exercise in the application of these fundamental laws, deducing the hydraulic characteristics of an aquifer, in terms of flow direction, rate of flow, and distribution of hydraulic potential, is a problem in mathematics. In practice, however, the problem is far from being straightforward, and it is rendered very difficult indeed, mainly because of the variable and oftentimes unknown permeability of the rocks comprising the aquifer and the uncertainty in defining the relevant boundary conditions.

Groundwater movement is a three-dimensional process, and the sediments, of which most aquifers are composed, vary in permeability in three principal coordinate directions. Changes in permeability with depth are not easily measured, but they have an important influence on underground water movement, if, for example, by pumping from an aquifer at shallow depth, an upward flow may be induced in the aquifer [4].

To a limited extent permeabilities may be determined from pumping tests, and boreholes can provide points of access for observation of rock characteristics and piezometric pressure. Geophysical techniques may also be applied to help locate pervious–impervious boundaries and to observe water-level changes indirectly, while considerable advances are being made in the application of radioactive tracer techniques for the direct observation of flow paths. Considerable assistance, too, may be obtained from model studies, using either liquid analogues in which the flow of groundwater is represented by the laminar flow of oil between narrowly spaced glass plates, or electrical analogues, which make use of the analogy between the flow of electrical current and groundwater under variable potential. Nevertheless, in the end the information obtained is liable to be costly, "spotty", and incomplete [5].

The problems involved in attempting to acquire the relevant data are no less difficult, although for different reasons, in the recharge and discharge zones. In the absence of any interference by man the input to the system is determined by geography and climate and consists solely of atmospheric precipitation. While precise measurements may be made over a limited time scale and limited area, quantitative measurements and forecasts over a longer time and over a wide area may only be roughly estimated. The hydrologist and meteorologist both talk in terms of average annual replenishment to the aquifer based on average rainfall and average snow depth, which in a balanced system must be equal to the average annual outflow. Actual measurements in a specific investigation may show figures that differ widely from average estimates, and the characteristics of variation are entirely random.

The proportion of precipitation input to a surface-water system depends upon evaporation, transpiration, shape and size of the drainage area, slope of the surface contours, and the length of the surface streamflow. The proportion of precipitation input to the underground system is indicated by progressive decrease in the percentage estimated runoff in the downstream areas compared with precipitation. This decrease is determined (a) by the direct infiltration of surface water into the underlying earth and its percolation to the groundwater table, (b) by surface water runoff in areas of high ground relief into the mountain-front alluvial fans, and (c) by percolation along the surface stream and river channels. In arid regions it is probable that the greater part of groundwater recharge is due to percolation along stream channels and in the mountain-front alluvial fans.

At first sight it might be expected that determination of outflow should be easier than that of inflow, because drainage sites are more localized than precipitation areas, and the flow of seepages, springs, streams, and rivers may be precisely measured and constantly monitored. Yet difficulties exist due to human, political, and social factors, and particularly in relation to outputs from wells and boreholes privately or individually owned. Political boundaries, subject to some overriding control authority, are not likely to coincide with hydrological boundaries, and private well-owners are generally loath to discuss how much water they actually use. They are likely to consider that this, like their bank balance, is a purely personal and private affair. In the absence of compulsory metering and recording at all points of outflow and water utilization it is difficult to see how this situation may be remedied. Until such time as it is remedied, groundwater hydrology, like rock mechanics, will remain very largely an intuitive art of making deductions, starting from a basis of incomplete and inadequate data. In applying this art, certain concepts are fundamental.

Underground Water Reserves

The general characteristics of the hydrological cycle are as depicted in Fig. 9.1. It is axiomatic that, over a long term, no aquifer should be exploited at a rate beyond that

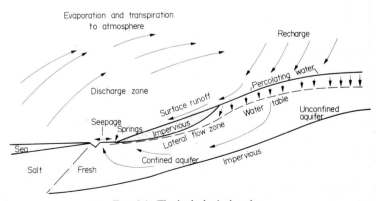

FIG. 9.1. The hydrological cycle.

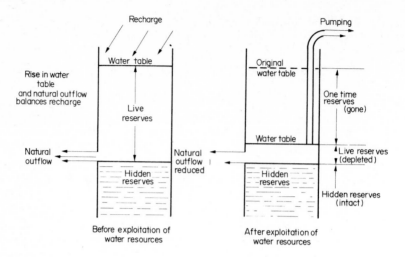

Fig. 9.2. One-time, live, and hidden reserves.

which can be made good by recharge, if the total reserves of water are not to be depleted. But such a simple balance does not permit of optimum utilization and effective exploitation of the reserves over a limited time. In considering such exploitation we should distinguish between three component parts of the total reserves. These are the fractions termed "one-time", "live", and "hidden" reserves, respectively, in Fig. 9.2. The live reserves comprise all that body of water that lies above the natural outflow or drainage level, and it is the gravitational force upon this mass which provides the energy of flow. Water in the aquifer below the natural drainage level forms a hidden reserve. The upper portion of the hidden reserves may lie in the flow path generated by the live reserves, but does not, in itself, generate flow.

The salinity of the hidden reserves increases with increase in depth and laterally towards the sea, with a diffused zone of mixing between fresh and salt water. In times of drought or excessive freshwater exploitation it is possible for this diffused zone to be drawn into the flow-stream pattern, so that the aquifer becomes contaminated.

The live reserves provide a volume of water that can be exploited at a rate that, when added to the natural outflow, exceeds the rate of recharge, but in so doing the level of the water table will fall. The original water-table level will then never be regained so long as the aquifer depends upon natural precipitation for its recharge. The excess volume of (exploitation plus natural outflow) over recharge can thus only be utilized once. Hence it is termed a "one-time" reserve.

Lowering the water table reduces the energy in the live reserves and this affects the flow characteristics and reduces the natural outflow, but the system responds sluggishly to change and there will be some time-lag between cause and effect. Obviously it is not wise to deplete the live reserves to the extent that natural outflow falls to unacceptably low levels, leaving no margin of safety to provide for unpredictable climatic drought, and the threat of salt water encroaching upon the freshwater supply. Not all the live reserves, but only a proportion of them, therefore constitutes the "one-time" reserves. What that proportion will be, in a given situation, 50%, 60%, or perhaps 70%, is very largely a matter of judgement. It depends upon the superficial extent of the watershed in relation to the storage capacity and extent of the aquifer, and the groundwater flow characteristics in relation to the projected exploitation and natural outflow. There must remain in the aquifer a sufficient volume of live reserves to provide for all contingencies, bearing in mind the relative seasonal probability of prolonged drought. Depletion of the live reserves to this level will then have no effect on the long term ability of the aquifer to function as a water-supply source provided that a balance is restored between recharge and (utilization plus natural outflow) before the danger level is reached. In the interim the one-time reserves may supply water over a period of many months, or even some years, during which time arid lands may be irrigated, farms built and brought into operation, communities and services established, pipelines and canals constructed, using water at a rate far exceeding that to which it must ultimately be reduced (in the absence of artificial recharge) to re-establish the hydrological balance.

Artificial Recharge

By artificial recharge is meant the importation of water from outside the natural hydrological basin. By utilizing the one-time groundwater reserves not only is it possible to provide the water necessary for immediate recovery of arid lands, but it is also possible to provide the time required for the construction of the engineering works that are needed to bring in water from outside—dams and reservoirs, pipelines, and canals.

In this connection the provision of adequate surface storage capacity presents problems, some of which are becoming increasingly difficult to solve with the passage of time. To begin with, most of the world's "easy" dam sites have already been used, and the sites that may have to be chosen in future are likely to present abnormal problems of relative inaccessibility, construction costs, or unfavourable geology in respect of either or both the impoundment of water in the reservoir and the rock abutments and foundations for the dam. And surface reservoirs are becoming increasingly unpopular anyway. The environmentalists hold that the world would be a much better place with fewer dams, not more. The stability of those that already exist is sometimes a matter of concern in that they may pose a threat to any communities living near them on the downstream side, particularly in the event of earthquake, landslide, or abnormally heavy rainfall or seepage.

Where the geology is favourable and no questions of stability exist, there may be dispute as to whether the new amenities for mass recreation that are often provided by the impounded water, together with the new roads which provide improved accessibility to the area, can fully compensate for what has been irretrievably submerged beneath the reservoir, while the preservation of the natural beauty and ecology of a native wilderness area becomes more difficult, and less likely to ensue, the more easily it becomes accessible to man. Even the utilitarian value of surface reservoirs can be disputed where other means exist, for in arid regions evaporation losses seriously impair the efficacy of a surface reservoir while in more densely populated humid regions the reservoir almost inevitably involves the loss of land which would otherwise remain for agricultural use.

Use of Aquifers in Artificial Recharge

The alternative to constructing surface storage reservoirs is to make use of underground storage capacity. The depletion of one-time reserves in a hydrological basin leaves a large volume of porous and permeable strata that are available for artificial recharge and storage. As yet we make relatively little use of this storage capacity. In wet and humid climates, such as Great Britain and western Europe, it is hardly used at all, and in these regions the community depends mainly on surface storage capacity coupled with groundwater utilization. The surface storage capacity is so limited, in proportion to the density of the population which it is required to serve, that a period of only two or three weeks without rainfall can threaten to pose water-supply problems to some, less favoured, communities. Yet the average annual rainfall is more than ample for every social, agricultural, and industrial need. In urban areas much of the rainfall is on asphalt, brick, and concrete, and passes directly into the town drains, to find its way to the sea, unused.

There are two methods of recovering runoff water. One is to channel it into low-lying "spreading grounds", where in wet periods it will form swamp areas from which percolation can find its way to the underground water table. This is a method best suited to areas of poor agricultural value in humid regions, or applied as a means of retaining flash flood water in arid regions. The other method is to channel it into settling ponds or "lodges" whence it may be injected, by pumping directly through boreholes, into the underground aquifer.

Injection of Recharge Water

Certain technical problems are liable to occur in connection with the injection of recharge water into an aquifer. The injection pressure required (which must exceed the piezometric head at any point) will build up progressively during pumping, due to the deposition of silt and the growth of algae on the walls of the borehole and in the interstices of the aquifer at and near the point of entry. Bacterial growth and algae can be inhibited by chlorination of the water, while the surface settling ponds must be of a size and depth that will allow silt to be deposited before the water enters the pump. The borehole walls can be cleared periodically by alternately putting the hole under high pressure and then suction, to flush foreign material off the wall and then draw it out. If water is injected at high pressure there will be considerable mechanical erosion of rock material around the point of entry and it should not be forgotten that any increase of fluid pore pressure in the rock will reduce its mechanical strength. This point is important if geological fault zones exist in the vicinity of the borehole, or if the aquifer contains open fissures or solution cavities in which air can be entrained. In the former event, not only is the fault plane subject to lubrication by water, so that its frictional properties are reduced, but the increased pore pressure reduces any stresses acting across the fault plane—two factors, either or both of which can lead to strata slip along the fault plane and hence result in a seismic disturbance.

The entrainment of air in the injected water should be prevented because any such air can find its way into fissures and solution cavities in the aquifer. Once there it will accumulate under pressure to blow back with considerable violence and hydraulic impact as soon as pumping stops. Besides being liable to cause mechanical damage to the pump system, in conjunction with the effects of increased pore pressure weakening the rock, this impact can be so great as to shatter the rock around the borehole and result in collapse of the rock walls.

Water injected at a location within the lateral flow zone will be available for recovery at a downstream point in the groundwater flowstream. Or the same well may be used alternately for pumping and injection, the "cone of depression" in the groundwater table around the well being restored by subsequently injecting water imported from outside the hydrological basin or recovered from what would otherwise run off during a wet season.

Groundwater Quality

All natural water contains some mineral matter in solution, the proportion being greater in groundwater than in surface water due to chemical interchange in the subterranean aquifers. The water becomes more saline with increase in the duration of contact with earth materials, and hence more saline in terms of distance and time. In the absence of interference by man the mineral matter contained in the natural outflow from the discharge zone, both surface and underground, is deposited in the sea. The effects of any utilization systems that may be imposed upon the natural hydrological cycle by man tend to deteriorate the quality of water, by reason of increase in salinity, to a progressively greater extent downstream.

This happens to surface water as well as to underground water supplies. For example, control of the natural drainage basins by the construction of dams and reservoirs in the Southwest region of the United States has reduced by a very considerable extent the volume of water that previously drained more quickly, carrying mineral matter into the Gulf of Mexico. Much of this water, being used to irrigate the region, is brought into intimate contact with soil and rock and picks up a much greater load of dissolved salts than formerly was the case. This water drains into the reservoirs impounded behind the dams of the Colorado and other rivers of the Southwest, whose lower reaches now have a serious salt problem. The potential consequences to such an area may be judged from the fact that a mineral content due to dissolved solids of about 800 parts per million, including about 160 parts per million of the sodium ion, is hardly noticed in drinking water but it is deadly to citrus crops.

Wells that pump from the groundwater in coastal regions should be carefully observed for possible contamination by seawater. In the undisturbed natural drainage system some portion of the aquifer's submarine outcrop will drain seawards, but the lower portion of the outcrop will be invaded by seawater. The density difference between saltwater and freshwater maintains a freshwater lens floating on the saltwater with about one-fortieth of the total thickness of the lens raised above sea-level. So long as natural drainage remains undisturbed, the slow seaward flow of freshwater maintains the zone of contact, in which mixing between saltwater and freshwater occurs, in a generally stable condition. In these circumstances the interface is more or less well defined and mixing is determined mainly by the process of molecular diffusion (Fig. 9.3).

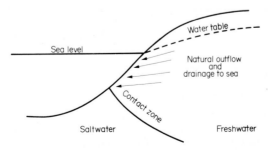

FIG. 9.3. Natural outflow and drainage to sea.

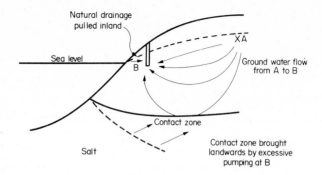

FIG. 9.4.   Excessive pumping of groundwater near saltwater contact.

But if groundwater is pumped from the coastal region, say from point *B* on Fig. 9.4, the contact zone is pulled towards the land, and if the rate of pumping is excessive the zone could be drawn into the flowstream, which can extend for some distance below the base elevation of the well-point. Besides bringing the zone nearer, the characteristics of mixing in the contact zone are changed. Kinetic energy in the drainage flow now plays a more important part, and in place of a well-defined interface controlled by molecular diffusion a much more diffuse interface results, in which hydrodynamic mixing occurs. As a result, the density differences between freshwater and mixed water are very much reduced and the mixed water is easily drawn into the flow-stream to the well. A mixture of 5 % seawater in freshwater is undrinkable in normal circumstances, while a 2 % mixture is fatal to all but the most salt-resistant herbage.

In urban areas, conservation, treatment, and recycling of water from industrial effluents, sewage works, and the like, all involve an increase in the residual salinity of the groundwater. Importation and artificial recharge systems can bring in freshwater that will dilute the groundwater salinity to acceptable limits. Conversely, where water with a high mineral content has to be used, either as a source of recharge water or disposed of as an effluent, it can be injected at an upstream point of the basin, so that it may mix and be diluted to acceptable salinity before being extracted for re-use at a point further downstream. In both cases, careful control, observation, and research on the processes of mixing are important factors on which success will depend.

Alternatives to Groundwater Management

From time to time other, more grandiose, schemes are advocated to solve regional water-supply problems, sometimes on an international scale. These include large coastal seawater desalinization plants, cloud-seeding to induce more precipitation, and the cross-continental transport of water in pipelines, e.g. from the Pacific Northwest and Canada to the American Southwest and Mexico. The transport of ice by towing large icebergs from the Antarctic to the desert countries of the Middle East is another scheme proposed. Whether any of these schemes will come to fruition on a large scale is very doubtful on grounds of engineering feasibility, practicality, and economics. More effective use of the world's groundwater supplies is another matter altogether, and one which should be regarded as a first priority, calling for urgent action at the highest national and international levels. One of the most immediately effective measures would be to require every user of water, individual and corporate, to account for what is used.

References and Bibliography: Chapter 9

1. BATISSE, M., *Problems Facing Arid-land Nations, Arid Lands in Perspective*, pp. 3–13, University of Arizona Press, 1969.
2. CHRISTIAN, C. S. and PERRY, R., *Arid Land Studies in Australia, Arid Lands in Perspective*, pp. 207–25, Tucson, Arizona, 1960.
3. FOGEL, M. M., Management of water resources, in *Mining and Ecology in the Arid Environment*, pp. 135–41, Univ. of Arizona Press, Tucson, Arizona, 1970.
4. MANDEL, S., Underground water, *Int. Sci. Technol.*, pp. 35–41 (1967).
5. HARSHBURGER, J. W., Source, movement, and developments of water in arid regions, in *Mining and Ecology in the Human Environment*, pp. 122–34, Univ. of Arizona Press, Mar. 1970.

FIG.
arra

The velocity of groundwater flow streams may be observed directly using radioactive tracers. These are injected at an upstream point and simultaneously monitored at various points downstream. Another instrument is the electronic liquid velocity evaluation system which applies an electronic heat source inserted into the flow stream at the base of a borehole. This generates a convection current which rises into an annular sensor. When no groundwater current flows the convection current is sensed uniformly around the annulus. In the event of underground water movement the convection current is deflected, its direction and magnitude being picked up by the sensors and recorded graphically on an *XY* plotter. The coordinate directions are established by a magnetic compass.

Recording Groundwater Flow Characteristics

The results of a study of groundwater characteristics during an engineering site investigation may be recorded on plans and sections. In general, plans on a scale of 1/25 000, or larger, are required, showing surface ground contours. On such a plan the observed depths to the water table at various points over the site are marked in, from which may be deduced the general contours of the underground water table, by joining points at equal level.

By combining the two families of contours, of surface level and water-table level, a third family of isopachytes, or lines of equal strata thickness, may be produced, showing the depths to groundwater over the site. The mode of construction of such a map is shown in Fig. 10.2. To avoid confusion it is helpful to plot the surface and water-table contours in two different colours and then draw the isopachytes on a separate tracing superimposed on the contour plan. From the combined plan, profiles may be produced showing surface slopes and groundwater table characteristics along specific directions over the site.

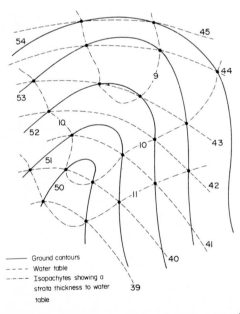

Ground contours
Water table
Isopachytes showing a strata thickness to water table

Fig. 10.2. Isopachytes showing strata depth to water table.

The water-table contours are lines of equal hydraulic potential which determine the general direction of groundwater flow along the water table. Since this will be in the directions of maximum hydraulic gradient, the flowstream is generally at right angles to the contour direction.

Similarly, if we consider flow at some depth within a saturated earth material, or in a confined aquifer under pressure, the total hydrostatic head at any point is the algebraic sum of velocity head, static pressure head, and gravitational potential head at that point, i.e. piezometric head or fluid potential (neglecting atmospheric pressure)

$$h = \frac{v^2}{2g} + \frac{p}{\gamma_w} + z.$$

In seepage problems the flow velocity is so small that the first term may be neglected, and

$$h = \frac{p}{\gamma_w} + z.$$

Flow between any two points A and B is then determined by Darcy's law $v \propto ki$, where i is the hydraulic gradient.

Hence
$$v = ki,$$

where k is the permeability, and

$$q = aki,$$

where a is the cross-sectional area of flow stream and q is the quantity flowing in unit time.

Flow Nets

The flow characteristics may be represented graphically in two dimensions, representing either a plan or a profile of the flow stream. In either case the two-dimensional flow pattern contains two families of curves: (a) equipotentials, representing lines of equal fluid potential, and (b) flow lines representing unit flow channels. Between each pair of adjacent flow lines representing a stream tube there is constant flow quantity. Such a plot is termed the flow net.

For many groundwater problems a graphical solution can provide results of practical value. Generally, such a solution can be obtained by drawing flow nets in one or more principal vertical cross-sections, conforming to the geometry and permeability of the earth medium and the known boundary conditions. Certain fundamental rules to be observed are:

1. Flow lines and equipotential lines cross each other at right angles.
2. Where a finite velocity exists only one flow line can pass through a given point at one and the same time.
3. The discharge along any stream tube is constant.
4. A stream tube cannot converge to zero. It can only terminate upon permeable boundaries by approaching zero as the cross-sectional area of the stream tube increases (Fig. 10.3a–c).

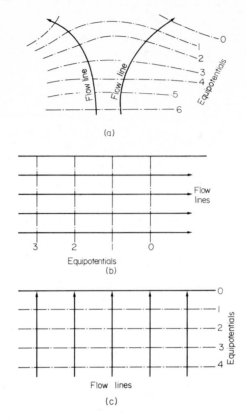

FIG. 10.3. Flow streams approaching permeable and impermeable boundaries: (a) boundary relatively permeable; (b) boundary impermeable; (c) boundary permeable.

5. The intensity of flow increases as the distance between equipotential surfaces decreases.
6. No equipotential surface can close completely upon itself.
7. No equipotential surface can terminate except upon a boundary of the field of flow.
8. When groundwater flows across a plane interface between rocks of different permeabilities, the flow lines and equipotentials refract by a tangent law (Fig. 10.4).
9. The tangential components of the potential gradient on opposite sides of any permeable surface are the same.
10. When two fields of flow are superposed the resultant flow vector is the vector sum of the vectors of the component fields, and the resultant potential field is the algebraic sum of the component potentials.
11. When static freshwater coexists with bodies of static air and saltwater, the interfaces are horizontal. If the freshwater flows the interfaces both tilt towards the freshwater layer in such a manner as to decrease its area of cross-section in the direction of flow.

Rule 10 sometimes has special implications, e.g. when uniform planar flow in an underground aquifer is disturbed by the extraction of water from a well-point, or when fluids are injected into an underground aquifer.

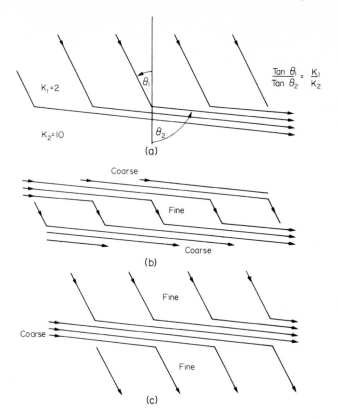

FIG. 10.4. Flow across interfaces of different permeabilities: (a) tangent law of refraction in fluid flow through porous media; (b) and (c) refraction across layers of coarse and fine material.

Groundwater Control and Slope Stability

The presence of groundwater in an earth or rock slope has a critical bearing on the stability of the slope. When measures to promote stabilization of a slope are being considered, first attention should be given to the drainage of seepage water. Figure 10.5a represents a cutting or embankment of permeable material resting on an impervious base. In the natural way drainage will occur towards the base of the slope and will appear as seepage for some distance up from the toe. The arrangement is simple and uncomplicated but has certain disadvantages. Erosion is liable to occur in the seepage zone. Seepage pressures, acting in the direction of flow, contribute to increase the pore pressures, which further weakens the material at the toe of the slope. In a cold climate a frozen ice wall can seal-off the seepage at and near the surface of the slope, and behind this barrier pore pressures will build up, to extend the volume of weakened material up the slope.

By inserting pipes horizontally into the toe of the slope as in Fig. 10.5b, the groundwater table can be lowered and material near the slope face drained completely. Not only are pore pressures reduced in the vicinity of the surface slope but seepage pressures are deflected towards the interior of the material. If the drain pipes are taken to a depth beyond the frost depth, the slope will continue to drain in winter and its stability will not

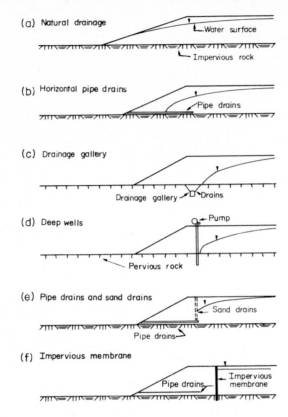

FIG. 10.5. Influence of groundwater on slope stability.

be impaired by freezing. Such drains may be constructed of perforated pipe installed in holes drilled horizontally into the slope face.

Alternatively, a drainage gallery could be constructed in the underlying strata to achieve a similar result (Fig. 10.5c). Methods (b) and (c) could be put into effect after the slope had been constructed. Their purpose would be either to promote the stability of a slope which threatened to slide or to make possible a steeper slope angle than could be maintained by method (a), with consequent economic benefits in excavation and materials handling. Method (d) employs deep wells drilled vertically from the surface and would be applied to de-water a site before beginning to excavate.

System (e) makes use of a permeable soil layer above the impermeable base. Vertical boreholes filled with coarse sand act as channels to intercept the seepage water and convey it to the base, where it is collected by pipe drains.

Seepage Pressures and Uplift Forces

Consider a mass of unconsolidated earth material, in which the void ratio

$$\frac{\text{volume of voids}}{\text{volume of solids}} = e$$

and let the specific gravity of the solids

$$\frac{\text{weight of solids}}{\text{weight of equal volume of water}} = \text{sp. gr.}$$

Suppose a depth of soil l below the drainage level be saturated with water with fluid potential of head h above the drainage level. The conditions of seepage through the earth material will then be as depicted in Fig. 10.6 with the base of the earth layer subject to a fluid potential of $l+h$ and an upward seepage flow through the material.

Consider a volume of the material, of depth l and horizontal cross-sectional area A. The stability of this material depends upon the upward forces due to hydraulic pressure not exceeding the downward forces due to gravity acting upon the material.

The downward forces are determined by the density of the submerged material, the bulk density of which is

$$\frac{\text{total weight}}{\text{total volume}} = \frac{\text{weight of solids} + \text{weight of water}}{\text{volume of solids} + \text{volume of water}},$$

$$\gamma_b = \frac{\text{sp. gr. } V_s \gamma_w + V_{\text{voids}} \gamma_w \, S}{V_s + V_{\text{voids}}},$$

where γ_b is the bulk density of earth material, γ_w is the density of water, V_s is the volume of solids, V_v is the volume of voids, and S is the per cent saturation.

Hence
$$\gamma_b = \frac{\gamma_w(\text{sp. gr.} + eS)}{1 + e}.$$

The density of this material when saturated is given by making $S = 100\%$ (or unity), from which

$$\text{saturated density } \gamma_{\text{sat}} = \gamma_w \left(\frac{\text{sp. gr.} + e}{1 + e} \right).$$

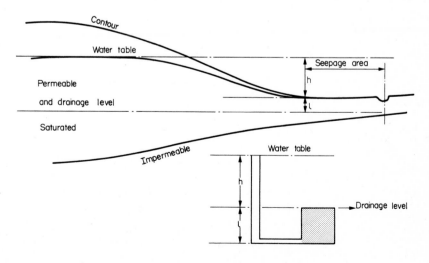

FIG. 10.6. Seepage through earth material.

When a material is submerged, its weight is reduced by the effect of buoyancy which exerts an upthrust equal to the weight of water displaced.

Therefore, considering unit volume,

$$\text{submerged density} = \text{saturated density} - \text{density of water.}$$

Hence

$$\gamma_{\text{sub}} = \gamma_w \frac{(\text{sp. gr.} + e)}{1 + e} - \gamma_w,$$

from which

$$\gamma_{\text{sub}} = \gamma_w \frac{\text{sp. gr.} - 1}{1 + e}.$$

Hence, in Fig. 10.6, the downward force equals submerged density × volume

$$= \gamma_w \frac{\text{sp. gr.} - 1}{1 + e} Al.$$

The hydraulic head causing flow is h and the upward force is $h\gamma_w A$.

Critical hydraulic gradient

Equilibrium will be maintained, so long as $h\gamma_w A$ does not exceed

$$\gamma_w \frac{\text{sp. gr.} - 1}{1 + e} Al.$$

That is, if h/l does not exceed

$$\frac{\text{sp. gr.} - 1}{1 + e}.$$

This value of h/l is termed the critical hydraulic gradient. Numerically it is around 1.0 for most soils, but it has particular significance in freely permeable materials such as sands and gravels where, if the hydraulic gradient is high enough, fluidization or "piping" can occur. This danger is less pronounced with finely grained silts and clays, in which the forces of cohesion may hold the particles together. In these circumstances the effects of excessive uplift pressure are more likely to be seen in plastic flow and upward "heave" of the foundation material.

The upward force $h\gamma_w A$ is the seepage force. Assuming it is distributed uniformly throughout the volume of material,

$$\text{seepage force} = \frac{h\gamma_w A}{lA} = i\gamma_w \text{ per unity volume of material,}$$

where i is the hydraulic gradient.

Seepage Through Earth and Rockfill Dams and Embankments

Observation and control of seepage through earth and rockfill dams is important from two aspects, (1) the effects of seepage in structural stability of the dam, and (2) downstream pollution due to seepage effluents from the impounded reservoir. The latter aspect applies particularly to tailings ponds from mines, mineral treatment plants, and chemical works.

The elementary method of estimating seepage in such a structure assumes isotropic permeability through homogeneous earthfill, resting on an impermeable base. The effects of water seeping through the dam to exude over an area of the downstream surface weakens the material in that region, and excessive seepage could lead to failure. It is customary, therefore, to reinforce the structure by the addition of rockfill at the tail, as shown in Fig. 10.7. This adds a permeable downstream section to the base of the dam. The downstream face of the dam is thus drained, and for a given slope angle will have improved stability. Conversely, for a given safety margin the downstream angle of slope may be made more steep. "Underdrainage" brings the flow lines and equipotentials to a parabolic form about a focus which lies at the junction between the impermeable base and the permeable tail. The graphical method of constructing a flow net in such cases was described by Casagrande [1].

Another method of obviating seepage through the dam is to construct it, not of homogeneous materials but in vertical layers of different materials, including an impervious core zone of clay. This method is generally applied in rockfill dams (Fig. 10.8). Alternatively, the dam may include an impervious membrane (Fig. 10.9). For example, the

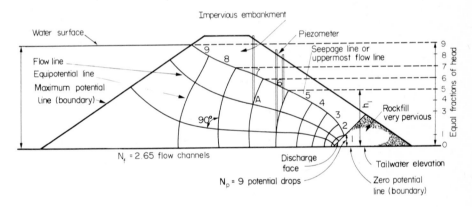

FIG. 10.7. Flow net for seepage through an earth dam.

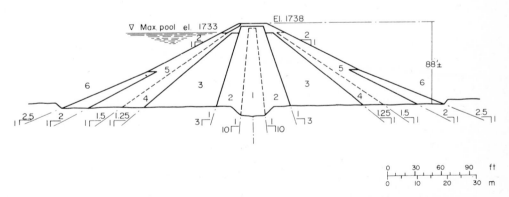

FIG. 10.8. Typical construction of a rockfill dam: (1) impervious clay; (2) impervious sands and clays; (3) weathered sandstone or shale; (4) sandstone rock rolled in layers to 225 mm; (5) rolled sandstone rock, to 0.5 m; (6) large-size rock boulders + 0.5 m.

Odiel Dam in Spain was constructed using a layer of polyethylene polymer roughly 1 mm thick held and protected by cushion zones of sand, in a rockfill embankment (Fig. 10.10).

Where alternative sites provide a choice between a permeable and an impermeable base for the dam, the impermeable foundation, together with an impermeable core or membrane in the dam, offers the best possibility of complete fluid impoundment on the upstream side. In general, however, the foundation materials are likely to be more or less permeable, in which case seepage through the foundations may require to be limited by the use of chemical or cement grouts injected into the rock pores and fissures through which water would otherwise flow more easily. Such methods would be combined with the provision of piped drainage systems, galleries, and pumps underneath the dam, by means of which the remaining seepage is controlled.

In the case of earthfilled dams, controlled drainage through and underneath the structure is achieved through the medium of a filter blanket of coarse, permeable material extending under the dam from the rockfill at the tail. If the foundation material and hydraulic gradient are such that the possibility of dangerous seepage pressures exists, then

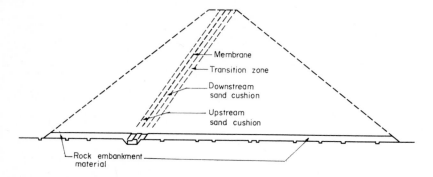

FIG. 10.9. Waterproof membrane in a rockfilled dam.

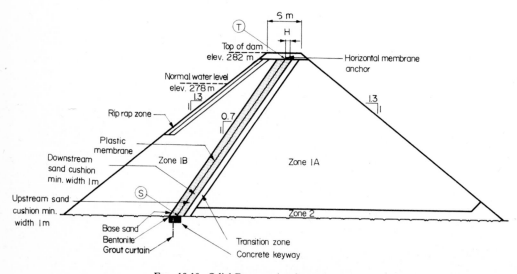

FIG. 10.10. Odiel Dam mode of construction.

a rockfill blanket could be thickened and also extended downstream of the tail. The weight of this blanket then helps to counteract the uplift pressures in this area.

The mode of construction of an earthfilled dam, by dumping and rolling, is conducive to stratification of the materials. As a result, the permeability of the material in a horizontal direction is likely to exceed that in a vertical direction. This distorts the pattern of flow from that deduced by means of an isotropic flow net and affects the area over which seepage must be collected in the tail filter (Fig. 10.11). Cedargren explores the subject of non-isotropic flow nets for a variety of structural dam forms [2].

While a graphical flow net can give results of practical value as a first approximation in design, precise determinations of seepage characteristics for specified boundary and permeability conditions are possible by the finite element technique [6].

In general it should be noted that the actual seepage in and underneath a dam should be constantly observed and recorded, together with the condition and possible movement of the dam material and foundations. It is possible that the observed seepage will be very different from design estimates and flow nets. The final provisions for seepage control, therefore, in the form of drainage facilities and pumping machinery, should provide an adequate safety margin to provide for all eventualities, such as abnormal seasonal water increase, and the arrangements should be capable of rapid expansion, if necessary, in the event of emergency.

Only in this way will it be possible to ensure safety of the structure, either in respect of stability of the earthfill, or the prevention of downstream water pollution.

Mine Hydrology

The exploitation of mineral resources in all the ancient civilizations, and in our own world up to a little more than 200 years ago, was limited to superficial earth materials and

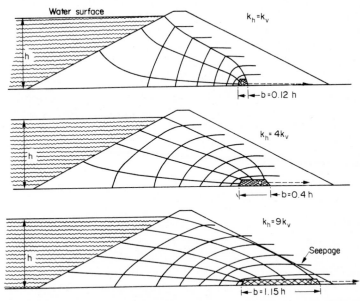

FIG. 10.11. Effect of embankment stratification on the required width of embankment drains (Cedargren [2]).

to the underlying rocks extending in depth only so far as the water table. Underground mine-workings were perforce limited to areas where the ground surface contour permitted of natural drainage into river valleys. It was the need to lift water up vertical mine-shafts which provided inventors with the incentive to produce reciprocating pumps and the first atmospheric and steam-engines with which to drive them, and thus helped to spark-off the industrial revolution.

Groundwater hydrology is an important factor in underground mining and tunnelling technology. A mine-shaft becomes a well as soon as it penetrates below the depth of the water table, and thereafter excavation is only possible if either the shaft is pumped at a rate that can overcome the influx of groundwater, or the area is protected by a strata wall that has been rendered relatively impermeable by grouting, freezing, or the construction of a coffer-dam.

Worley describes a method of estimating the influx of groundwater into a vertical mine-shaft from homogeneous and permeable strata [7]. The method uses data obtained from pumping tests and a modification of the Theis equation [8]. A system for the analysis of shaft water problems is described, using published nomographs, by means of which calculations of influx may be made, based on draw-down estimates.

With extension of the underground workings laterally and in depth, pumping of water from the mine-shaft generates a cone of depression in the water table, with the shaft at its apex forming a central sump, towards which lateral groundwater flow is induced. Le Grand has described a typical situation with progressive extension of the influence of the cone of depression as the mine extends in depth. To keep the workings drained the rate of pumping must increase as the mine extends if new sources of recharge water are tapped by the widening radius of influence [9].

If new recharge sources are not tapped by the expanding limits of the cone of depression, the required rate of pumping to de-water the mine may not have to increase because the rate of inflow generally decreases with increase in depth. Also, it is to be expected that permeability will also decrease due to the higher strata pressures at greater depth. Excavations at depth in impermeable strata can be expected to be dry, but a sudden influx of water can result if the excavations run into fissured ground or intersects a fault zone making contact with saturated strata.

Old and abandoned mine workings generally fill with water, and their presence poses a constant hazard to active mines in their vicinity because the boundary limits of the respective excavations may not be precisely known. Consequently any approach from the active side to intermediate "barrier zones" of unmined strata, which common prudence requires, should be made with extreme caution.

Tunnelling Under Saturated Areas

Similar precautions are required when tunnelling at relatively shallow depth underneath waterlogged areas such as peat or moss, or beneath a river bed or the sea. Any underground work should be preceded by a detailed site investigation from which an isopachyte contour map may be prepared showing the strata thickness between the projected underground working and the base of the saturated zone all over the area. In the case of a civil engineering project, such as a vehicular tunnel, driving the main bore would be preceded by the construction of a pilot bore or exploratory tunnel of small dimensions.

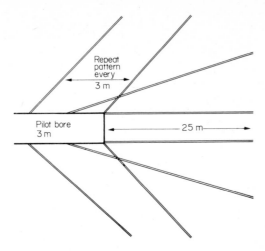

FIG. 10.12. Protective borehole scheme in advance of pilot bore when in the vicinity of waterlogged ground.

This would give complete and detailed information as to strata characteristics, including permeability and seepage, along the line of the proposed work.

Particular care must be exercised where buried river channels may exist, e.g. in areas subject to past glaciation. Such buried channels, usually filled with permeable material, can provide direct access for water to penetrate any impermeable cover above the tunnel.

Where underground work is known to be approaching a saturated area, boreholes should be drilled in advance of the heading, two directly ahead and others inclined as shown in Fig. 10.12 to form a protective exploratory screen. To protect the existing tunnel in the event of mishap, a temporary dam should be erected in the tunnel, having a steel door which can be rapidly shut and sealed. One or more ranges of steel pipes laid through the door can then provide for controlled drainage after the door is sealed.

The procedure of drilling boreholes through a rock or concrete barrier to drain a waterlogged area calls for expertise and care. If the site is at depth underground, high hydraulic pressures must be expected. The procedure usually adopted calls for drilling through a pipe and valve system. The pipe is firmly anchored into the rock or dam, and the valve controls any water flow that may be tapped by the drill.

Sealing-off Groundwater

Grouting

The process of grouting consists of injecting fluid substances or mixtures into the walls of an engineering excavation. The grout materials are deposited or solidify inside the interstices of the soil and rock wall, and thus reduce its permeability. The object of the operation is usually to seal the walls against percolating water, but the process may also be applied with the primary object of strengthening the walls.

Typical applications of the process include:

(a) Grouting to provide an impermeable "grout curtain" in the foundation rocks of a dam.
(b) Sealing the foundation materials and perimeter dykes of a storage reservoir.
(c) Sealing the walls of a tunnel or mine-shaft in water-bearing strata so that work can proceed with the aid of pumps to drain off any residual percolation.

Materials injected as grouts include cements, bentonite clay, chemical solutions, plastics, and asphalt emulsions.

Suspensions

The cement and clay grouts are solids in suspension. For these solids to be able to pass into soils and rocks, without bridging at the entrance to pores, the average pore diameter must be at least three times the maximum grain diameter in the grout material. This means that suspension grouts are only suitable for the more permeable rocks, cement grouts being coarser than clay suspensions. The range of grain sizes in grout materials and their range of applicability to various soils is described by Kravetz [10] (Fig. 10.13).

Mixtures of cement, sand, and clay are sometimes used, the function of the sand being to act as a filler, while the bentonite clay, added up to about 20% by weight of the total mix, gives viscosity to the mixture and helps to keep the sand in suspension during the pumping process. If the bentonite were not used, the sand would tend to settle out of the mix in the pipelines and valves, and so block the pump system.

The progress of the sealing operation is indicated by the build-up of pressure in the injection pipeline as observed on a pressure gauge. The injection pressures required range

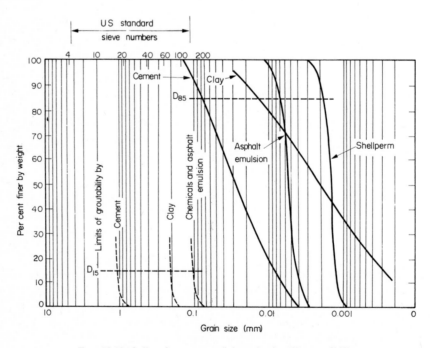

FIG. 10.13. Soil and grout materials grain size (Kravetz [10]).

from around 0.14 MPa up to a maximum ranging from 0.7 MPa to 7 MPa when sealing coarse voids, and up to around 18 MPa when sealing-off seepage through the pores of a permeable rock. If the rock contains large cracks and crevices it is necessary to add to the grout materials which swell in water, such as sawdust, grain meal, chopped straw, and wood shavings. Examples of grout mixes and pressures used on tunnel projects are quoted by Woodruff in Table 10.1 [11].

TABLE 10.1. *Examples of pressure grout mixes used on tunnel projects*

	Low pressure	High pressure
Delaware Aqueduct (New York)	100 psi (0.7 MPa) Neat cement 1:(6)[a] Cement:sand 1:1:(10) Cement:sand 1:1:(7)	600 psi (4.1 Mpa) Neat cement 1:(10) to Neat cement 1:(20)
Waikaremoana Tunnel (New Zealand)	80 psi (0.5 Mpa) Cement:sand 1:1:(3) 2% bentonite added	
La Cienega and S. F. Sewer Tunnel (Los Angeles, Cal.)	90 psi (0.6 MPa) Cement:sand 1:1:(10)	300 psi (2.0 MPa) Neat cement 1:(10)

[a] Figures in parentheses indicate gallons of water per bag of cement.

Chemical grouts

Some materials will resist penetration by suspension-type grouts due to their fine porosity. In such cases very high injection pressures may fail to seal the rock against water percolation. Or effective penetration might only be possible using very thin slurries, so that very large volumes of grout mixture would be required to produce relatively small volumes of sealed rock.

In such circumstances chemical grouts may be used. There are several of these, some of which are proprietary substances, but in general all employ water-soluble chemicals, usually two reagents which, when they encounter one another upon injection into the earth material, form a gel which fills the pores and interstices in which they stand. The method was first employed using sodium silicate as the base solution to which was added a reagent such as hydrochloric acid, sodium aluminate, calcium chloride, or salts of iron and copper. By diluting the base and the reagent, or possibly both, to various concentrations of solution, the characteristics of the gel and the setting-time can be varied. Kravetz [10] records the setting-times of various reagents added to a 50% saturation solution of sodium silicate in water, in Fig. 10.14.

There are two methods of application. In the first method the base solution and the reagent are mixed and this mixture is injected. In the second method a full-strength solution of sodium silicate is first injected and this is followed by a separate injection of reagent using adjacent alternate pipes for each injection. Contact of the two solutions then produces almost instantaneous flocculation, and the resulting gel has maximum rigidity. This method is used when rapid sealing-off is required.

In recent years proprietary grouts such as AM-9 (American Cyanamid Company) have come into widespread use. The materials used here have the advantage that a stiff gel can be produced from dilute solutions in water, so that the grout has low viscosity. This makes possible the effective penetration of finely porous materials at low injection pressures.

Since grouting was first introduced considerable advances have been made, not only in the development of techniques of application, but in understanding the fundamental processes of fluid flow through porous media. With the concurrent advance of theory and practice, new grouts having a wide range of viscosities and capable of penetrating finer materials than previously could be sealed are periodically introduced. Ischy and Glossop have reviewed the application of grouting processes to the problems of sealing alluvial earth deposits against percolating water, and of increasing the mechanical strength of the earth material [12].

Grouting Procedure

The general arrangements for pumping cement grouts are illustrated in Fig. 10.15 and for chemical grouting in Fig. 10.16. The grout is applied to the surrounding strata by

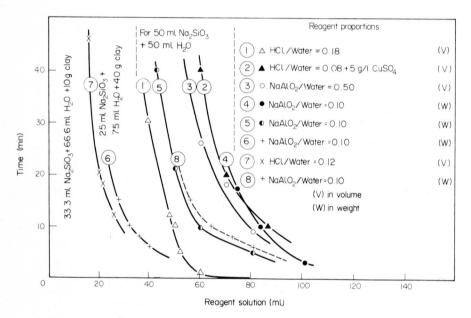

FIG. 10.14. Typical setting times and reagent curves for chemical grouts (Kravetz [10]).

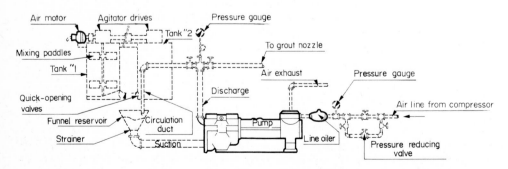

FIG. 10.15. Typical cement-grouting plant.

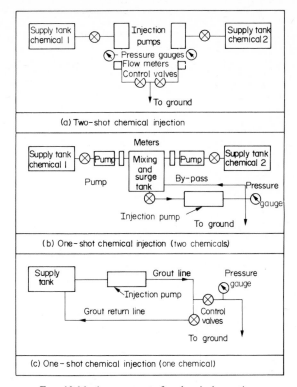

Fig. 10.16. Arrangements for chemical grouting.

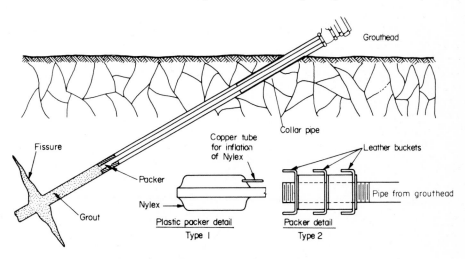

Fig. 10.17. Grout packer arrangement (Rankin [14]).

means of pipes connected to "groutheads", which are inserted into boreholes 25–50 mm in diameter (Fig. 10.17). The injection holes are spaced at intervals, determined by trial and error, such that grout injected from one hole penetrates the ground to overlap the next hole. A solid screen of impermeable strata is thus created. Injection of grout is usually preceded by injection and circulation of water to test the injection pressure required and also, in fissured rocks, to flush out loose grains and sediment from the nearby fissures as much as possible [14].

Sealing-off the strata around an excavation is effected in stages, and to a depth depending upon the circumstances to be served. For example, when forming a "grout curtain" in the foundations of a dam, the holes would be drilled from a drainage tunnel running parallel with the dam. The grout-holes would be aligned to form a fan extending to a depth of 6–12 m from the horizontal to the vertical at each horizontal interval along the drainage tunnel (Fig. 10.18).

A vertical shaft could also be protected and sealed against water by "fanning" grout-injection-holes from the base of the shaft and deepening the shaft in stages, each stage being excavated to about half the depth of the sealed strata. This would leave a solid plug from 3 to 5 m thick at the base of the shaft, which would have to be penetrated by the drill-holes of the next stage in turn (Fig. 10.19).

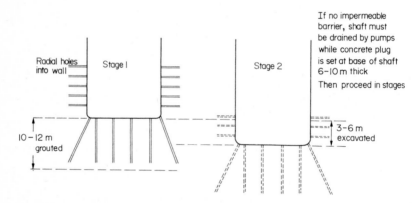

Fɪɢ. 10.18. Grouting procedure for sinking a vertical shaft through water-bearing ground.

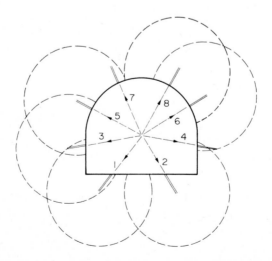

Fɪɢ. 10.19. Grouting round a tunnel in porous rock. Inject grout in segments in order indicated.

Sealing-off by Freezing

It has long been the practice in cold regions to take advantage of the winter climate to excavate and construct earthworks in alluvial materials which might be completely unconsolidated and fluid during the summer. The method of artificially freezing saturated and unconsolidated earth materials to facilitate engineering work such as the construction of foundations, shafts, and tunnels originated in Germany nearly 100 years ago. Since that time it has been widely applied in the mining industry and, to a lesser extent, in civil engineering. In more recent years it has become very important in relation to the development of deep underground works, particularly potash mines, in the Southwest United States, Canada, and Britain. There is also a strong possibility that recent modification of the technique will become increasingly important in connection with ground stabilization and earthworks in civil and geological engineering generally.

The need for freezing arises when materials are encountered that cannot be sealed by grouting, and where the hydraulic pressures, or the depth to solid ground, may be too great to permit the use of coffer-dams and caissons. Such conditions exist when thick surface deposits of peat or alluvial sands and gravels are saturated and subject to fluidization or "piping". Similar sand deposits may be encountered at considerable depth from the surface. For example, when sinking mine-shafts in the concealed coalfields of Western Europe, the coal measures must be approached through considerable thicknesses of water-bearing ground, including fissured limestones and sandstones. Sometimes these strata can be successfully penetrated with the aid of grouting techniques, but in other cases freezing becomes necessary either because the fissures are too large to be filled by grout or beds of "quicksand" or fluidized sand exist which grout cannot consolidate.

A notable example occurred in the north-east of England at Seaham, where mine-shafts were sunk which had to penetrate a considerable thickness of water-bearing strata. This included fissured limestones for which a cement-grouting process might have succeeded. However, beneath the limestone lay a quicksand bed. This stratum outcrops underneath the North Sea, and when the shaft reached it an inrush of water filled the shaft up to sea level and thereafter rose and fell with the tide.

The shaft was eventually recovered and extended in depth to penetrate the water-bearing limestones and the running sand by applying the freezing process.

In more recent years, the extensive development of potash mining in Saskatchewan would have been impossible without the freezing technique. Here the economic minerals to be won lie at a depth of from 975 to 1100 m below the surface. They must be approached first by penetrating about 100 m of glacial drift, which is incompetent and saturated. Below the surface alluvium is a thickness of Cretaceous shale, competent and dry, for a thickness of around 380 m. This is underlain by the Blairmore formation, which consists of loose sand with lenses of clay and shales from 50 to 100 m thick, and saturated with brine, at a pressure around 4 MPa (600 psi). Below the Blairmore there are from four to ten further water-bearing zones with hydraulic pressures up to 7.6 MPa (1100 psi) [16].

When the Grand Coulee Dam was under construction in the United States, an earth dam was constructed and frozen to prevent the intrusion of landslide debris on to the site [17]. On a smaller scale, freezing was used in the construction of a 3 m diameter shaft through 40 m of ground in connection with a drainage and water pollution control problem in an area of Manhattan. The area was underlain by several layers of incompetent material, and it was feared that any movement of ground into the excavation while it was in progress would disturb the foundations of important structures nearby [18].

Layout of Freezing System for Shaft Sinking

The process of excavating and construction in unconsolidated ground by the aid of freezing consists of several component operations.

1. Boring vertical holes, each about 200 mm in diameter, at horizontal intervals of from 1 to 2 m apart on the central axis and to the full depth of the ground to be frozen. In the case of a circular shaft the holes are drilled on a circle whose diameter is from 2 to $2\frac{1}{2}$ times the required diameter of the finished shaft.
2. A cold brine solution is pumped down these holes and circulated from a refrigeration plant erected at the site to freeze the ground around each hole so as to form an ice wall. There are two possibilities: (a) if the freeze-holes extend from the surface to an impermeable base, excavation can begin when the ice wall has closed to a cylindrical annulus of sufficient thickness to withstand the hydraulic and strata pressures behind the wall; (b) if the holes do not extend to an impermeable base, then freezing must continue until the central core of the ice cylinder is also solidified.
3. When the ground is frozen then excavation can proceed and a strong watertight lining can be constructed around it, inside the ice wall.
4. The ice wall is allowed to thaw and the refrigeration plant and circulating brine pipes are removed.

The general arrangements are shown schematically in Fig. 10.20. The boring and freezing operations are conducted from a "freeze cellar" constructed around the collar of the future shaft. The progress of the boreholes is carefully surveyed, and they are maintained precisely in alignment. They are generally vertical but in some cases may be inclined slightly outwards from centre if it is desired to increase the thickness of the ice wall outside the excavation at depth, so as to counter increasing external pressure.

Ammonia evaporator–condenser refrigeration plants are used, usually circulating a calcium chloride brine of sp. gr. 1.24 to the freeze-holes. The required refrigerator capacity is substantial, ranging from around 120 tonnes, circulating from 3600 to 4600 dm^3/min of

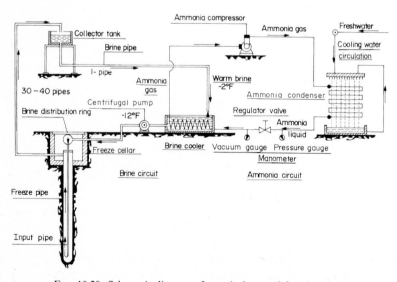

FIG. 10.20. Schematic diagram of a typical ground-freezing plant.

brine through 21 –25 holes on a freeze circle around 10 m in diameter and from 40 to 110 m deep, and up to around 400 tonnes refrigeration capacity in a Saskatchewan potash project, with 30–40 freeze-holes to a greater depth.

The freeze-holes are plugged at their base so that the brine coolant can be circulated through two steel pipes, the outer pipe 150 mm in diameter, "mudded in" to the hole by bentonite, and enclosing the inner pipe which is 50 mm in diameter. These are connected to ring mains at the freeze cellar. The brine circulation is usually down the inner pipe and rising up the annulus of the outer pipe. This promotes freezing to build the ice wall from the base upwards. Occasionally circulation of brine down the outer annulus is used if it is desired to freeze around the upper part of the shaft, in alluvium, first. This will then permit of excavation and construction at the surface while the lower sections are freezing. The temperatures in each pipe are monitored. In a typical system brine leaves the refrigerator at $-24°C$, rises up the pipe at $-20°C$ and returns to the refrigerator at $-19°C$.

It is usual to drill an observation and pressure-relief well-hole at the centre of the freeze-circle, with additional observation boreholes at several radial intervals extending across the area. These are used to facilitate determination of the groundwater flow conditions before the project begins, and subsequently to observe the rate of growth and limits of the ice wall (Figs. 10.21 and 10.22).

In locations where the underground water is saline it is necessary to flush the area with freshwater until the water tested at the observation holes is observed to be diluted to a salinity that can be frozen. This may also necessitate the circulation of a coolant other than calcium chloride brine, and either magnesium chloride or lithium chloride is sometimes used.

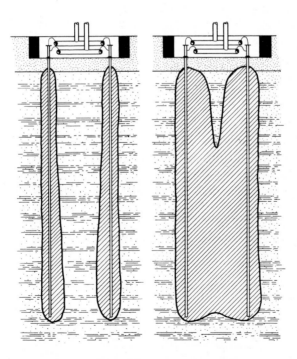

Fig. 10.21. The formation of an ice wall.

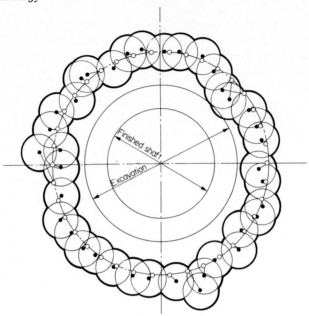

F IG. 10.22. Typical plan of an ice wall.

The success of the method depends upon control of the freezing process to build up an ice wall of adequate strength that will permit safe excavation and construction to proceed as soon as possible. Observation of the ice wall limits is made by taking temperatures using thermocouple probes in the observation holes. Temperature–time plots for specific locations will enable a forecast of the time of ice-wall closure to be made. These measurements may be supplemented by geophysical observations using an ultrasonic probe to determine the limits of the ice wall. A pronounced change in wave-propagation velocity is observed at the transition from unfrozen to frozen material.

Design criteria in the freezing process

There is a general lack of information on the design details of the freezing process and a great need for fundamental research on the subject. Contractors who specialize in freezing are relatively few in number and, having gained their expertise largely as a result of their own experience, are naturally reluctant to broadcast what they have learned.

The required thickness of ice wall depends upon the strength of the frozen ground and the external pressures to be withstood. Domke's formula is commonly used,

$$t = R[0.29K + 2.3K^2],$$

where t is the minimum required thickness of ice wall, R is the radius of excavation, and K is the ratio

$$\frac{\text{(strata pressure + hydrostatic pressure)}}{\text{compressive strength of frozen ground}}.$$

The required freeze-circle diameter is then $2(R + t)$ [20].

The refrigeration capacity determines the rate of heat removal by the circulating brine.

$$C = BWSTt,$$

where C is the heat removed, B is the brine circulation, W is the weight of brine, S is the specific heat of brine, T is the time, and t is the temperature difference in brine circuit.

One tonne of refrigeration capacity will remove 3517 J/s (288 000 Btu/day).

The total heat to be extracted from the ground is determined by the mass and thermal properties of the materials concerned, both solid and liquid, and the temperature difference that must be encompassed in order to promote freezing.

As a first approximation, if t_1 is the average initial ground temperature and t_2 is the final freezing temperature,

$$Q = Q_1 + Q_2 + Q_3 + Q_4,$$

where Q is the total heat to be extracted and Q_1 is the quantity of heat removed from the ground solids in reducing their temperature from t_1 to t_2,

and
$$Q_1 = W_s C_s (t_1 - t_2),$$

where W_s is the weight of solids and C_s is the specific heat of solids;

Q_2 = quantity of heat removed from the groundwater in reducing its temperature from t_1 to $32°$ F

and
$$Q_2 = W_f (t_1 - 32),$$

where W_f is the weight of groundwater to be cooled;

Q_3 = quantity of sensible heat removed from the frozen groundwater

and
$$Q_3 = W_f C_f (32 - t_2);$$

Q_4 = latent heat of freezing the groundwater

and
$$Q_4 = 144 W_f.$$

However, in practice the problem is complicated by heat flow into the cooled region from the outside earth material, and in the case of a deep shaft the geothermal gradient will have an appreciable effect. Another complicating factor is that, in all probability, there will be groundwater movement and consequent heat transfer in the groundwater. These are matters that call for a rigorous theoretical treatment.

Tsytovich and Khakimova have described the process of ground freezing as applied to mining and civil engineering foundation works in the USSR [20]. They quote formulae by means of which may be determined the rate of formation of the ice wall and the time required for this to close. Mankovski describes the use of electrical analogue computers to solve the heat-flow problems that are involved in shaft sinking by the freezing process. The models provide solutions for the temperature field around a freeze ring of boreholes in earth materials of known thermal conductivity, using circulating fluid at specified temperatures [21].

Excavation and support of the frozen ground

When the freeze-holes bottom in impermeable ground, excavation may commence when the ice wall has grown to the finished limits of the proposed excavation. Excavation

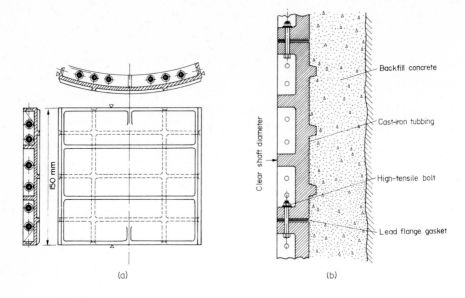

FIG. 10.23. (a) Cast-iron tubbing segment; (b) cast-iron tubbing system with backfill concrete.

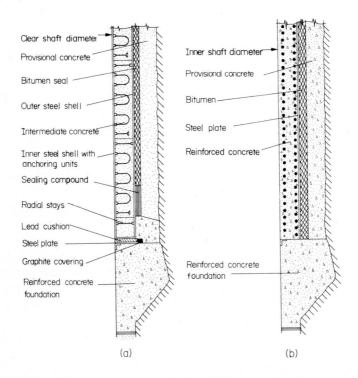

FIG. 10.24. (a) Double steel cylinder watertight lining, with inner and outer concrete; (b) reinforced concrete with watertight sheet steel lining, backfilled with concrete and sealed with bitumen.

should be by mechanical means other than drilling and blasting, since it is important not to weaken the ice wall nor to fracture the freeze pipes by impact or concussion. The excavation is lined to provide a strong watertight structure, capable of withstanding the external pressures that will build up as the strata subsequently thaw. Cast-iron segments, bolted together then caulked with lead and backfilled with concrete, provide a strong, rigid lining for situations where maximum hydraulic thrust must be resisted (Fig. 10.23). If some flexibility in the lining is required, to conform to possible strata movement, and at the same time preserve a watertight seal, the lining may be constructed of sheet-steel plates embedded in concrete and sealed with bitumen, as shown in Fig. 10.24.

Alternatives to brine systems

Brine circulating systems are slow, taking from 10 to 15 weeks to freeze the ice wall. Their expense is likely to be heavy, depending upon the prevailing power costs. Some freezing operations might be conducted more economically with the aid of liquid nitrogen or carbon dioxide. Liquid nitrogen can freeze the wall in 2–3 days, but the associated drilling and casing of holes, together with the general assembly of apparatus, would probably occupy about 1 week. Freezing the wall by circulating liquid CO_2 would require about 2 weeks after drilling the rock and assembling all equipment which, in this case, would have to operate at a working pressure around 2 MPa (300 psi).

References and Bibliography: Chapter 10

1. CASAGRANDE, A., Seepage through earth dams, *New England Water Works Ass. Proc.* **51** (2) (1937).
2. CEDARGREN, H. R., *Seepage, Drainage and Flow Nets*, Wiley, New York, 1967.
3. BARNES, J. M., Design and construction of Summersville Dam, *Proc. 8th Int. Conf. on Large Dams, Edinburgh, 1964*, **3**, 763–80.
4. JOHNSON, C. E., Construction of a dam with a plastic interior membrane, *Pacific Southwest Mineral Industry Conference*, AIME, Las Vegas, April 1972.
5. HUBBERT, M. K., The theory of ground water motion, *J. Geol.* **48** (8, Part I) 785–944 (1965).
6. KEALY, C. D., and BUSCH, R. A., *Determining Seepage Characteristics of Mill Tailings Dams by the Finite Element Method*, US Bur. Min., Rep. Invest. No. 7477, 1971.
7. WORLEY, M. T., Groundwater influx into a vertical mine shaft, *Trans. AIME* **223**, 428–31 (1962).
8. THEIS, C. V., The relation between the lowering of the piezometric surface and the rate and duration of discharge of a well using groundwater storage, *Am. Geophys. Uni. Trans.* **16**, (2) 519–24 (1935).
9. LE GRAND, H. E., An overview of problems of mine hydrology, *Am. Inst. Min. Engrs AGM, San Francisco*, preprint No. 72–AG–5, New York, Feb. 1971.
10. KRAVETZ, G. A., Cement and clay grouting of foundations, *Proc. Am. Soc. Civil Engrs Soil Mech. Found. Div.* **84** (1), paper 1546 (Feb. 1958).
11. WOODRUFF, S. D., *Methods of Working Coal and Metal Mines*, Vol. 2, pp. 322–37, Pergamon, New York, 1966.
12. ISCHY, E., and GLOSSOP, R., An introduction to alluvial grouting, *Proc. Inst. Civil Engrs, London* **21**, 449–74 (1962); and **23**, 705–25 (1964).
13. ANON., Chemical grouting, *Proc. Am. Soc. Civil Engrs, J. Soil Mech. Div.* **83** (SM4), paper No. 1426, Nov. 1957.
14. RANKIN, I. A. R., Underground grouting practice, *Chem. Engng Min. Rev.*, pp. 358–63 (July 1954); pp. 411–15 and 449–55 (Aug. 1954).
15. GLOSSOP, R., The rise of geotechnology and its influence on engineering practice, *8th Rankine Lecture, Geotechnique* **18**, 105–50, 1968.
16. SCOTT, S. A., Shaft sinking through Blairmore sands and Paleozoic water bearing limestones, *CIM Bull.* **56** (610) 94–103 (1963).
17. FROAGE, L. V., Frozen earth dam at Grand Coulee, *Mech. Engng New York*, pp. 9–15 (Jan. 1941).
18. GAIL, C. P., Stabilization of excavations by freezing, *Proc. Am. Soc. Civil Engrs New York* (Oct. 1961), Winston Bros. Co., Minneapolis, Minnesota.

19. WALLI, J. R. O., The application of European shaft sinking techniques to the Blairmore formation, *CIM AGM, Edmonton, April 1963*, Associated Mining Construction Ltd., Regina, Saskatchewan.

20. TSYTOVICH, N. W., and KHAKIMOVA, K. R., Ground freezing applied to mining and construction, *Proc. 5th Int. Conf. Soil Mech. and Found. Engng New York, 1955*, pp. 737–41.

21. MANKOVSKI, G. I., The use of models to solve mining problems, *Min. Mag.* **115**, 212–75 (Oct. 1966).

22. KITAMURA, I. (ed.), Tunnelling under difficult conditions, *Proc. Int. Tunnel Symp., Tokio, May–June 1978*, Pergamon Press, Oxford, 1979.

23. WEST, G., and O'REILLY, M. P., Methods of treating the ground, *Tunnels Tunnelling*, **10** (7) 25–29 (1978).

24. CAMBEFORT, H., Principles and applications of grouting, *Q. Jl Engng Geol.* **10** (2) 57–95 (1977).

25. SCOTT, R. A., Fundamental conditions governing the penetration of grouts, in *Methods of Treatment of Unstable Ground*, pp. 69–83, Newnes–Butterworth, London, 1975.

26. LENZINI, P. A., Ground stabilization—a review of grouting and freezing techniques for underground openings, *Underground Space* **1** (3) 227–40 (1977).

27. BROWN, D. R., and WARNER, J., Compaction grouting, *J. Soil Mech. Found. Div.* **99** (SM8) 589–601 (1973).

28. PLAISTED, A. C., The application of chemical grouts to ground consolidation, *Ground Engng* **7** (4) 420–44 (July 1974).

29. DU BOIS, H. L. E., High pressure grouting, in *Proc. World Congress on Water in Mining and Underground Works, Granada, Spain, September 1978*, **1**, 391–407, Asociación Nacional de Ingenieros de Minas, 1978.

30. ANON., Freeze it, *Civil Engng London*, pp. 53–55 (Dec. 1977).

31. PEKHOVICH, A. I., *Calculating the Speed with which Water-permeable Ground Is Frozen by a Row of Columns before Frozen Ground Cylinders Join*, US Army CREEL Trans. TL 590, Feb. 1977.

32. PEKHOVICH, A. I., *Method of Calculating the Freezing Rate of Single Column Water Permeable Soils*, US Army CREEL Translation TL 594, Feb. 1977.

33. BRAUN, B., and MACCHI, A., Ground freezing techniques at Salerno, *Tunnels Tunnelling*, pp. 81–89 (Mar.–Apr. 1974).

CHAPTER 11

Underground Storage of Oil and Gas

WITHIN recent years the need for increased storage facilities for oil and gas has grown markedly. This has come about as a result of two major forces: (1) the need to provide for peak seasonal demands for gas when consumption may exceed pipeline distribution system capacity, and (2) the growth of international and intercontinental transport of oil and of liquefied petroleum gases and liquefied natural gas.

To a limited extent, peak seasonal demands for fluid reserves may be met from stocks stored in old and depleted oil and gas reservoirs, or in porous-rock aquifers, where suitable geological rock structures exist close to the major market areas. More extensive storage facilities are required, however, to provide and maintain strategic national reserves of crudes, fuel oils, and petroleum products, and also to provide for the day-to-day needs of traffic in liquefied natural gas. Natural gas, composed mainly of methane, becomes a clear liquid with about half the density of water when it is cooled to $-162°C$ at atmospheric pressure. The advantage of transporting and storing methane in this form is that each unit volume of liquid yields some 600 volumes of gas at normal temperature and pressure. However, the cryogenic temperatures necessary for transport and storage of the liquid create problems in the choice of site for, and mode of construction of, suitable storage facilities.

The fluids stored may be water and compressed air, but these are considered in another context. We are here concerned mainly with the hydrocarbons, methane, propane, butane, and ethylene, both in gaseous and in liquid phases, and with the less-volatile crudes and fuel oils. The confining storage arrangements may include aquifers, salt cavities, mined caverns, wide boreholes, pressure tunnels, shallow covered surface quarries and pits, and underground receivers or pre-stressed pressure vessels. The methods of underground containment may be classified as being either high-pressure or low-temperature storage (Fig. 11.1).

Storage at High Pressure

Any one of three general procedures may be followed: (a) storage where the maximum gas pressure is equal to or less than the hydrostatic head of groundwater above the storage cavity; (b) storage such that the maximum gas pressure is greater than the hydrostatic head of groundwater above the storage cavity but less than the *in situ* strata pressure; (c) storage at pressures higher than the *in situ* strata pressure. Method (a) is more common than the others. The rock walls are generally unlined and the percolating groundwater around the cavern acts as a seal to contain the stored fluid. In method (b) the cavity would need to have an impermeable lining, and in method (c) the constructed pressure vessels would be, in principle, similar to those used in above-ground storage.

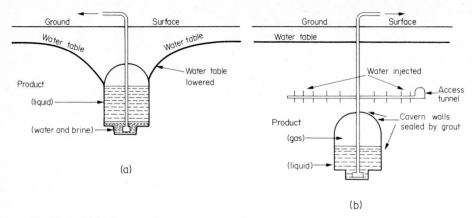

FIG. 11.1. (a) Underground storage at atmospheric pressure; (b) underground storage at high pressure.

Storage at Low Temperature

Low-temperature storage systems include: (a) insulated storage units in which the insulation is designed to maintain the soil temperature at its normal level; (b) systems that utilize the earth materials at the site to provide the necessary insulation; and (c) frozen-ground techniques where the containing earth materials are frozen to a thickness and depth sufficient to encase the liquefied gas inside an ice wall. The formation of the ice wall and the rate of increase of its thickness with time will be dependent on the groundwater flow characteristics and the permeability of the earth materials involved. The ice wall is gradually formed by artificial freezing during the initial construction, and it will continue to build up in thickness when the excavation is subjected to the lower temperatures introduced by the stored product, e.g. liquid methane. Ultimately a steady state is reached. The existence of the ice wall then acts as an insulator and helps to prevent unlimited extension of the ice wall after the storage chamber has been filled with liquid gas at low temperature.

General Classification of Gas- and Oil-storage Systems Underground

A wide variety of systems and arrangements for gas and oil storage exist. They include:

Aquifers

Usually water-bearing sand or sandstones sealed by a capping of impermeable clay (Fig. 11.2). For use as a natural gas reservoir the geological structure should be that of a dome, or anticline, or syncline, such that the containing rocks are provided with an adequate gas seal. The minimum gas pressure in the aquifer is the relevant groundwater pressure, and the maximum pressure must not exceed the *in situ* strata pressure [1, 2].

To use such a reservoir, gas is pumped down a cased borehole into the reservoir rock in which the gas fills the available pore space, being confined by the impermeable cap-rock. Before the injection of gas, the *in situ* strata pressure is determined by residual tectonic

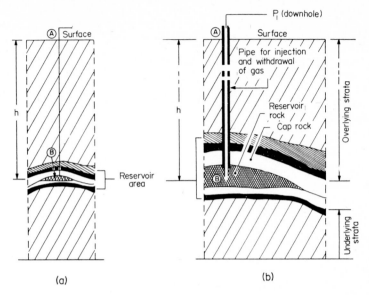

FIG. 11.2. Simplified gas-storage reservoir: (a) typical anticlinal aquifer; (b) enlarged view of reservoir area (Hardy [6]).

forces and the weight of overburden, but after gas injection and pressurization the maximum pressure to which the reservoir may be subjected and still remain mechanically stable is termed the "optimum pressure". In broad terms the optimum pressure is dependent on the *in situ* strata pressure, the geometry and geology of the reservoir, and the mechanical properties of the reservoir, cap-rocks, and surrounding strata.

If the optimum pressure is exceeded, the reservoir becomes mechanically unstable, and fractures develop, which may eventually lead to the escape of the stored gas.

Salt Cavities

Salt cavities are formed by pumping circulating water to leach out underground chambers in salt domes or in bedded salt deposits (Fig. 11.3). Such cavities have been used to store liquid hydrocarbons, such as propane, butane, and ethylene, for many years. Salt domes are massive vertical intrusions of salt, thrust upwards from very deep sediments, into the surrounding rocks. Small oil- and gas-fields commonly occur around the edges of the domes, where the overlying porous rocks are tilted and folded to form natural reservoir traps. These domes, of which more than 300 occur in the Gulf Coast area of the United States, are ideal for the construction of storage cavities.

Salt cavities may also be formed in bedded salt deposits that are sufficiently thick and deep enough to sustain the pressures under which the product oil or gas must be stored. The shape of the leached cavity is controlled during construction by protecting the roof from solution by injecting gas or light emiscible liquid [3, 4].

In general terms, either of two methods of procedure is adopted—wet and dry storage. In wet storage the product floats on a base of water or brine and is collected by displacement. The pressure within the cavity must be maintained higher than the brine hydrostatic pressure but less than the *in situ* strata pressure. In dry storage the product is

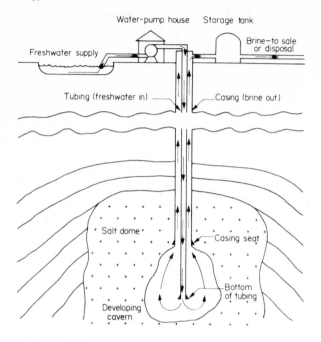

FIG. 11.3. Schematic diagram of a salt-cavern storage development (Slater [3]).

not displaced, but water or brine is pumped from the cavern as gas is injected into it. This forms a pressurized underground tank of gas. The strength of the cavity must sustain the *in situ* strata pressure plus the internal pressure of the product [5, 6].

The salt cavity may be created by drilling a well in which a pipeline is cemented from the top edge of the salt deposit to the ground surface with a line of smaller pipes housed inside the outer pipe. Water is pumped down the inner pipe into the salt formation, where it dissolves the salt. The resulting brine is pumped up the pipe annulus for disposal at the ground surface.

In salt deposits which are interbedded with non-soluble rock materials more than one well will be required. Two wells may be driven and connected by the process of hydrofracture at depth. Freshwater can then be injected into one of the wells and brine extracted from the other [7].

Mined Caverns

Mined caverns for the storage of gas and oil may be considered in relation to (a) the use of existing mines, either in active production or abandoned, and (b) the construction of new mined excavations especially designed for the purpose.

Existing Mines

The urgent need for stores of petroleum products and of fuel oil, to meet sudden and periodic international crises, has focused attention on the possible use of existing mined cavities capable of immediate utilization. Davies and Willet [7] have described the identification and preliminary assessment of existing mines at selected sites in the United States as possible repositories of strategic oil and gas reserves. Simpson catalogues the possible utilization of deep underground mines in Canada to store solid and liquid wastes, and to provide live storage of oil and gas. The possible storage of crude oil in disused mine-workings beneath the sea-floor off the coast of Newfoundland is included [9].

In Britain, possible existing storage sites are limited to small-scale limestone and gypsum mines at shallow depth, while more extensive salt, anhydrite, and potash mines are in active production. There are many abandoned coalmines, some of which supply stored natural methane gas for local industrial use. One major installation, at Point of Ayr Colliery, supplies methane collected from beneath the Dee estuary in North Wales.

If an existing or abandoned mine is to be used for gas storage, several conditions must be met. The various interconnected passages to be utilized should be intact, preferably not flooded, and completely isolated from other workings by adequate solid barriers. All subsidence should be complete, with no possibility of further subsidence or collapse. Buttiens describes the conversion of collieries in Belgium to low-pressure gas storage, with particular reference to sealing-off old shafts and workings by gas-tight plugs, around which the ground was frozen, to provide a secure base for the construction of the seals [10].

Old and abandoned metalliferous mines may offer more favourable sites for possible adaptation than do coal-mines. They are frequently to be found more easily accessible from the surface than are deep coal-mines, and the surrounding rock strata are likely to be stronger than carboniferous sediments. Many old mines contain caverns that have remained open and unsupported for many years, and these remain available for conversion to storage.

Wettlegren quotes two examples from Sweden. One consists of an abandoned open pit, originally worked from the outcrop (Fig. 11.4a). The necessary construction for conversion to oil storage included (1) covering the pit by a reinforced concrete dome, and (2) boring a vertical shaft from the surface pump-room to intersect an inclined tunnel providing a water pipeline between the ground surface and a sump constructed at the bottom of the pit.

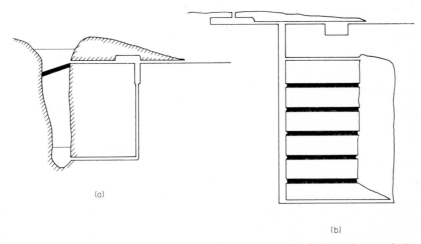

(a)

(b)

FIG. 11.4. Use of abandoned mines for gas and oil storage: (a) open pit; (b) underground mine.

An example of the conversion of an abandoned underground mine is shown in Fig. 11.4b. In this case the ore-body had been mined from a vertical shaft, with levels and cross-cuts intersecting the ore at several horizons between the ground surface and the shaft bottom. The necessary conversion work included sealing off all the intermediate levels other than the top and bottom horizons by concrete plugs. The oil product pipeline was connected to the top horizon level and the water pipeline connected to the bottom sump, using the vertical shaft to house product and water pipeline connections to the pump-house [11].

Layout of Mined Caverns

The layout of a mined underground storage system may generally follow room and pillar principles or, alternatively, may be provided by a tunnel network. The use of rapid tunnel-boring techniques would favour the latter layout (Figs. 11.5 and 11.6) [12]. Liquefied petroleum gases require a storage cavern in unlined rock at least 60 m below ground surface level, and preferably around 100 m. It is important that the storage cavern be at a sufficient depth below the water table that there should be no migration of fluid outside the cavern walls. If there is any movement of fluid, this should be such that groundwater enters the cavity rather than have gas escaping from the cavern.

Pressure Tunnels

Cavities of circular cross-section can be readily produced by modern tunnel-boring machines. Such tunnels are often constructed to serve as pressure tunnels in civil engineering, and they are eminently suited to contain liquefied gas under pressure. Tunnels in massive strong igneous and metamorphic rocks can withstand pressures up to 300 atm. In weaker sediments the lining and support system can include a plastic impermeable membrane. The cavity in this case might withstand pressures around 50 atm.

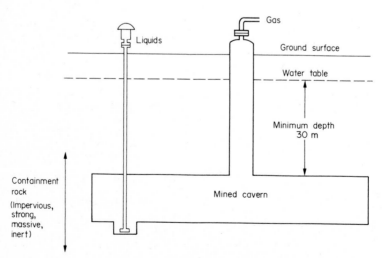

FIG. 11.5. Mined cavern storage.

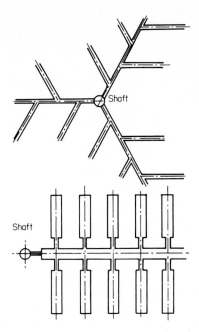

FIG. 11.6. Layout of underground storage in mined caverns.

Covered Surface Pits

For liquids stored at low temperature, excavated storage units may be cut into bedrock, insulated at the sides and base, and covered by a suitably designed insulated roof with a gas seal (Fig. 11.7) [13–15].

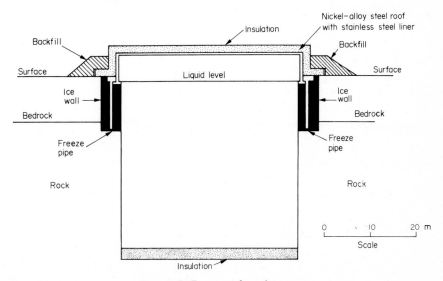

FIG. 11.7. Frozen surface-pit storage.

217

General Principles of Underground Storage of Petroleum Products

Storage of petroleum products in unlined caverns, in which the product is contained in direct contact with the rock walls, with no interposed containment structure or lining, is described by Jansson[13]. In massive igneous and metamorphic rocks, permeated by cracks and discontinuities such as joints and cleavage, the movement of groundwater and of the petroleum product is relatively easy. In these circumstances, if a cave is mined out below the water table and the water table lowered by pumping, a sump forms at the base of the cavern.

If petroleum products are then introduced into the cavern they will float on the groundwater in the sump, and if pumping is stopped then the petroleum products will float up and out into the surrounding rocks, above the groundwater. Two methods of procedure may be followed.

Fixed Water-bed

The water-bed is maintained at constant level near the cavern base, its upper surface being controlled by means of a retaining wall constructed around the sump. Groundwater seeps into this sump and is pumped from it. The petroleum product being stored floats on the water-bed, the level of its upper surface being controlled by a product pump. The space in the cavern above the product level is filled with a mixture of air and petroleum gas (Fig. 11.8).

Mobile Water-bed

In this system the cavern is filled with the petroleum product to ceiling height, and is maintained at this level. The volume of water in the water-bed will therefore be variable to control the product level. This is done by pumping water in as the product is removed, and, alternately, removing water when more product is pumped in for storage.

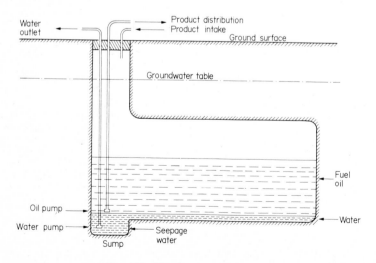

FIG. 11.8. Fixed water-bed storage.

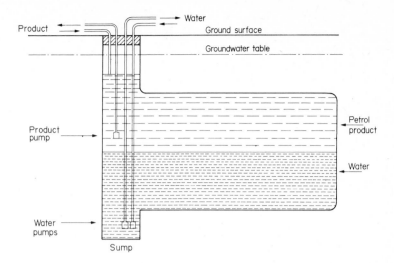

FIG. 11.9. Mobile water-bed storage.

Fixed water-bed storage is more commonly used when storing heavy crudes and the less-volatile fuel oils, since it requires less pumping and adjustment of fluid levels. Only seepage water needs to be removed. Mobile water-bed storage is favoured for more volatile petroleum products because the volume of free space above the product is smaller (Fig.11.9).

Product Migration

Lowering the groundwater level should effectively prevent the migration of product into the cavern walls; any movement of fluid should then be towards the cavity. If two or more caverns are constructed in close proximity to one another it may be necessary to include a protective curtain between adjacent chambers. This can be achieved by injecting water at high pressure into the intervening strata from an intermediate water-tunnel.

Prevention of Leakage

Liquefied petroleum and other gaseous products can be stored in unlined rock caverns at temperatures above $0°C$ provided that due precautions are taken to prevent leakage. The caverns must be situated at an adequate depth below the groundwater table. Aberg makes the point that it is not sufficient merely that the hydrostatic pressure of the groundwater surrounding the cavern should exceed the internal gas pressure, but the groundwater must flow in the downward direction with a hydraulic gradient not less than unity if the upward passage of gas bubbles through the pore spaces of the rock is to be prevented.

If an adequate downwards percolation cannot be achieved by the natural groundwater forces, then water should be injected from a system of boreholes situated in the capping rock. A water seal can thus be maintained in the fine fissures and pore spaces. Wider fissures, such as joints, etc., will need to be sealed with cement grout [16].

Applied Geotechnology

If the storage caverns are at a depth insufficient to provide the flow of groundwater that is necessary to prevent leakage, an alternative to water injection could be to so design the storage cavern to include a reinforced concrete lining to give mechanical strength, incorporating also a plastic membrane as an impermeable lining. Tabary describes tanks constructed in this manner to store liquid butane and propane. A feature of special interest is the provision of a means of collecting any fluid that may penetrate behind the plastic lining in a collection sump with a monitoring system to detect and measure any leakage hydrocarbons [17].

Engineering Properties of Frozen Ground

The storage of liquefied products at cryogenic temperatures in unlined caverns is complicated by the uncertainties attributable to the effects of low temperature on the rock walls. Thermal cracking may occur, leading to leakage and to boil-off problems.

Investigations into the engineering properties of frozen rocks and soils have acquired considerable importance today, mainly as a result of the strategic importance of the polar regions and their economic development. Much of the work involved has been of a classified nature, and the results of this have not been accessible for the general reader, but in recent years a growing volume of unclassified material has accumulated and is available for reference.

We are here concerned with the properties of mechanical strength, compressibility, and permeability. Frozen ground consists of solid mineral particles, ice particles, water, and air, and a proportion of water vapour. A measured value of the compressive strength of frozen gravel at the test site of the Engineering Standards Institute in the United States is quoted as approximately 1500 psi (10 MPa) [18]. The uniaxial compressive strength of water-saturated sandstone and limestone rocks was observed by Lindblom to decrease with the lowering of temperature to a minimum for sandstone at a little below 100 MPa at $0°C$, rising to a value approaching 200 MPa at $-160°C$. The corresponding strength of limestone ranged from a minimum of around 50 MPa at $0°C$ to around 160 MPa at $-160°C$ [19]. Dreyer observed that rock-salt and gypsum became brittle at cryogenic temperatures with a uniaxial compressive strength around 33–35 MPa [20].

Both Tystovich and Mellor explain the increase in stiffness and strength of water-saturated rock specimens on freezing to be due to the influence of mobile water in the pores and cracks of the rock matrix. All of this water does not solidify at once. Due to the increasing volume at the change of phase, liquid to solid, the residual water is squeezed into progressively smaller cracks and pores until ultimately it solidifies. There is thus a progressive reinforcement of the weaker portions of the rock matrix as the temperature decreases [21, 22].

There is also a generated tendency for the rock material to expand significantly at cryogenic temperatures, which could lead to potential tension failure. This tendency will be further intensified by changes in the volume and shape of ice crystals in the rock matrix.

The Mechanical Properties of Ice

Ice normally forms as a polycrystalline aggregate with grain sizes varying from less than 1 mm to more than 1 m in diameter.

Deformation of Ice Under Load

When a constant load is applied to an ice specimen the deformation–time curve is characterized by an instantaneous elastic deformation followed by a creep curve which eventually reaches a steady state. When the load is removed an instantaneous elastic recovery takes place of the same magnitude as the initial elastic deformation. This is followed by an elastic after-effect, which corresponds to the recovery of the transient creep. The creep recovery is not complete and the residual deformation represents the plastic flow which took place during the time that the stress was applied. Figure 11.10 is a typical deformation–recovery curve for polycrystalline ice with random *c*-axis orientation.

The elastic component of the deformation–time curve is due to the distortion of the crystal lattice, while the transient creep in the viscoelastic portion is due to intergranular slip or glide hindered by grain boundaries, and this is reversible. The plastic portion is due to intragranular slip or distortion of the grains.

Single crystals of ice show quite a different behaviour from that demonstrated by polycrystalline ice. The deformation rate in this case increases with time, and is almost entirely plastic with very little recovery taking place after the load is removed. Figure 11.11 is a strain–time curve for single crystalline ice oriented with the optic axis at 45° from the direction of application of the load.

Among others, Jellinek and Brill have shown that for ice Newtonian flow takes place only in stress regions below 2 kg/cm^2 [23, 24]. At higher stresses the minimum creep rate vs. stress curves are no longer linear. Butkovich and Landaur have shown that the minimum creep rates may be related to the shear stress by a power law [25].

Viscosity coefficients can be calculated in the region of Newtonian flow. Jellinek and Brill quote values averaging 5×10^{14} Ps for polycrystalline ice at $-5°$ C, and about 5×10^{12} Ps for glacial single crystals. For ice the rate of strain under load is dependent on the applied stress and its duration. For all practical purposes ice has no definable yield point since it creeps at extremely small stresses.

Ice may be treated as an elastic material provided the rate of stress application is fast enough or that low stresses are involved only for short periods of time. Values of Young's modulus, as measured by different observers, using bending, compression, and extension

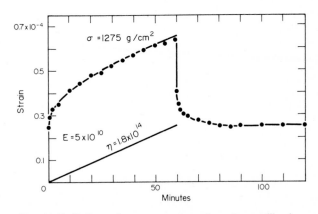

Fig. 11.10. Deformation–recovery curve for polycrystalline ice.

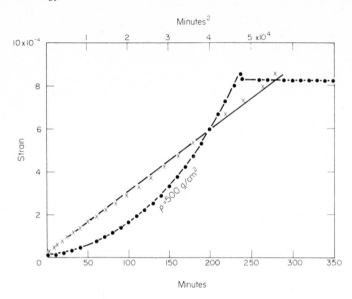

Fɪɢ. 11.11. Deformation–recovery curve for single ice crystal.

methods, show wide variation, probably due to the complications caused by creep. Gold made careful measurements of the static modulus of elasticity ranging from 5200 MPa at $0°$ C to 8300 MPa at $-40°$ C [26].

Ultimate Strength of Ice

Several investigators have made a study of the strength properties of ice, the reported results again showing wide divergence. They apparently depend upon two sets of conditions. The first are those imposed objectively, such as grain size, orientation, and impurities. The others are those imposed subjectively by the observer. Many observers choose test techniques apparently arbitrarily. Oftentimes these depend on tools, materials, and facilities available. Butkovich has recommended standards for small-scale ice-strength tests, specifying specimen size, shape, and loading rates, together with descriptions of various test techniques [27].

The compressive strength of ice, as a function of the time to failure, was observed by Hawkes and Mellor. Figure 11.12 brings out the high sensitivity of measured strength to the rate of loading for test durations less than approximately 100 s at $-7°$ C. It also shows that the measured strength is relatively insensitive to changes in the rate of loading for test durations more than about 10 min. The apparent strength for tests lasting about 30 min was only about 35% of the apparent strength for rapid tests over a few seconds [28].

The risks inherent in using rock quality indices based on compressive strength in frozen ground where tension failure is a major hazard is apparent from Fig. 11.13. This shows the failure strain as a function of the average strain rate. The extent of strain at failure decreases with increasing strain rate, both in tension and in compression, but the failure strain in tension is about one order of magnitude less than that in compression.

Standard test methods usually measure unconfined compressive strength, tensile strength, and shear strength. For unconfined compressive strength of lake ice the values

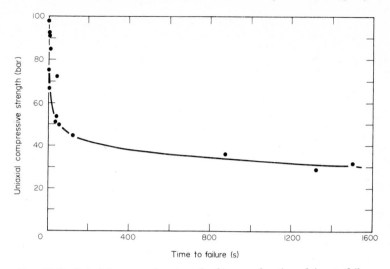

FIG. 11.12. Uniaxial compressive strength of ice as a function of time to failure.

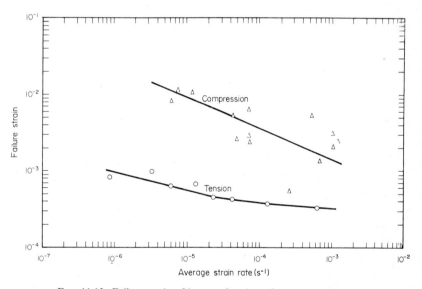

FIG. 11.13. Failure strain of ice as a function of average strain rate.

range between about 3.5–5.5 MPa in the temperature range − 5 to − 15° C depending upon orientation. For tensile strength a commonly accepted value is between 1.4–1.7 MPa in the same temperature range. Shear strength values, as measured in torsion, show a relationship similar to that of tensile strength values (Fig. 11.14).

The Effects of Impurities

The presence of impurities in the ice changes its mechanical properties. There exists, in nature, many types of ice in the impure state. For example, ice formed from seawater

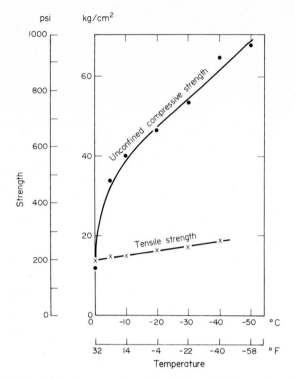

FIG. 11.14. Unconfined compressive and tensile strength of ice vs. temperature.

contains solubles that are excluded from the ice crystals and from brine pockets, or contain solid salts, depending upon the temperature. Glacial ice very often has included foreign matter, in size from fine silts to large pebbles, usually concentrated in stratified layers. The extreme case of ice containing impurities is permafrost, formed from water-saturated soils [29].

Figure 11.15 compares the unconfined compressive strength of sea ice and saturated silty clay with pure ice as a function of temperature. Both of the samples of ice with impurities show a reinforcement of strength. Observations in an ice-tunnel by Butkovich and Landaur show that ice layers containing glacial debris extrude from the clear ice, which seems to indicate that the impure ice had a creep rate greater than that of clear ice.

Livingston describes explosive blasting tests in frozen Keeweenaw silt. The frozen soils with the finest gradation had the highest moisture content and the lowest density. The moisture content of the silt was, in many cases, in excess of their water-holding capacity, which implies the presence of ice lenses. The test data showed that in nearly all cases failure occurred in the plane of the ice-lens stratification, regardless of the orientation of that stratification to the direction of loading. Hence the strength of the frozen soil was largely determined by the volume of ice within each specimen [30, 31].

Stresses Around Underground Storage Caverns

In storage caverns, where the temperature of the cavity surface is higher or lower than the ambient strata temperature, thermal stresses will be induced. If the storage generates

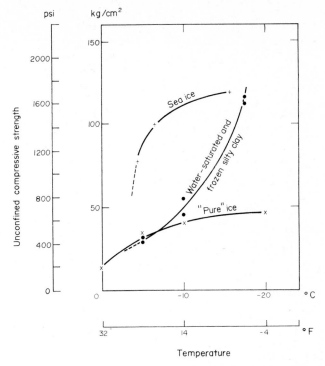

psi   kg/cm²

Fɪɢ. 11.15. Comparison of the compressive strength of ice showing the effect of impurities.

higher temperatures, those stresses will be compressive, and if at lower temperatures, tensile. The effects of thermal shock to the strata in cryogenic storage can be minimized by controlled, slow filling when the tank is first brought into commission. Thereafter, an effective control can only be exercised if suitable monitoring systems are included in the design to detect the leakage of hydrocarbons and to observe the state of stress in the cavern walls.

Theoretical and laboratory model studies of pressurized underground storage cavities have been conducted by Siskind *et al.* in an investigation of their loading and failure characteristics [32]. Thompson and Potts found that thermal stresses can be critical in changing a stable cavity in an elastic material into one where either a tensile or shear failure is possible. Tension failures may be a problem in shallow salt cavities used for high-pressure gas storage, due to the thermal variations that occur as a result of pressure changes. Tensile failure is less likely at greater depths because of the increased compressive gravitational load. Increased creep rates at higher ambient temperatures, however, will be important at depth [33–35].

Siskind *et al.* used cylindrical models in which cavities were fabricated to scale to simulate storage reservoirs. These were loaded in controlled stress fields and subjected to internal pressurization. Tangential and radial stresses on the cavity walls were inferred from theoretical solutions. The observed reservoir failures were tensile in form and occurred at predicted locations. However, the internal pressures that were necessary to produce reservoir failure were approximately three times those that were predicted on the expectation that failure would occur when the tangential tensile stresses exceed the tensile strength of the rock surrounding the cavity. The most critical reservoir parameters were found to be the values of tensile strength and Poisson's ratio for the rock surrounding the cavity [32].

Monitoring the *in situ* State of Stress

Details of the principles and methods of *in situ* stress measurement in rock mechanics will be found in *Geotechnology*, pp. 171–201. In practical terms, as applied to underground storage cavities, two methods are of prime interest—the Hawkes vibrating wire stressmeter, and indirect monitoring by microseismic technique.

Hawkes Vibrating Wire Stressmeter

The vibrating wire stressmeter, an exploded view of which is shown in Fig. 11.16, is designed to measure stress changes in rock. When inserted into a 38-mm diameter borehole and prestressed the cylindrical gauge can sense changes down to 14 KPa (2 psi). The changes cause minute alterations in the diameter of the steel body of the gauge, and they are measured in terms of the period of resonant frequency of a highly tensioned steel wire clamped diametrically across the gauge.

Because the stressmeter is rigid compared with the surrounding rock material, conversion of the frequency readings to stress change does not require accurate knowledge of the rock modulus.

The stressmeter is coupled to a solid-state meter with digital read-out. Since the signal is a frequency contact the electrical resistance of the leads does not affect the gauge readings, so that it may be read at remote distance (up to 1.5 km) without special cables or boosters.

The stressmeter can be inserted into boreholes in rock drilled to a depth of up to 30 m, and it is set either by a manual or by a hydraulically operated setting tool. The setting procedure tightens a wedge which is interposed between the gauge body and a metal platen, the width of the platen being selected to suit the rigidity of the rock concerned. Its range in hard rock is up to 70 MPa (10 000 psi) and in softer rock 40 MPa (6000 psi) [36].

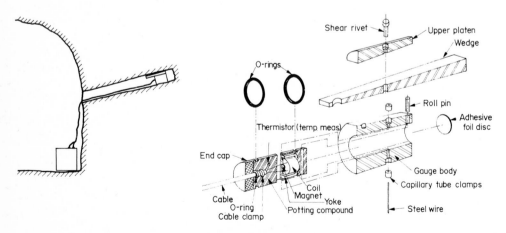

F IG. 11.16. Hawkes vibrating wire stressmeter.

Monitoring by Microseismic Technique

Since it is economically desirable to store gas at the highest possible pressure it is important that we should be able to determine with positive assurance what is the optimum pressure, i.e. what is the maximum pressure to which the reservoir can be pressurized while still remaining mechanically stable. In approaching this problem Hardy has employed a seismic method, coupled with analytical studies, to develop suitable criteria.

The analytical studies were conducted by use of finite element and basic mechanics techniques, in which various cavity shapes, spherical, cylindrical, and ellipsoidal, located in a gravity field, were studied.

The microseismic technique is of more practical application. The technique is based on the fact that many materials, including rocks and ice, emit transient vibrations in the audible and subaudible range, when subjected to stress under load. The origin of these "rock noises" appears to be related to processes of deformation and failure, which are accompanied by a sudden release of strain energy.

From the viewpoint of practical application in the present context the importance of the phenomenon is that microseismic pulses originate at localities where the rock material is mechanically unstable, and they can be detected at some considerable distance from their source. By using an array of stations the location of the source may be inferred, and the rate of occurrence of the pulses, their magnitude, and frequency spectra, all provide indirect evidence of the type and the degree of instability.

The technique has been used for many years to monitor the stability of the rocks around mining excavations, in slope stability investigations, and in underground mines to give warning of the possible imminence and location of rock-bursts and of rock and gas outbursts [37, 38].

Environmental Factors in Oil and Gas Storage

The environmental impact of underground oil and gas storage may be considered in respect of (i) ground surface utilization and amenity, and (ii) safety and health aspects. Given the absolute necessity of storage provision, then the only alternatives are whether to store in tanks above ground or in tanks and caverns underground.

Underground storage leaves the surface virtually free for other development or for agriculture or amenity use, apart from the necessary valves and distribution pipework. Difficulties may be presented by the proximity of urban or rural residential areas, the population of which may well object to the idea of living adjacent to or on top of gas- and oil-storage tanks. The risks must not be ignored. Nor should they be exaggerated.

After all, most of our villages, and all of our towns and cities, have had gas-holders in their midst for the better part of a hundred years and more, with little consequence other than damage to the aesthetic quality of the surroundings. Underground storage facilities, situated as they can be in an environment capable of being closely controlled, can manifestly be made safer against the risks of explosion and fire than surface works and gas-holders can possibly be. They will certainly be far less obvious and less ugly. Indeed, in not a few cases it is possible that the general public is unaware of their existence.

This is not to say that precautions are not required, or that the procedures at present being followed are satisfactory in all respects. Accidents have been known to happen, and even experts can make mistakes. In the end it is a question of probabilities. The risk of mishap must be brought down to acceptable proportions.

So far as the risk of ignition, explosion, and fire is concerned, various principles of action are possible:

(i) To interpose and maintain between the stored materials and the outside atmosphere a blanket of inert gas which is incapable of supporting combustion.

(ii) To allow the vaporized products to accumulate so as to build up an extinctive atmosphere containing insufficient oxygen to support combustion.

(iii) To ventilate and maintain a non-incendive atmosphere, below explosive limits, in all localities where possible sources of ignition may exist in proximity to stored products and the passages connected thereto.

(iv) To eliminate all possible sources of ignition.

(v) To install monitoring systems to observe the stability of the containment systems and to detect and measure any leakage of hydrocarbons.

Principle (i) offers positive protection and is of proved efficacy in maritime tanker transport. It is, however, costly. Principles (ii) and (iii) are more commonly followed, but they cannot remove the risks altogether, since there must be, between a non-incendive atmosphere due to lack of oxygen and a non-incendive atmosphere below explosive limits, a fringe zone in which explosive mixtures must exist. This means that principle (iv) is always necessary, the main hazard coming from open sparking in electrical equipment and from static electricity. Similarly, principle (v) should be a universal requirement, preferably incorporating automatic alarm and control systems.

Nevertheless, experience in the chemical industry and in mining shows that engineering operations are possible in circumstances where explosive atmospheres may exist provided that they are carried out in accordance with approved procedures and codes of practice. These procedures should be governed by the appropriate legislative authorities, by acts and regulations carefully formulated in the light of research and experience, strictly supervised by a vigilant safety and health inspectorate, intelligently applied by educated and responsible operatives, and acceptable to an informed public at large.

References and Bibliography: Chapter 11

1. EVANS, B. M., Design and construction of underground storage units for gaseous and liquid products, *Inst. of Gas Engrs 29th Autumn Res. Meeting, London, November 1963.*
2. ANDERSON, P. J., and KHAN, A. R., Storage of liquefied natural gas, *ASCE Transport Engng Conf., Minneapolis, May 1965*, reprint 216.
3. SLATER, G. E., *Salt Caverns—Multipurpose Storage Vessels*, Res. Rep. Penn. State Univ. Dept. of Mineral Engng, 1974.
4. KATZ, D. L., and COATS, K. H., *Underground Storage of Fluids*, Lurich Inc., Ann Arbor, Mich., 1968.
5. JACOBY, C. H., Storage of hydrocarbons in bedded salt deposits, *3rd Symp. on Salt, Northern Ohio Geol. Soc., Cleveland, Ohio, 1968.*
6. HARDY, H. R., Model studies associated with the mechanical stability of underground natural gas storage reservoirs, *2nd Cong. Inst. Soc. Rock Mech., Belgrade, September 1970*, paper 4–42; and Stability studies on gas storage reservoir models, *Proc. Am. Gas Ass. Transmission Conf., Denver, Colo., April 1970.*
7. DAVIES, R. L., and WILLET, D. C., The identification and preliminary design of existing mines for the US Strategic Petroleum Reserve Program, *Rockstore 77, Stockholm* (Bergman, ed.) **1**, 19–23, Pergamon Press, 1978.
8. KATZ, D. L., *et al.*, *Handbook of Natural Gas Engineering*, McGraw-Hill, New York, 1959.
9. SIMPSON, F., Potential for deep underground storage in Canada, *Rockstore 77, Stockholm* (Bergman, ed.) **1**, 37–41, Pergamon Press, 1978.

10. BUTTIENS, E. F. J., Conversion of abandoned collieries in Southern Belgium to low pressure gas storage, *Rockstore 77, Stockholm* (Bergman, ed.) **3**, 651–5, Pergamon Press, 1978.
11. WETTLEGREN, G., Underground oil storage in disused mines, *Underground Space* **3**, 9–17 (1978).
12. McCARTHY, D. F., Underground mined type LPG storage, *Underground Space* **3**, 243–51 (1979).
13. JANSSON, G., Storage of petroleum products in unlined caverns, *Underground Space* **2**, 27–37 (1977).
14. AISENSTEIN, B., BRAWN, N., FLEVER, A., and VERED-WEISS, J., Oil storage in dry chalk, *Rockstore 77, Stockholm* (Bergman, ed.) **3**, 675–9, Pergamon Press, 1978.
15. MAURY, V., Underground storage in soft chalk, *Rockstore 77, Stockholm* (Bergman, ed.) **3**, 681–9.
16. ABERG, B., Prevention of gas leakage for unlined reservoirs in rock, *Rockstore 77, Stockholm* (Bergman, ed.) **2**, 399–413, Pergamon Press, 1978.
17. TABARY, J., Monitored impermeability for petroleum storage, *Rockstore 77, Stockholm* (Bergman, ed.) **2**, 507–12, Pergamon Press, 1978.
18. ANON., *Investigation of Description, Classification, and Strength Properties of Frozen Soils*, Arctic Construction and Frost Effects Laboratory, US Army SIPRE Rep. No. 8, 1951.
19. LINDBLOM, U. E., Research related to LNG storage in rock caverns, *Rockstore 77, Stockholm* (Bergman, ed.) **2**, 487–93, Pergamon Press, 1978.
20. DREYER, W. E., Cold and cryogenic storage of petroleum products, *Rockstore 77, Stockholm* (Bergman, ed.) **2**, 467–79, Pergamon Press, 1978.
21. TYSTOVICH, N. A., The fundamentals of frozen ground mechanics, *Proc. 4th Inter. Conf. Soil Mech. and Found. Engng, 1950,* **1**, 116–19.
22. MELLOR, M., *Phase Composition of Pore Water in Cold Rocks*, US Army CRREL Rep. No. 292, Hanover, NH, 1970.
23. JELLINEK, H. H. G., and BRILL, R., Viscoelastic properties of ice, *J. Appl. Phys.* **27** (10) 1198–1209 (1956).
24. BUTKOVICH, T. R., The mechanical properties of ice, *Proc. 3rd Symp. Rock Mech., Colo. Sch. Mines* Q. **54** (3) (July 1959).
25. BUTKOVICH, T. R., and LANDAUR, J. K., *The Flow Law for Ice and the Creep of Ice at Low Stresses*, US Army Corps Engrs SIPRE Rep. No. 35, 1959.
26. GOLD, L. W., Some observations on the dependence of strain on stress, *Can. J. Phys.* **36** (10) 112–15 (1958).
27. BUTKOVICH, T. R., Recommended standards for small scale ice strength tests, *Eng. Inst. Can. Trans.,* **2** (3) 112–15 (1958).
28. HAWKES, I., and MELLOR, M., Deformation and fracture of ice under uniaxial stress, *J. Glaciology* **11** (61) 103–31 (1972).
29. SWINZOW, G. K., Certain aspects of engineering geology in permafrost, *Engng Geol.* **3**, 177–215 (1969).
30. LIVINGSTON, C. W., *Excavations in Frozen Ground*, US Army Corps Engrs Spec. Rep. SIPRE 30, 1956.
31. TYUTYUNOV, I. A., *An Introduction to the Theory of Frozen Rock*, Pergamon Press, Oxford, 1964.
32. SISKIND, D. E., HARDY, H. R., and ALEXANDER, S. S., A study of pressurized underground storage cavities: theoretical and laboratory investigations, *Int. J. Rock Mech. Min. Sci.* **10**, 133–40 (1973).
33. THOMPSON, T. W., and POTTS, E. L. J., The influence of thermal stresses on the stability of underground storage cavities, *Int. J. Rock Mech. Min. Sci.* **16**, 117–25 (1979).
34. KRENK, S., Internally pressurized spherical and cylindrical cavities in rock salt, *Int. J. Rock Mech. Min. Sci.* **15**, 219–24 (1978).
35. OTTOSEN, N. S., and KRENK, S., Nonlinear analysis of cavities in rock salt, *Int. J. Rock Mech. Min. Sci.* **16**, 245–52 (1979).
36. HAWKES, I., and HOOKER, V. E., The vibrating wire stressmeter, *Proc. 3rd Congr. Int. Soc. Rock Mech. Denver, Colo., 1974,* **2A**, 439–44.
37. KNILL, J. L., FRANKLIN, J., and MALONE, A. W., A study of acoustic emission from stressed rock, *Int. J. Rock Mech. Min. Sci.* **5**, 87–121 (1968).
38. HARDY, H. R., Microseismic techniques—basic and applied research, *Rock Mech.* **2**, 93–114 (1973).

CHAPTER 12

The Disposal of Waste Materials

THE production of solid and fluid wastes is associated with many industrial processes and their disposal can be a particularly severe problem. Dumps of earth and rock debris are a characteristic feature of the landscape in mining areas, the debris coming both from open-pits and from underground workings. Tunnelling operations in civil engineering also create a waste-earth and rock-disposal problem. So far as mining is concerned, some of the waste can be returned and stowed underground, but this is not always feasible.

Dumping Solid Wastes

The disposal of earth and rock wastes may be considered to be solely a matter of materials handling, and the main problem one of finding a suitable site, close to hand, on which to dump the materials. But the matter is seldom one of engineering economics alone. Questions of aesthetics and of public health and safety must also be borne in mind. Where space is limited or expensive there is a natural temptation to dump the materials until their maximum slope angle is attained, and then to move the tippler point laterally until the site is completely covered. The inevitable result is the creation of a high dump that, in addition to being an offence to the eye, stands at about its limit equilibrium slope condition only for so long as it and the ground upon which it stands remains dry and unyielding.

In many communities today legislative control of urban and rural development imposes restrictions as to the manner in which waste materials may be disposed of. Even where such controls do not exist, the mounting pressure that is generated by public attention towards the provision of environmental quality provides an incentive for planned and careful disposal of industrial wastes even for no other reason than to improve the public image of the industry concerned. It is fair to say that today no industrial concern that is conscious of its public image would fail to include in its systems and economic analyses on any important new project the procedures and cost of waste disposal in such a manner as to cause the minimum disturbance of environmental quality. Indeed, some may be designed so that in the long run the environment is improved.

So, instead of the high mounds that disfigure the landscape of many of our older industrial areas, good contemporary practice limits the dump height, with the intention of moulding it into the natural contour as much as possible. In a flat area this inevitably involves a substantially greater lateral spread, which can be expensive in terms of land values. The prior removal of top soil for subsequent replacement and recovery of the land so used is today regarded as standard practice in such cases, as it should be also for open-pit strip mining. Where the relief of the landscape is suitable, solid wastes should be

230

dumped below the natural contour so as to preserve the natural drainage characteristics. Special care is required on the downslope face of the dump to maintain its stability, limit its erosion, and to collect and regulate all seepage and run off water. These aspects become progressively more important in regions of high relief and where the safety and convenience of downslope communities must be preserved. The National Coal Board's code and rules appertaining to the construction of colliery spoil-tips in Britain requires that dumps on sloping ground must be designed to a slope safety factor at least 0.25 greater than that required on flat ground.

Dumping Materials Liable to Fire

Certain earth materials are liable to spontaneous heating when they are exposed to oxidation. Such materials include carbonaceous sediments and some sulphides, and when both these materials occur in combination they may constitute a particularly dangerous fire risk. Two procedures are applicable as protective measures. One is to ventilate the material, using pipes 250–150 mm in diameter to let in air and to drain off the heat as fast as it is generated. The other procedure is to seal off the material by covering it with a coating impervious to air, so as to prevent oxidation. The object in both procedures is to limit the rise in temperature below that which might result in the occurrence of objectional fumes, smoke, or active fire.

To be effective, either procedure must be accompanied by constant surveillance and measurement to monitor the location, extent, and rate of temperature rise so that corrective measures can be put into effect should the need for them arise. The former procedure is applicable to dumps of limited size, such as coal-storage areas, but probably would not be attempted at a large waste dump where the process of attempted sealing would be more practicable. Methods of restricting the access of fresh air to materials at depth include mechanical compaction concurrent with dumping and the provision of a final coating of fine, rolled, topsoil.

The suppression of spoil-bank fires by blanketing and trenching was described by Carr [1]. Blanketing materials included a depth of approximately 225 mm of limestone dust. Trenching to a width and depth of 2×2 m was undertaken to limit the lateral spread of fire. The base of the trenches was flooded with a limestone dust and water slurry, which was soaked into the material to the limits of saturation. The application of water in the form of a fine mist spray has often proved to be effective in promoting consolidation and sealing of the surface layer, fine solids being carried into the interstices of the spoil, in suspension. If water is sprayed on it is important not to apply this in heavy jets the runoff from which will form erosion channels and so open up passages for the easy ingress of air. There is considerable risk, too, if water jets are played directly on to an incandescent carbon mass, for this then forms a water–gas generator, carbon monoxide and hydrogen being generated. These gases form a highly explosive mixture with air, and an explosive eruption could occur in the spoil heap.

Close attention should be placed at all times to the following precautions when dumping carbonaceous materials.

(1) Remove all vegetation from the ground around the periphery of the spoil heap.
(2) Compact the spoil material as it is laid.
(3) Streamline the surface of the heap to a smooth contour.

(4) Prohibit the inclusion of any inflammable material in the spoil.

(5) Ensure that all hot materials such as boiler ashes are thoroughly quenched before dumping them.

(6) Prohibit the use of braziers on the heap.

(7) Seal all holes and excavations and all erosion channels with fine backfilling material.

(8) Set up constant inspection systems and monitoring to detect the emission of fumes and to locate hot spots [2].

Temperatures may be monitored by means of thermistors distributed at various depths within the material. Infrared photographs scanning the surface can identify the precise position of localized temperature increase at a very early stage of heating. It is important that early detection should be made so that protective measures can be effective. Such measures could consist of cement and lime grouts and of plastic foam injected into and sprayed over the areas concerned. Should these measures not succeed in halting the temperature rise, then direct attack to expose and quench the heated material is imperative. Failure to do this will almost inevitably result in active fire, which if allowed to take firm hold within the heap will be extremely difficult to eradicate.

Dumping Dry Dusty Materials

The proper disposal of fine dry materials requires that dust-suppression techniques be adopted to minimize health risks and nuisance in neighbouring down-wind areas. Water sprays should be positioned at all material transfer points and chutes on the transportation system. The efficacy of water is substantially improved by the addition of chemical wetting agents to the solutions sprayed.

In dry climates wind-blown dust, lifted from the dump, frequently poses a problem. Again, water sprays aided by wetting agents and followed by mechanical compaction may be applied. A more lasting solution may require that the dump material be fastened by vegetation, and much research is at present in progress to determine suitable vegetation and optimum modes of treatment for different geographical and climatic conditions [3].

Protection of Waste Dumps Against Surface Erosion and Runoff

Besides giving protection against wind erosion the vegetation also affords some protection against erosion by rain. However, it is of little avail to attempt to seed a slope which is inherently unstable. It is also important to grade the upper surfaces of the dump so as to promote controlled drainage away from the peripheral slope, or towards the headwall, to prevent water lodging in pools, and to prevent the opening-up of erosion channels that will provide easy access paths for moisture into the underlying slope material [4].

Rainfall seeping into the dump in quantity will reduce its mechanical strength and hence reduce its stability. On the other hand, the retention of a limited amount of moisture will give a loose unconsolidated material added cohesion with which it can resist surface erosion. In desert and semi-arid regions a steep slope may be constructed with contour trenches to control excessive runoff during wet seasons, when deep channels would otherwise be eroded. Such trenches are constructed along zero grade at vertical intervals of from 7 to 15 m, with check dams at about 30-m intervals. The trenches are intended for storage of peak flow and the cross-dykes channel-off excess water. The angle of slope of the waste dump in these circumstances should not exceed $2\frac{1}{2}$ to 1.

Open-pit Strip Mines

A typical general layout of a modern open-pit strip-mining operation is illustrated in Fig. 12.1. Surface material or overburden, which overlies the mineral body, is removed in

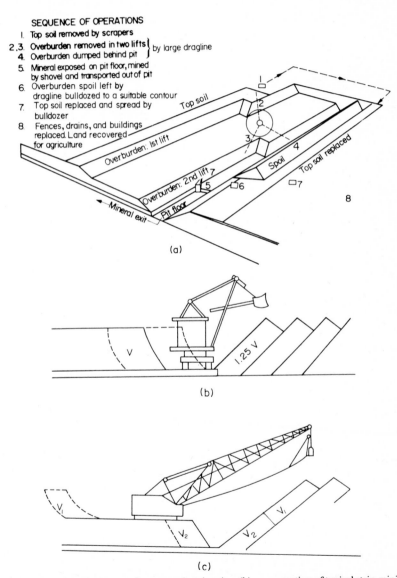

SEQUENCE OF OPERATIONS
1. Top soil removed by scrapers
2,3. Overburden removed in two lifts
4. Overburden dumped behind pit } by large dragline
5. Mineral exposed on pit floor, mined by shovel and transported out of pit
6. Overburden spoil left by dragline bulldozed to a suitable contour
7. Top soil replaced and spread by bulldozer
8. Fences, drains, and buildings replaced. Land recovered for agriculture

FIG. 12.1. (a) General layout of an open-pit strip mine; (b) cross-section of typical strip-mining operations (bucket-shovel excavator); (c) cross-section of a walking dragline operation.

stages. The first stage involves the removal of topsoil which is placed temporarily at the side of the area to be worked. Next, the remaining overburden is stripped off to expose the mineral. A cut in the base of the trench so formed then provides a working face along which the mineral can be excavated and loaded into a transport system running along the base of the trench. The overburden is dumped behind this trench by means of mobile draglines, bucket-arm excavators, or rotary wheel-diggers, positioned on top of the exposed mineral. If the excavation is deep the overburden is removed in stages by means of an excavator positioned on top of the next lower stage. The waste material is also dumped in stages, replacing each lift in appropriate sequence behind the excavated area.

The final stage is one which in earlier times was, and sometimes even today is, omitted, with the result that much opprobrium is often directed by the general public towards the minerals industries. In many quarters strong efforts are made to try to prohibit strip mining altogether, on grounds of aesthetics, ecology, the destruction of surface land, and loss of public amenity. But this need not always be so if proper restoration procedures are followed. Such procedures include bulldozing the spoil back to a contour suitable for agriculture, relaying the original topsoil, constructing access roads, drains, and buildings, replanting trees and other vegetation to promote, in the long run, a general recovery of the area to standards of agricultural value, ecology, and aesthetics, no less desirable than those which originally existed.

Solid–Liquid Waste Materials

The surface disposal of mixed solid and liquid wastes forms an essential feature of the minerals and metallurgical industries. Few natural raw materials occur in a form suitable for immediate transport to market. Most have to be separated from waste, or concentrated, in the process of which they are crushed and subjected to various forms of hydro-physical and hydro-metallurgical treatment. The waste material from such operations is usually in the form of a liquid–solid mix, which must be disposed of. Problems of allied but differing character arise in connection with the debris from coal-mines and the wastes from hard-rock mines and from metalliferous plant.

Coal-mine Debris

In October 1966 there occurred a disastrous slide of colliery debris which flowed down the side of the valley at Aberfan, in Wales, to engulf part of the village below, including much of the school in which the children were gathered at assembly. This disaster, which shocked the world, was followed by a rigorous tribunal of inquiry, the report of which makes clear the causes and mode of occurrence of the disaster, and includes recommend-ations that are aimed to prevent the future occurrence of similar disasters in Britain [5]. Some years later, in February 1972, a coal-refuse dam failed in West Virginia. The resulting flooding of the Buffalo Creek Valley had national ramifications in the United States. The immediate consequences of the flooding included the deaths of 118 persons, with 7 more reported missing, the loss of over 500 homes, and extensive flood damage. This disaster was also followed by several investigations and inquiries [6].

It is easy to be wise after the event, and the disasters at Aberfan and Buffalo Creek differ in that in Wales a refuse tip collapsed, whereas in West Virginia the failure of an upstream dam triggered a chain reaction of subsequent failures extending progressively downstream. Yet the official reports on the respective disasters contain many parallels and provide us with much salutary reading which should help to prevent future occurrences of a similar nature. In both cases there was a previous history of limited failures, the significance of which was either ignored or not recognized by the respective engineers and supervisory personnel. Neither at Aberfan nor at Buffalo Creek were the operations conducted in accordance with the requirements of good engineering practice. The Aberfan inquiry brought to light no evidence of an investigation to establish the suitability of the site before dumping commenced, and no systematic observations and measurements to monitor the stability of the material subsequently being dumped. At Buffalo Creek the dams were not designed or engineered. They were built largely by end-dumping of refuse from trucks, followed by grading with bulldozers to create a relatively smooth surface.

The foundation materials of the dams were thick deposits of fine-grained coal-sludge waste, having a low unit weight and a high susceptibility to erosion and piping. Neither in Britain in 1966, nor in the United States in 1972, was there any significant documented information on the engineering properties of colliery waste materials that could have formed a rational basis for design. Although in the intervening period between 1966 and 1972 much work was done and published in Britain, a review of this was not generally available in the United States until January 1972 [7]. Both at Aberfan and at Buffalo Creek the season was wet, but not abnormally so. There were, in each case, inadequate provisions for conveyance of runoff water from the embankments and no adequate means of controlling internal seepage.

The Wimpey Laboratory Report [8] presents details of typical discard materials from coal-treatment plants, comprising fine discards in the form of slurry and tailings of coarse reject solids. Particle size analyses and consolidation characteristics of material from various localities are given, together with representative shear stress envelopes, as determined by shear box and triaxial tests. In general the shear strength envelopes show some departure from linearity, so that a tangent to the envelope at a given value of normal stress indicates the friction angle and the cohesion constant for that particular condition. Hoek and Bray describe the mode of application of this type of information to the stability analysis of mine-waste dumps, assuming circular shear slide failure and various drainage conditions [9], while Bishop discusses in authoritative detail the stability of tips and spoil heaps in general [10].

Colliery Slurry Lagoons

Disposal of the slurry effluent from coal-treatment plant involves ponding the slimes behind embankments usually constructed of coarse solid wastes. It was such embankments that failed at Buffalo Creek. In these ponds or lagoons the fine solids in the slurry are allowed to settle and the clarified surface water is removed. The embankment may be constructed as a drainage bank that permits percolation and forms a boundary drain around the lagoon. This promotes drainage from and consolidation of the deposited fines that, after the lagoon becomes filled and the deposit drained and consolidated, ultimately might be excavated and used as fuel, or, if of no economic value, would probably be covered with later spoil. Such a lagoon embankment would have relatively low strength.

If an adequate safety factor is to be achieved its height must be restricted and a low angle of slope maintained.

Alternatively, the lagoon embankment may be constructed as an earthfill dam, incorporating an impermeable membrane or core. Such a construction might be necessary to ensure the stability of an embankment that is required to impound a large volume of water. However, it would considerably prolong the time required for the deposited fines to consolidate, and it would also increase the amount of residual moisture in the deposit. A compromise mode of construction may therefore be adopted if the waste materials are suitable. This requires that the main embankment structure is constructed of material with a low permeability. To prevent seepage and erosion problems on the downstream face, either an internal drainage system is required or else a "drainage berm" of permeable material may be placed so as to reinforce the downstream slope, to a height equal to at least half the depth of the lagoon. Another mode of construction for an impermeable embankment is to lay a blanket of low permeability and not less than 3 m thick along the lagoon contact face. If erosion of this face by wave action and weathering is a possibility, then the blanket could itself be faced with a layer of resistant strength, as for a conventional rockfill dam.

Lagoons should be constructed on firm foundations at ground level. Any seepage from the lagoon, into and through any underlying spoil, will greatly influence the stability of the spoil slopes. The construction of a slurry lagoon on top of a spoil heap is therefore a practice not to be recommended. However, it may be that circumstances may render the utilization of such a site necessary. Much then depends on the characteristics of the spoil heap material. If the lagoon is filled rapidly, seepage will occur through the underlying spoil and through the retaining embankment. Possible danger may then arise in two ways: (a) if the spoil contains coarse granular material then high seepage pressures may promote piping and consequent failure at the foot of the embankment; (b) if the spoil contains fine-grained argillaceous sediments these will be affected by water and they will lose much of their strength. If they remain saturated and subjected to seepage pressures for a prolonged period they may deteriorate to the stage of plasticizing and beyond this to eventual liquefaction and mud-flow.

On the other hand, if the lagoon is filled slowly more time will be allowed for the deposit of slimes to coat the bottom of the pond with a relatively impermeable layer and, in consequence, seepage through the underlying spoil will be reduced. The rate of percolation will decrease as the thickness of the relatively impermeable layer increases. The ideal situation would be that the quantity of fluid fed into the lagoon would be insufficient to maintain a continuous flow through the spoil, which would then drain off. To achieve this condition the slurry should be fed into the pond slowly, so as to extend an apron of deposited silt gradually from the inlet pipe and across the base of the lagoon to the retaining embankment, controlling the level of the fluid at all times by an overflow system (Figs. 12.2, 12.3, and 12.4).

Tailings Dams

The effluent or tailings from metalliferous ore treatment plant commonly consists of two fractions—denoted "sands" and "slimes". As in the case of colliery debris, there has hitherto been little published information as to the critical design parameters of such

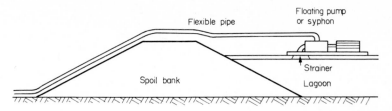

FIG. 12.2. General arrangements for drawing-off water from coal slurry lagoons (National Coal Board).

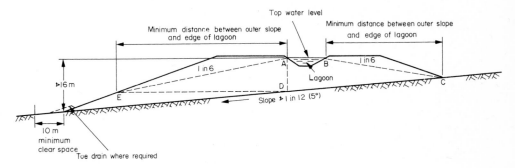

FIG. 12.3. Minimum requirements for constructing a lagoon on an existing spoil tip (National Coal Board).

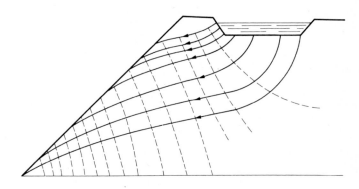

FIG. 12.4. Flow net for a slurry lagoon on a spoil base.

material, in particular, data concerning density, shear strength, and permeability. However, in view of the enactment of control legislation in recent years concerning the design, construction, and maintenance of tailings dams, these deficiencies are being remedied. It is now most unlikely that any new tailings dam of consequence would be constructed without first commissioning both a detailed site investigation and a materials study (Fig. 12.5).

In the construction of a tailings dam either one of two general methods of procedure is usually adopted: (a) the mill-tailings constitute the structural medium, or (b) a primary retaining structure is built from imported materials against which the tailings are impounded. There may be some variation of method (b) in that sometimes the original retaining dyke may be built of borrowed materials, later extensions being constructed of

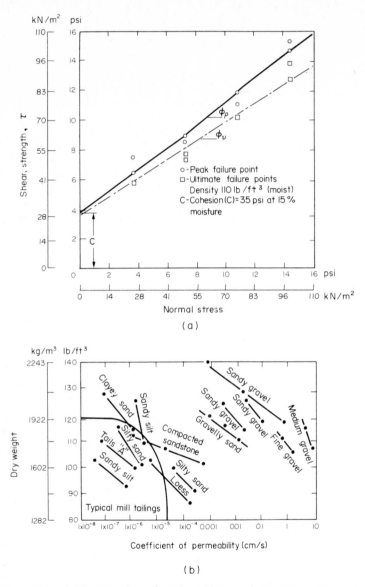

FIG. 12.5. (a) Typical failure envelopes for mine-tailings spoil; (b) density of typical mill-tailings related to permeability [11].

mill-tailings. In other cases the retaining dyke or berm is constructed of imported materials throughout.

Upstream Method of Construction

The upstream method of construction is illustrated in Fig. 12.6a. After preparing the foundations for the dam on a firm base, cleared of all organic material, the original dyke is constructed of pervious material such as coarse rock and gravel, with some sand admixed,

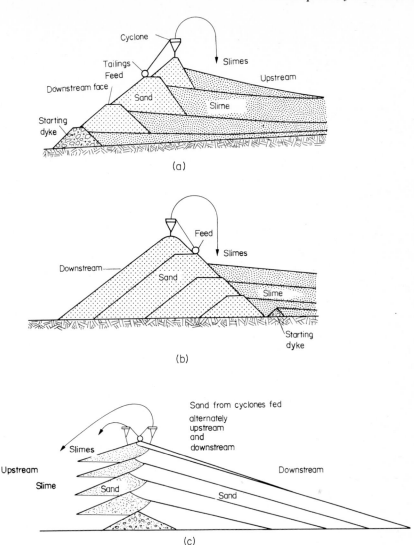

FIG. 12.6. (a) Constructing a tailings dam by the upstream method; (b) constructing a tailings dam by the downstream method; (c) constructing a tailings dam by the centreline method.

with a layer of sand on the upstream side to prevent the subsequently impounded tailings piping through the rockfill. The tailings are then discharged from pipes set along the dyke at intervals from 3 to 15 m apart. From these the coarse material settles out first, to form a beach, while the finer material is carried in suspension into the pond. The separation of coarse and fine fractions may be aided by the use of cyclones at each discharge point, so that the slimes can be conveyed separately from the sands and well into the interior of the pond. This provides a more stable construction because the slimes settle into a much weaker material than the sands, and it is undesirable to have them deposited near the berm. After the initial dyke has been completed and filled, subsequent stages are added, progressing upstream, in such a manner as to present a continuous slope to the downstream face. If the dykes are not built of imported rock throughout, subsequent berms after the first are scraped up from the sand beach by dragline or bulldozer, which can also be used to compact the sand.

Downstream Method of Construction

In the downstream method of construction the coarse material is deposited on the downstream side of the dyke and the slimes are deposited in the pond. A pervious triangular-shaped dam results, the volume of which increases progressively as the dam gets higher (Fig. 12.6b). Comparing the upstream and downstream methods of construction, the latter procedure results in a more stable dam, which is sustained by the progressively increasing mass of the structure as it gets higher. Provided that the downstream slope angle is properly designed, in accordance with the shear strength, density, and permeability of the sand, the height of the structure is limited only by the volume of sand that is available.

The stability of a dam that is constructed by the upstream method, however, is less easily assured, with extension in height, because in this case it depends on the shear strength of the impounded material rather than on that of the dam. In this type of structure the impounding dykes form a relatively thin shell as compared with the gravitational mass of a dam constructed by the downstream method. Nevertheless, an acceptable factor of safety can be designed into the structure if the strength parameters of the impounded sands and slimes are known, and assuming particular modes of failure.

Centreline Method of Construction

In the centreline method of construction a rockfill starter dam is constructed, extending both upstream and downstream of the centre line. The dam is subsequently raised by cycloning sand alternately upstream and downstream. All the slimes are deposited in the upstream pond (Fig. 12.6c) [11].

Failure of Mill-tailings Dams

Three possible modes of failure must be considered: (a) circular shear slide failure, (b) liquefaction under seismic shock, (c) piping failure.

Shear Slide Failure

In the early stages of construction, with a low dam, the critical failure surface is likely to lie close to the dyke shell, since at this stage the sands are not compacted. They are saturated, with very low cohesion and friction. As the dam grows in height the lower layers of sand become compacted and stronger, so that the stability of the structure depnds more on the strength of the deposited slimes. The critical failure surface then will lie further within the impounded material [12].

Liquefaction

In the event of seismic shock the strength of both the sands and the slimes may be destroyed by liquefaction. Instead of a circular shear slip, liquefaction of the tailings occurs behind the approximate contact between consolidated and weaker material, which forms a relatively thin retaining wall. This wall is then likely to be breached by the high hydraulic pressures generated within it [13].

In his evaluation of the stability of tailings dams under seismic shock, Brawner quotes Dobry and Alvarez, who maintain that sand with a relative density of 50% or more cannot liquefy, regardless of the constituent size gradation [14]. Seed and Idriss determined the potential liquefaction for sands of various consolidation characteristics, as indicated by standard penetration resistance [15]. It follows that the design of tailings dams in earthquake-prone regions must take into account the probability of seismic accelerations and provide solutions, either by compacting the tailings to a sufficiently high density so that they will not liquefy under the anticipated acceleration, or, alternatively, they must be adequately drained so that they cannot remain saturated. If high accelerations are possible it may be necessary to adopt either the downstream method of construction or the centreline method, despite their greater space requirements and higher construction costs.

Piping Failure

Piping and internal erosion occur as a result of excessive seepage pressure. Fine particles of soil are impelled into and through the voids between coarser fragments. A progressively less-restricted channel for fluid flow is thus generated until ultimately the material may "boil" to form a large failure cavity, sink-hole, or pipe. The phenomenon is liable to occur at the downstream toe of a retaining dyke, and it may be obviated by controlling the seepage through the permeable foundations and dyke materials. Klohn describes a typical piping failure on a tailings dam. The cause of the failure appeared to be the lack of a proper filter zone between the rockfill and the tailings, which allowed internal erosion of tailings into the rockfill [12].

Methods of controlling seepage through permeable foundations and earth dams include:

(1) Construction of a vertical impermeable cut-off through the permeable deposit.
(2) Inclusion of an upstream impermeable blanket.
(3) Inclusion of a drainage blanket beneath the outer downstream zone.
(4) Construction of relief wells close to the outer toe.

Relief wells are normally used where the foundation strata are subject to artesian pressures in order to reduce the head of groundwater near the toe of a slope, heap, or dam. Seepage from the wells is then collected and channelled off.

Underground Disposal of Waste Materials

However desirable it might seem to return underground all the waste materials that are produced in geotechnical operations, the cost of so doing is such as to render it most unlikely that this will often be done on aesthetic grounds alone. But there may be sound technical reasons to justify the procedure. These include: (1) where the excavations require the support afforded by solid-packed or "stowed" wastes, either to promote the effective control of strata movements or to limit the extent of subsidence of the ground surface, due to the underground workings; (2) where there is a risk of spontaneous combustion underground and it is desired to restrict the passage of leakage air from the ventilation system; (3) where poisonous or radioactive wastes have to be buried at depth. Basically, the adopted procedures are matters of materials handling, but the implications that frequently arise in relation to strata control, support of the excavations, and subsidence of the surface make it necessary for the geotechnologist to study the techniques applied. Depending upon the method of transportation employed, the various alternative systems may be classified as (a) mechanical, (b) pneumatic, and (c) hydraulic stowing.

Mechanical Stowing

In mechanical stowing systems the waste earth and rock material to be stowed is transported in trucks or on belt conveyors to the disposal point, where it is tipped. It is then moved to each side of the disposal point by some means, such as by scraper buckets pulled by wire ropes. Alternatively, the material is tipped on to a short, high-speed, conveyor belt, or on to a mechanical centrifugal impeller, which throws it into the underground void being filled. The scraper method is more common, although it leaves a poorly filled pack. Belt and rotary vane impellers, being subject to high impact, abrasion, and wear, are relatively expensive to maintain (Figs. 12.7 and 12.8) [16, 17].

Pneumatic Stowing

In pneumatic stowing the waste rock is crushed to below 75 mm in size, and then transported underground. At a transfer point near to the stowing machine the material is transferred into a feed-pipe 150 mm in diameter which passes it into a hopper and thence into a rotary valve. This valve passes the material into a chamber which is pressurized by compressed air, so that the stowing material forms a fluidized stream along the stowing-pipe, from which it is blown into the area being filled. A deflector at the end of

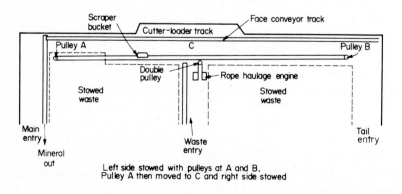

FIG. 12.7. Solid stowing by scraper packer.

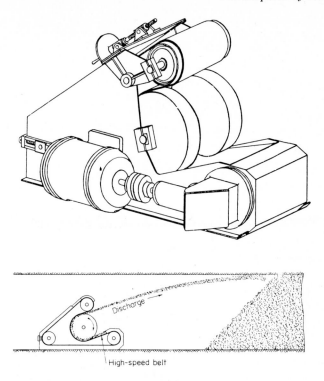

FIG. 12.8. Mechanical stowing by high-speed belt.

the pipe controls the direction of the ejected stream (Figs. 12.9 and 12.10) [18, 19]. Considerable research has been conducted on the fundamentals of pneumatic stowing. Whetton and Broadhurst determined the pipe friction coefficient, used in calculations of the drop in air pressure for compressible flow of air in stowing pipes, to be

$$f = 0.055V^{-0.384}$$

where V is the quantity of free air flowing per minute.

An air velocity of 24 m/s in a pipe 150 mm in diameter was found to be adequate to fluidize material of sp. gr. 2.6 and 75-mm size. This should provide an exit velocity of the ejected material around 30 m/s. To keep energy losses low, the radius of bends in the pipeline should not be less than 1 m and ideally should be about 2 m [20].

Hydraulic Stowing

Hydraulic stowing has been practised for many years, both in coal and in metalliferous mines. The stowing material commonly consists of sands from the mill-tailings plant or, if quarried and imported, sands and fine gravels up to 50-mm size. Coal-mine debris up to 75-mm size may be employed. The stowing material is fed from a bunker into a mixing cone, where a suspension of solids in water in the proportion of 1 volume of solids to 1 to 3 volumes of water is prepared and fed into the stowage-pipe. This may be done either at the surface of the mine or at some convenient location underground. After leaving the

During terminal storage the containment of radioactivity is further aided by interposing various radiation-resistant and insulating materials around the canisters and between the canisters and the rock walls of the repository. All these barriers are engineered, and they are further reinforced by the design, layout, and mode of construction of the underground chambers, the arrangements for handling and transportation of the waste containers, and the methods for controlling dust, ventilation, lighting, and air conditioning in the underground environment.

Protective Geological Barriers

The ultimate safety of a repository will depend not on the engineered barriers but on the natural barriers interposed by the geology, rock mechanics, and hydrology of the geological strata in which the repository is constructed. The time factors are so great that the ultimate safety of a high-level repository cannot possibly be demonstrated realistically today. We must therefore assume that ultimately the engineered barriers may fail and that radioactivity may gain access to the underground hydrology. We will then depend upon the natural geological barriers, which must be capable of retarding the passage of radioactivity over the very long time involved to a level which is acceptable in everyday life within the human environment. This means that the added risk that is presented by the storage of radioactive waste must be comparable with all the other risks to which humanity is inevitably subject in modern society.

Alternative Geological Formations

Various geological formations are at present being investigated, in many parts of the world, as possible sites for underground repositories. They include salt, anhydrite, crystalline igneous rocks (mainly granites but also some basalts), clay and shale sediments, and calcareous limestones and dolomites. Much attention has been given to the saline deposits, due to their widespread distribution, relative accessibility, and ease of excavation. They also have the advantage of low water content and favourable thermal insulation properties. On the other hand, under some conditions salt and clay shales can be regarded as being plastic, so that mined openings in them may progressively creep and close under normal overburden loads.

The main disadvantages of salt as a radioactive waste repository are its solubility in water, its low adsorption capacity, and its relative attractiveness to future generations as an economic deposit and as a possible storage medium for other materials. The clay sediments generally contain 15–20% water, which could create problems in relation to the effects of heat-generating sources on the surrounding walls of rock chambers. On the other hand, they have extremely low permeability and they are less corrosive than salt. Their main disadvantages lie in their low thermal conductivity, while mining in clay sediments can sometimes present problems.

The hard igneous rocks and massive limestones demonstrate negligible creep, but they may be fractured and hence possess high hydraulic permeability. However, in certain favourable localities masses of hard rock, free from circulating ground water, may exist. Fractures may be filled by secondary mineralization, and at great depth (7000 m or more) such fractures are likely to be kept closed by tectonic processes. In some arid desert localities, large masses of hard rock may exist above the water table.

Possible Thermal Effects

The possible effects of heat, generated by the radioactive waste, must be borne in mind. These effects include the weakening of rock pillars due to intensified creep, particularly in salt deposits. The rock fabric may also be weakened by intergranular crack promotion. The rock walls may be subject to thermal spalling, while the expansion of moisture and gases entrapped within the rock substance, possible chemical reactions and phase changes—all will affect the integrity of the rock mass. All these matters must be studied, coupling the thermal and chemical properties with rock mechanics in the formations concerned.

A suitable geological repository site for radioactive waste would be one free from any possible seismic shock or vulcanism. It would be free from fault dislocations, the stability of which might be affected adversely by the injection of fluids into the strata at high pressure. The rocks should be impermeable and free from open fractures. They should have the ability to absorb radioactivity, and any groundwater present should flow at a rate slow enough to ensure that any contained radioactivity decays to a harmless level before it enters the biosphere.

Access to a repository for the purposes of mining and transportation would be eased by limiting the depth to about 500 m, but it must be borne in mind that over the time scale involved there is a need to provide for possible denudation by geological erosion, or possible meteoric impacts, that might remove 300 m or more of surface cover. It would be an advantage if the repository were constructed in rocks that are likely to have little or no economic value, either now or in the future, while, to minimize the possibility of accidental human interference, the site should preferably be in a locality remote from centres of population and difficult of access [24–26].

Repository Design for Long-term Storage

Considerable attention has already been given to the design of suitable underground repositories for radioactive waste. In general these are basically mines that are excavated on a room and pillar layout in which the excavated rooms form the storage chambers into which the waste-containing canisters may either be stacked, or possibly inserted into holes drilled into the floor, the top of each hole then being sealed by a plug. Access to the repository would be by vertical shafts and down inclines, using shielded vehicles to carry the containers of high- and medium-level wastes, possibly handled by remote control. One alternative to the room and pillar layout plans to use spherical rock caverns surrounded by clay barriers, into which would be inserted the high-level waste containers mounted in large concrete balls [27, 28].

Safety Considerations

However difficult the technical, mining, engineering, and geological problems may be, it is probable that the most difficult aspects to resolve will be those that are concerned with obtaining public acceptance of the risks involved. For this to be possible the issues must be identified and described in terms that the representatives of the general public can understand, will accept, and can recommend. These issues are too important to be left workings. So far as coal-mining is concerned there is a wealth of richly documented experience upon which to draw, going back more than two centuries from the present day. While the results of this experience can undoubtedly help engineers to plan their current activities in carboniferous sediments, there must still be some doubt as to the extent to which the observed data can safely be extrapolated into the future, to excavations in non-carboniferous rocks, possibly at much greater depths than those so far encountered in coal-mines.

Evolution of Room and Pillar Practice

During the early stages of evolution of the room and pillar system in Britain it was not until the early nineteenth century that the method was developed to the stage when complete extraction of the pillars could be attempted. Table 13.1 gives details of typical pillar dimensions and extraction percentages during the period before the systematic extraction of pillars was usual in a second working. The percentage of the seam that was taken out in the whole working varied from 60 to 80 % at the shallowest depths to 30 or 40 % in the deeper workings. The practice of pillar robbing may have resulted in a slightly

TABLE 13.1. *Early room and pillar workings. Pillar dimensions and extraction percentages (whole workings only)*

Date	Depth of working (m)	Thickness (m)	Width of rooms (m)	Dimensions of pillars (m)	Percentage extracted	Location
—	14	—	2.7–3.6	2.7 × 2.7	81	Durham
—	18–36	1.5	1.8–3.6	22 × 2	61.5	Northumberland
17th century	—	—	—	1.3 × 1.8	—	Lancashire
1708	0–110	—	2.7	18 × 3.6	48	Northumberland
18th century	90	1.8	2.7–3.6	24 × 6.4	41	Northumberland
1750	150	—	3.6	12 × 12	41.5	Cumberland
1754	—	—	2.7	3.6 × 3.6	67	Berwick
1765	302	—	2.7–4.5	13.7 × 13.7	38	Cumberland
1778	180	—	1.8–3.6	18 × 7.3	39.5	Northumberland
1800	250	—	1.8–3.6	18 × 7.3	39.5	Northumberland
1800	303	—	3.6	24 × 7.3	30.5	Cumberland
1812	—	—	1.8–3.6	18 × 18	37	Northumberland

Reference sources for data in Table 13.1.

GALLOWAY, R. L., *Annals of Coal Mining*, Griffin, 1920.
ANON., *Historical Review of Coal Mining*, Institution of Mining Engineers, London, 1926.
ASHTON and SYKES, *Coal Industry of the 18th Century*, OUP, 1930.
DUNN, M., *Winning and Working of Collieries*, and *View of the Coal Trade*, both published at Newcastle upon Tyne, 1848.
BUDDLE, J., Coal mining in North East England, *Trans. Natural History Soc. of Northumberland* **2**, 333 (1850) and *Proceedings Archeological Institute Newcastle* **1**, 199 (1852).
BARKUS, W., Ventilation of coal mines and working of pillars, *Trans. NE Institute Mining Engrs* **1**, 239 (1852), and **8**, 31 (1859).
WALES, J., Working and ventilation of coal mines, *Trans. NE Institute of Mining Engrs* **6**, 204 (1857), and 7, 9 (1858).
CRONE, S. C., Pillar working in Northumberland and Durham, *Trans NE Institute of Mining Engrs* **9**, 17 (1866).

TABLE 13.2. *Percentage extraction in whole workings. Broken workings to follow or surface to be supported*

Date	Depth of working (m)	Thickness (m)	Width of rooms (m)	Dimensions of pillars (m)	Percentage extracted	Location
1860	106	1.2	4.5	36.5 × 14.5	34	Northumberland
—	110	1.2	4.5	27.5 × 14.5	34	Northumberland
—	136	1.4	4.5	27.5 × 27.5	26	Northumberland
1900	154	0.9	4.5	18 × 18	36	Durham
—	207	1.5	4.5	27.5 × 16	33	Northumberland
1920	215	0.9	4.5	22 × 18	33.3	Lancashire
1928	228	1.7	7	16.5 × 16.5	40	Lancashire
1944	252	1.7	4.5	40 × 40	19.5	Durham
1900	274	0.9	4.5	30 × 20	20.4	Durham
1848	305	—	1.75–3.6	45 × 9	23	Lancashire
1900	320	1.1	4.6	60 × 30	19.5	Durham
1860	329	2.1	4.6	46 × 30	21	Northumberland
1930	329	0.8	4.6	36.5 × 36.5	21	Northumberland
1944	366	1.2	4.6	40 × 40	19.5	Durham
—	427	1.8	4.6	50 × 50	16	Durham
1848	549	1.2	1.75–3.6	36.5 × 26.5	15	Durham
1930	549	2.1	3.6	46 × 46	14.2	Lancashire
1944	549	1.2	4.5	46 × 46	14.1	Durham
1933	610	1.7	3.6	55 × 55	13.8	Yorkshire

higher percentage extraction, but up to the turn of the eighteenth century the maximum extraction that was considered to be possible in the deepest workings of the period was around 45 %.

Table 13.2 gives similar details for the period from the early nineteenth century onwards. In this table the maximum extraction percentage in the shallowest workings is from 40 to 50 %, and in the deepest workings it is as low as 14 %. In these cases, however, it was probably the intention to leave pillars of an ample size to support the strata, so that a second working could be made to extract the pillars completely. A noted mining engineer of the time, Matthias Dunn, gave a rule for determining the size of pillars at various depths as "at 180 feet a width of 5 yards, to be increased by 1 yard for each additional 60 feet of depth". Dunn's data were subsequently published as listed in Table 13.3, being recommended pillar sizes for various depths, with rooms 4 m wide connected by cross-cuts 2 m wide.

The information in Tables 13.2 and 13.3 is plotted in graphical form in Fig. 13.5, from which it can be seen that for depths greater than 150 m Dunn's figures follow a linear relationship, but at shallower depths the line comes to a maximum of 60 % extraction at 36 m. It should be borne in mind that at the time when Dunn published his figures only one or two mines were working at depths beyond 300 m or so, and the bulk of his information must have been obtained from seams lying at much shallower depths. It is therefore probable that the curved portion of Dunn's graph refers to observed data while the straight line from 150 m onwards is, in the main, conjectural.

The extraction percentages from Table 13.2, when plotted, do not give a series of points on a straight line. If they follow any law at all it is more in the nature of the curve D to B, and it is of interest that the curved portion of the line plotted from Dunn's figures follows smoothly on DB from B to A. The observations from seams below 366 m lie much nearer

TABLE 13.3. *Scale of pillars for first working with the design of afterwards taking out the pillars (M. Dunn)*

Depth (m)	Dimensions of pillars (m)	Percentage extracted
36.5	18 × 4.5	59
73	18 × 5.5	50
110	20 × 6.5	48
146	20 × 7.25	43
183	20 × 8.25	41
219.5	20 × 9	39
256	22 × 10	37
292.5	22 × 11.75	34
329	24 × 17.75	31
366	24 × 14.5	30
402.5	25.5 × 16.5	27
439	25.5 × 18	24
476	27.5 × 19	23
512	27.5 × 20.5	20
549	27.5 × 22	18

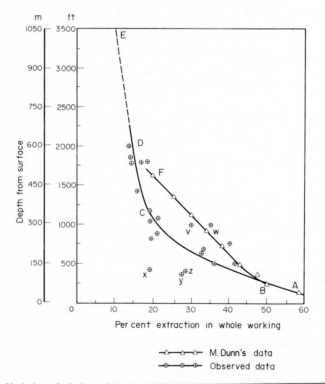

FIG. 13.5. Variation of whole working extraction percentage with depth in British coal-mines.

to the curve *DC* than do those of the shallower workings from *C* to *B*. This may be due to the effect of depth, the increase in depth from 300 m onwards rendering all the other variables relatively unimportant.

258

Effect of Depth on Pillar Dimensions

The advantage of making the pillars square and of maintaining them square when extracting the broken workings has long been recognized. In the late 1940s, when the system had probably reached its maximum technological development in Britain, at a depth of 300 m pillars were made from 30 to 40 m square, depending upon the thickness of seam and the character of the associated strata, while at depths greater than 300 m the size of the pillars was increased to about 60 m square under 600 m of cover.

At shallower depths, say less than 300 m, many factors other than depth may have an appreciable effect on the conditions underground. For example, pillars will need to be larger if the floor is soft than they would if the floor is hard, other factors being equal, while the thickness of the seam being extracted will have an effect on the strength of long, narrow pillars, or pillars of small dimensions, such as are often left in seams at shallow depth. But at greater depth, where pillars must necessarily be large, the thickness of the seam extracted can have little significance. For example, the strength of a pillar 30–60 m square will be, for all practical purposes, about the same no matter whether the seam is, say, 1 m or 2 m thick.

The strength of the associated strata is also more apparent in shallower seams than it is at depth, because the pillars are smaller. Any local failure which results in small pillars being overloaded will cause a high intensity of loading on the floor, which may then creep and flow, or on the roof, which may then crush. In deeper workings the greater pillar dimensions give a measure of protection against this, because serious overloading of a large pillar would be required to produce an intensity of loading equivalent to that which could result from a slight overload on a small pillar.

We can therefore expect observed data relating to pillar sizes and extraction percentages to show considerable variation in shallow seams, depending on the relative importance of the many factors involved. In Fig. 13.5, for example, the points *V* and *W* refer to seams in Cumberland and Lancashire at depths of 300 m and 366 m respectively, with strong roof and floor. Extractions of 30 and 35 % were made in the whole working. On the other hand, points *X, Y,* and *Z* refer to seams in Northumberland less than 150 m deep, which could only be worked satisfactorily with a whole working extraction of 20–30 %, due to a friable roof and a weak floor. Between these extremes, observations from seams of more general character approach the curve *ABC* and at depths beyond 366 m; although observations are included from seams of varied character, all the points

Fig. 13.6. Retreat room and pillar with mobile loader served by conveyors.

approach the curve from C to D. If the curve is extrapolated smoothly to E it becomes apparent that at depths of 1000 m and beyond, the probable maximum extraction in room and pillar whole workings in carboniferous sediments is no more than 10 or 12%.

These figures assume that it will be possible to drive and to maintain whole workings at the depths stated. As we shall see later, this is by no means certain, but, assuming that the whole workings could be supported and worked at depths beyond 760 m, the subsequent extraction of pillars by a systematic broken working, using any of the layouts so far described, would be likely to prove very difficult. At the present time there are very few examples of room and pillar workings in carboniferous sediments below more than about 600 m of covering rock, and it became widely accepted that 600 m to 760 m or so represented the probable limit of depth at which it was practicable to apply the conventional room and pillar system as had evolved in British coal-mines up to the year 1945.

During the past 30 years, however, there have been some important developments that have changed the picture considerably. Probably the most important of these has been the contribution of extensive rock mechanics research on the fundamentals of pillar support systems. The incentive for this has come, not only from the need to mine coal at greater depths, but from the increasing need for excavations in saline evaporites, mainly rock-salt and potash, in the case of the latter mineral at depths of 1220 m or more.

Room and Pillar Coal-workings Outside Britain

If corroboration of British experience in the application of room and pillar workings at depth were needed we could find it in some overseas areas, particularly in Australia and North America. The room and pillar system was adopted in those areas, the method being carried by emigrants from Britain. In these new coalfields, almost without exception, methods that had been successful in shallow mines were applied to deeper seams without making adequate provision for the greater strata pressures which were inevitably encountered at depth. As a result, many millions of tons of the world's mineral resources have been irretrievably lost by crush and creep on small pillars. In some cases, statutory action has been necessary to restrict wasteful methods of working, the percentage extraction in the whole workings being limited by law.

In the coalfields of New South Wales restricting clauses in the Crown leases laid down minimum pillar and maximum room dimensions for specified depths of working. On the south coast of Australia, in the Illawarra Coalfield, the percentage extraction in the whole working ranged from 50% in seams 60 m or less in depth to 8% in the deepest seams, under 450 m or more of cover. In the latter case the pillars were made about 100 m square.

Extraction of the submarine coal of Sydney, Nova Scotia, began in 1867 by the room and pillar method, the rooms being 5 m wide, with pillars 12 m wide by 37 m long giving an extraction of 38%. In later developments the width of the rooms was increased to 5.5 m, with 4 m cross-cuts. Up to the year 1900 it was not contemplated to attempt complete extraction, but as the seaward limit of workable coal increased pillar extraction began. Complete extraction was not possible under less than about 213 m of cover, but was attempted at greater depths. In submarine coal under solid cover of between 122 m–213 m rooms and cross-cuts 5 m wide were adopted, the size of pillar varying with thickness of cover and the height of the seam.

Many features of the room and pillar layout generally adopted in the United States bear a strong resemblance to the mining system that evolved in the north-east of England during the early nineteenth century. In particular there was the adherence to the panel system, with large barrier pillars to take crush, and an extraction of 40–50 % within these barriers. Concern as to the consequent waste of natural energy resources in recent years has occasioned a trend towards change in the established character of underground coal-mining in the United States. The introduction of new continuous mining techniques in Europe, allied to methods of strata control by caving behind strong hydraulic supports, has focused attention of American mining engineers on the possibility of adopting longwall methods in preference to room and pillar methods in the United States. Complete extraction of the mineral is more closely attainable, and the associated problems of ventilation and transport are easier in longwall working.

As yet, however, it is probably still true to say that the room and pillar system is the predominant method of underground coal-mining in North America, often with extraction percentages little, if any, better than those described by H. H. Otto in 1925 [3]. An examination of existing workings then enabled Otto to produce curves showing the effects of depth and thickness of bed on the percentage extraction (Fig. 13.7). He also gave the figures listed in Table 13.4 relating to extraction percentages attained in some anthracite coal-mines.

Strata Control in Narrow Excavations

It is frequently observed that some rocks are far from isotropic in their geological and physical characteristics. They may show pronounced differences in certain preferred directions and contain planes of weakness related to stratification, cleavage, jointing, and cleat. Sometimes these differences are so pronounced as to have a major effect upon the techniques adopted for the excavation and support of tunnelling operations within the rocks. The orientation of an excavation may be arranged in some cases to facilitate driving the tunnel by making use of the weakening effect of major discontinuities at the

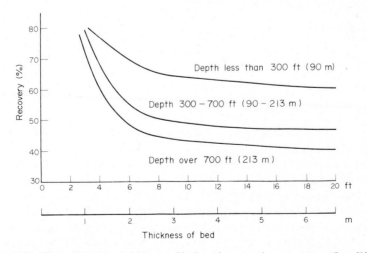

FIG. 13.7. Effect of depth and thickness of bed on the extraction percentage (Otto [3]).

TABLE 13.4. *Average extraction in early room and pillar workings (Otto [3])*

Bed thickness (m)	Percentage removal	Bed thickness (m)	Percentage removal
0.9	72.3	3.4	60.8
1.2	75.1	4.0	68.9
1.5	71.0	4.6	73.4
1.8	71.5	5.2	65.6
2.1	70.6	5.8	72.0
2.4	68.6	6.7	67.7
2.7	72.8	8.2	62.5
3.0	72.1	9.1	57.5

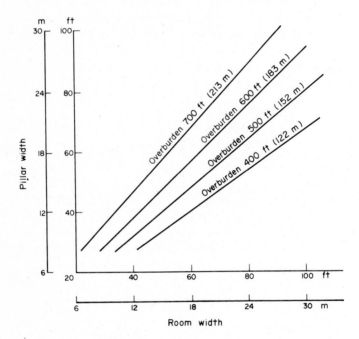

FIG. 13.8. Pillar and room widths for square pillars in Colorado oil shale (Wright and Bucky [24]).

working face or, in other cases, taking into account the existence of such discontinuities when designing the support system.

For example, in a room and pillar layout the direction of advance of the rooms may be aligned across the major planes of weakness or cleat, in which case the cross-cuts or "walls" will be aligned at right angles to that direction (Fig. 13.9). In such a layout at depth it is very likely that the support of the rooms may be much more difficult than that of the walls, due to the greater rigidity of the pillar edges along the sides of rooms, being across the cleat, as compared with the weaker sides of a wall, running parallel to cleat. Observation shows that the opening of an excavation in rock results in a reduction in the

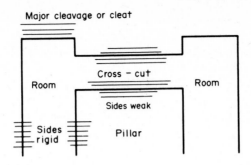

Fig. 13.9. Effect of cleat direction on the relative strength of pillar sides in rooms and cross-cuts.

strength of the strata and in an intensification of the forces around the excavation. The effects of greater rigidity in the immediate surroundings is to produce concentrations of stress around the excavation perimeter, while a less-rigid wall is more likely to yield and so deflect stress concentrations deeper into the pillars.

If the strata are relatively weak, the stress concentrations may be so pronounced as to cause local failure of the periphery rocks. This is particularly likely to occur in shales, where the roof in cross-cuts or walls is likely to remain in better condition than that in rooms of equal width, the difference being most pronounced at depths beyond 300 m of cover. Under these circumstances it is quite common for the roof of walls to remain unbroken, while the roof of rooms may be extensively fractured. This difference is only slight at shallow depths, and it is not likely to be very noticeable even at greater depth if the roof beds consist of strong rock. Walker *et al.* [2] report the occurrence of three distinct types of fracture in carboniferous shales, to some extent related to depth and attributed to bending stresses, shearing forces, and lateral pressure, respectively.

Fracture Control

Where fracturing attributable to lateral pressure occurs, improved roof conditions in the rooms is sometimes attained by a process of fracture control in which the planes of weakness are systematically broken into by an advance cut into the rock face, always at the same side of the room. This induces the fracture always at that side, and the support system can then be designed accordingly. The depth to which the incision need be made, ahead of the general line of advance, depends upon the rigidity of the rock, some few centimetres being sufficient in strong rock, but up to 1 m or so being required if the rock face is weak and yielding.

Pressure Relief

The effects of strata pressure due to depth, intensified by the stress concentrations around zones of high resistance, are evidenced by the occurrence of strata failure and may be countered to some extent by the techniques of pressure relief, some aspects of which have been described. The two possible modes of failure, colloquially termed thrust and creep in coal-mines, may also be seen in other strata.

Sudden failure, in the nature of brittle fracture, is sometimes experienced around narrow excavations, being evidenced as rock outbursts from the high-stressed zone. Protection against such outbursts may be obtained by the technique of de-stressing, or stress relief, by means of which the extent of the relaxed zone of deformed or fractured rock around the excavation is deflected deeper into the solid. This is done by boring or by blasting into the high-stress zones in advance of the main working or into the pillars on either side.

Strata Control in the Broken Workings

The layout of the broken workings, to facilitate extraction of the pillars, calls for care, and its safe execution requires considerable expertise. Some basic principles are:

(1) The pillars cut out by the whole working should be of ample size to take the abutment loads that will subsequently be generated by the broken working.
(2) Large pillars will be much stronger and better able to accommodate the abutment loads than would a greater number of smaller pillars of the same total area. Hence, as a general rule, pillars should not be split before being extracted. The extraction of pillars by means of narrow roadways driven through them is good practice only if the pillars are originally much larger than needs be for the loads they have to carry.
(3) Following from (2), pillars should be extracted by taking slices or "lifts" driven along the pillar edges.
(4) An irregular line of extraction produces local zones of stress concentration near the edge of the caved area. Pillars should therefore be extracted on a straight line of retreat so as to maintain an even stress distribution over the working area.
(5) The system should be so organized that all operations are completed without delay to minimize deterioration of the strata due to creep. High rates of advance and retreat should be maintained so that a room can be opened up and taken rapidly to its full length and then the neighbouring pillars extracted swiftly in a series of short lifts. In this way the working place is finished before it becomes untenable.

The Yield-pillar Technique

With increasing depth of working the conventional practice of increasing the size of the pillars and reducing the extraction percentage in the whole workings becomes progressively more difficult to apply with success. Not only is this due to the increased strata pressure, which affects the support of narrow excavations, but it is also due to greater difficulties in the associated problems of ventilation, materials handling, and equipment disposition. These are all factors that together limit the depth of cover under which it is possible to apply pillar support systems relying on pillars of large dimensions in a conventional room and pillar layout.

However, if the main strata pressures can be transferred on to abutment zones, to bridge the working area over a span wide enough to include both the advancing rooms and the retreating pillar extraction lifts, over a limited area, then the depth of cover ceases to have its overriding significance. The yield-pillar technique is intended to achieve such a result (Fig. 13.10).

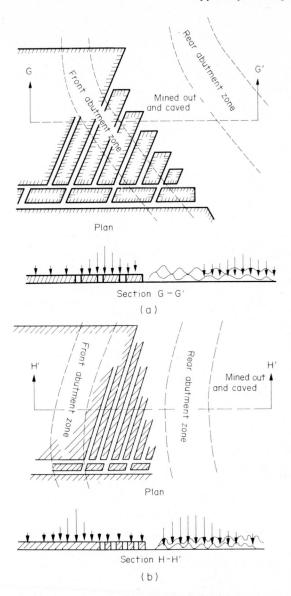

FIG. 13.10. The yield-pillar technique: (a) Pillars too wide to yield. The front abutment will damage the development openings. (b) Pillars narrow enough to yield. Main roof load is transferred to the abutments. The development area is narrow, so that it lies within the abutments (Woodruff [31]).

The idea is to have, interposed between the major abutment zones, a region in which the roof loads are carried by pillars too large to crush but weak enough to yield slightly, and so cause a transfer of superincumbent load on to the abutments. Whether such an arrangement can be designed will depend mainly upon the distance to which the abutment pressure zone exists over the solid and in advance of the workings. Walker *et al.* suggest that there is a general relationship between the depth of cover and the distance between

TABLE 13.5. *Width of the maximum pressure arch for various depths* (*Walker et al.*)

Depth (m)	Width of maximum pressure arch (m)
122	40
183	50
245	60
305	70
306	80
427	90
488	100
549	110

the abutments of the maximum pressure arch that will develop above the broken workings. On the assumption that the front and rear abutments are symmetrically disposed about a mean at the working face, they quote the figures listed in Table 13.5. In practice, however, there is considerable doubt as to where the abutments will fall. The position of the front abutment is of major importance, and the only wise course is to locate it by methods such as those described in *Geotechnology*, pp. 227–9.

The Design of Pillar Support Systems

Determination of Pillar Loads

In the absence of more precise information it is common to assume that the pillars will be required to carry all the weight of the overlying strata, evenly distributed upon them. The maximum pillar load in a large horizontal excavation may then be calculated from the vertical succession of strata, as proved in shafts and boreholes. These will supply data relating to thickness, density, and weight of each bed, and the load equivalent of the total weight. Pillar stresses at that cover load may then be calculated for various extraction percentages.

Assuming that the strata behave as an isotropic elastic material, the distribution of pillar stresses may be calculated both as to the vertical weight stress and the corresponding lateral stress that will be generated by Poisson's ratio for the material and the degree of constraint that is imposed upon the pillar.

At the edge of a pillar, in a direction normal to that edge, the horizontal stress will be zero. The imposed load will cause the pillar sides to yield, so that they tend to fracture and creep towards the excavation. Friction is generated between the yielding pillar material and the strata in contact with it, above and below. The restraint that is imposed upon the pillar material increases in intensity, from zero at the pillar edge to a maximum somewhere within the pillar. If the pillar is a large one its centre will be completely restrained, so that it will be able to support any load that it may be called upon to bear. Hence the reason for preferring a smaller number of large pillars rather than a greater number of small pillars of equivalent total area. Similarly, a square pillar is stronger than

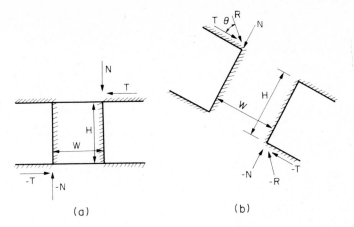

FIG. 13.11. Forces acting upon a pillar in inclined strata: (a) horizontal deposits; (b) inclined seams.

a rectangular pillar of the same area. This assumes that the support of the rooms can be safely achieved over the span concerned.

In inclined strata the effect of gravitational forces acting down the dip must be considered. Figure 13.11 shows that an inclined pillar suffers a net thrust down-dip, in the roof, and a net thrust up-dip, in the floor. Hence an overturning couple will be created at the top of the rise side of the pillar, and this is opposed by a couple acting at the base of the pillar, on the dip side. In order to counter this effect of the down-dip gravitational forces the pillars in inclined strata should be made rectangular, with the short axis horizontal.

Stress Concentrations

The act of driving a tunnel through an elastic medium disturbs the natural stress field and produces stress concentrations, higher than the original stress, in the strata around the opening. However, at a distance of about two diameters from the surface of the opening the intensity of the stress field is virtually the same as the original value. Hence the stress concentrations around multiple openings driven more than two diameters apart may be considered as being identical with those around separate single openings, whereas openings that are driven at closer intervals to one another will produce interacting stress concentrations in the surrounding rock.

The stress concentrations around a single opening will be generated almost as quickly as the working face advances, but those produced by multiple openings will take time to interact. They will progressively increase in intensity as the surrounding excavations are extended.

The complexities around multiple openings make theoretical treatments difficult and questionable, and this has caused many investigators to seek the aid of models and to attempt to make direct measurement of *in situ* stress. Photoelastic models have been widely applied. Bodonyi [7] used gelatine models in order to observe the stress distribution due to gravitational loading. Dixon built models of pillars with photoelastic coatings on

TABLE 13.6.   *Stress concentration factors round mine openings*

Openings			Opening width / Pillar width (m)	Ratio of stresses	Work by	Technique	Maximum stress con- centration
Shape	H/W ratio	No.					
Ovaloid	0.5	5	1.03	1:0	Duvall	Photoelastic	4.05
Ovaloid	0.5	5	4.28	1:0	Duvall	Photoelastic	6.10
Ovaloid	0.5	2	1.01	1:0	Duvall	Photoelastic	4.00
Ovaloid	2.0	5	1.04	1:0	Duvall	Photoelastic	2.81
Ovaloid	1.0	1	—	1:1	Duvall	Theoretical	2.10
Circle	—	2	2.00	1:0	Panek	Theoretical	3.26
Circle	—	2	0.50	1:0	Panek	Theoretical	2.99
Circle	—	2	0.53	1:0	Duvall	Photoelastic	2.97
Circle	—	5	1.07	1:0	Duvall	Photoelastic	3.29
Ellipse	0.5	1	—	1:0	Inglis	Theoretical	5.00
Ellipse	0.5	1	—	1:1	Eynon	Theoretical	4.10
Ellipse	1.0	1	—	1:0	Inglis	Theoretical	2.90
Ellipse	1.0	1	—	1:1	Eynon	Theoretical	2.00
Rectangular	0.5	1	—	1:0	Holland	Theoretical	2.50
Rectangular	0.5	1	—	1:1	Holland	Theoretical	1.40
Rectangular	1.0	1	—	1:0	Holland	Theoretical	2.00
Rectangular	2.0	1	—	1:0	Holland	Theoretical	1.75

simulated equivalent materials. Biron made a photoelastic model study which showed stress concentrations of 3.2 times the overburden stress at the corners of pillars, and 2.3 times the overburden stress along the pillar edges. Caudle and Clark [10] reviewed much of the work that has been done, both theoretically and experimentally, on models, with openings of various size and configuration, while Nair and Udd [11] showed how alternative solutions for different force vectors can be obtained by the method of superposition. Berry [13] suggested that stratified rocks may be represented by a transversely isotropic elastic model, and Salamon [14] showed how this can be applied, with the aid of an analogue–numerical technique, to solve two-dimensional problems. Crouch [15] described a digital computer method for calculating the stresses and displacements in a single flat-lying seam, while Dhar and Coates [16] developed a three-dimensional approach to predict the pillar stresses, with irregular room and pillar geometry, in elastic ground. Tables 13.6 and 13.7 were compiled by Moore in a review of work conducted by means of models, and theoretical treatment, in addition to *in situ* measurements. Figure 13.12 shows the results obtained on a photoelastic model compared with measurements made on an underground pillar using a strain-gauge stress-relief technique.

Tributary-area Theory

Duvall [18] has suggested a method of predicting pillar loads by a tributary-area theory. If it is assumed that the pillars take the entire weight of the superincumbent rock, uniformly distributed, then the average pillar stress is related to the applied stress by

$$S_p A_p = S_v(A_e + A_p),$$

where S_p is the average pillar stress, S_v is the vertical stress, A_e is the excavated area, A_p is the pillar area, and $A_t = A_e + A_p$ = total area.

TABLE 13.7. *Maximum stresses measured in mines*

| Openings | | Material | Work by | Technique | Normal stress | | Maximum stress | | Concen-tration factor |
Shape	H/W ratio No.				(lb/in^2)	(MPa)	(lb/in^2)	(MPa)	
Rectangle	0.5 Many	Coal	Van Heerden	CSIR cell	580	4.	2 200	15	3.8
Rectangle	0.75 1	Hard rock	Leeman	CSIR cell	6 500	45	12 000	83	1.85
Rectangle	– 1	Lava	Leeman	CSIR cell	7 000	48	15 000	103	2.14
Stopes	– –	Hard rock	Leeman	Maihak	10 384	72	20 000	138	1.92
Rectangle	1.0 Many	Trona	Panek	Deformation gauge	1 250	8.6	3 500	24	2.80
Rectangle	0.47 7	Coal	Moore	Photoelastic stressmeter	1 280	8.8	2 880	20	2.25
–	– 1	Quartzite	Moxon	Photoelastic overcoring	600	4.1	3 500	24	5.80

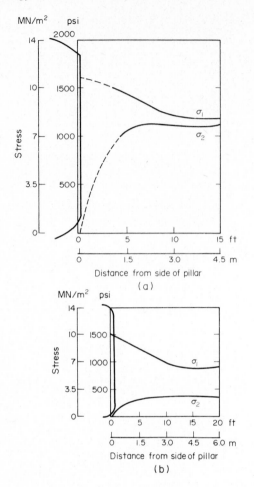

FIG. 13.12. Comparison of stresses observed on models and *in situ* underground: (a) stresses in mine pillar *in situ*; (b) stresses determined from model (Leeman [12]).

For a series of parallel openings separated by rib pillars, the average pillar stress

$$S_p = \frac{S_v(W_p + W_o)}{W_p},$$

where W_o is the width of the openings and W_p is the width of the pillars.

Extraction ratio R_a = excavated area/total area

and $\qquad R_a = A_e/A_t = I - A_p/A_t \quad \text{or} \quad I - S_v/S_p.$

By replacing S_p/S_v by $C_p/F_s S_v$, where C_p/F_s is the safe pillar load,

$$R_a = 1 - \frac{F_s S_v}{C_p}.$$

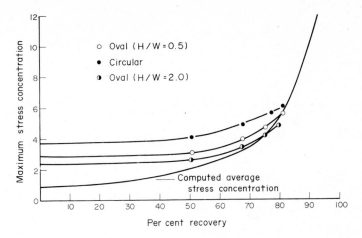

Fig. 13.13. Stress concentrations for oval- and circular-shaped openings (Duvall [18]).

In Fig. 13.13 Duvall plots the computed average stress concentration together with the maximum pillar stress concentrations for circular-shaped openings and two oval-shaped openings of height/width ratio 0.5 and 2.0, respectively, for various extraction percentages.

The Strength of Rock Pillars

Some aspects of the effects of scale on the strength of a rock sample are discussed in *Geotechnology*, pp. 147–9. The strength of a rock specimen increases as the ratio height/width decreases, chiefly because the restraint that is imposed by the platens of the testing machine also increase at the same time, and this produces an effect on an incipient fracture plane similar to that which results from increased lateral constraint. However, if the proportions of the specimen are always such that the ratio height/width is always greater than 2.0, then the influence of end-restraint will not be likely to extend over a fracture plane near the middle of the specimen, and changes in the proportions height/width will have little significance.

Effect of the Width/Height Ratio

Eynon [19] reports on the relationship between the ratio H/W and the rock strength (Fig. 13.14), while Holland [20] discussed the strength of coal pillars. A realistic value for the compressive strength of a rock requires that the specimens tested have a ratio H/W greater than 2.0, and as the pillars in an underground support system usually have a ratio much less than this the compressive strength, as determined by a conventional laboratory test, is of little value in pillar design except for use on a comparative basis.

Holland gave an empirical rule for determining the strength of a coal pillar as:

$$S \propto \frac{\sqrt{W}}{H},$$

where S is the strength of the pillar, H is the height, and W is the width.

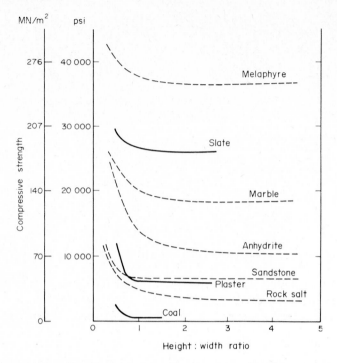

FIG. 13.14. Variation of compressive strength with H/W ratio in different rock materials (Eynon [19]).

K is a material characteristic, obtained from uniaxial compression tests on rock cubes, and

$$K = S \sqrt{D},$$

where D is the edge dimension of the cube.

For rib pillars in massive elastic rock Obert and Duvall [6] quote

$$S = C(0.778 + 0.222 \, (W/H)),$$

where C is the compressive strength of specimens having $W/H = 1.0$.

Effect of Roof and Floor Strata

It is well known that the conditions of end restraint on specimens under load on a compression test have a pronounced effect on the results obtained. This comes about as a result of the loading platens and the specimen having different elastic properties. Vertical deformation of the specimen under load then generates different lateral expansions between the specimen and the material above and below it. As a result, we may expect, for example, that a rock pillar situated between a hard-rock roof and floor will be stronger than a similar pillar interposed between weak shales and clay sediments.

Serata [21] records observations of the compressive strength of specimens representing model pillars, for different end restraint conditions, as shown in Fig. 13.15. Type A restraint simulated low-friction end conditions, while types B and C had the larger lateral

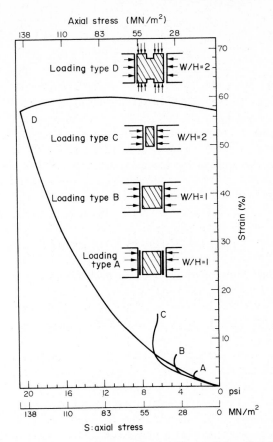

Fig. 13.15. Behaviour of pillars in various loading conditions (Serata [21]).

friction developed between steel platens. This doubled the fracture strength, as compared with the uniaxial condition, type A. In the type-C specimen the W/H ratio was doubled, and this trebled the fracture strength. Finally, when the ends of the specimen were confined by rings, to simulate the confined constraining medium of the roof and floor on a model pillar the fracture strength was increased to eight times the uniaxial value.

Modes of Pillar Failure

Creep

We may expect creep to be an important factor in the support of excavations made in evaporite rocks, and, to a lesser extent, also in clay sediments. The rational design of pillar support systems in rocks that exhibit creep must be based upon keeping the stress distribution sufficiently low that the onset of tertiary creep is delayed until such time as the support given by the pillars is no longer required.

While the onset of tertiary creep can be delayed indefinitely by keeping the stresses low, in practical terms it is more realistic to think in terms of a limited span of time, which, indeed, may be lengthy. For example, Wynn reports that the creep characteristics of the Jefferson Island salt dome are such that present openings in the salt will probably remain open for approximately 900 years, presumably before the onset of tertiary creep.

When reporting on an investigation into the design of room and pillar workings in rock salt, Potts [23] describes the concept of "time–safe–strength", which is based upon observations of strain, rate of strain, and rate of loading, in creep tests on pillars of various width/height ratio. The deformation is measured with time until the change in deformation approaches zero, and the test is then repeated at higher levels of stress. Stress is then plotted against the estimated ultimate strain, so that an estimate may be made of the stress at which the curve is asymptotic. This value is defined as the time–safe–strength.

A logical corollary is the concept of "time–safe–strain", which suggests that a pillar will support a given load for a defined period of time, instead of attempting to define that the pillar will support a given load indefinitely. In order to apply this concept, observations must be made of the constant vertical strain rate during the secondary stage of creep and the critical vertical strain rate which a pillar can withstand before it enters the accelerating stage of tertiary creep which precedes failure.

Wright and Bucky [24] investigated pillar loads and time effects on the Colorado oil shales, and estimated that the pillars would stand for many years if they were not stressed beyond 75 % of their maximum observed compressive strength. Based on a safety factor of 3, applied to a minimum observed compressive strength, reduced by a factor of 0.75 to include the time effect, they produced relationships for pillar and room widths at various depths of cover, as shown in Fig. 13.8.

The creep behaviour of pillars is a critical factor when mining in the potash deposits of Saskatchewan, Canada. In the absence of full understanding of the creep properties of evaporites, the producers have been reluctant to increase the extraction ratio above 30 % when mining by the room and pillar method. For the same reason they have been deterred from conventional mining at depths beyond 1040 m. The problem is further complicated by the presence of water-bearing strata above the evaporites and the occurrence of clay sediments adjacent to the ore.

King [25] reports on the results of a study of the deformation of model pillars related to changes in stress and temperature. From this study some empirical laws governing creep in potash have been obtained. In Fig. 13.16 the extraction ratios are plotted for temperatures and stresses chosen to be representative of those that exist in Saskatchewan. The temperatures of 80° F and 110° F correspond to depths of working of 1040 m and 1400 m of cover, respectively. For pillars designed to take the full weight of the overburden, King concluded that a width/height ratio not less than 4 is sufficient to preclude tertiary creep failure in Saskatchewan potash, but the required ratio is doubled if clay seams are present in the roof and floor. However, Baar [27] counsels extreme caution when comparing the results of laboratory model studies with field conditions in potash. Salt rocks possess mechanical properties which make it dangerous to base pillar design on laboratory-observed parameters for excavations in deep salt and potash deposits. In particular, many laboratory tests show an apparent strain-hardening process on salt specimens subjected to load cycling, whereas, in reality, underground the same rocks may not always exhibit strain hardening. The laboratory tests may therefore show salt rocks to be apparently stronger than they really are.

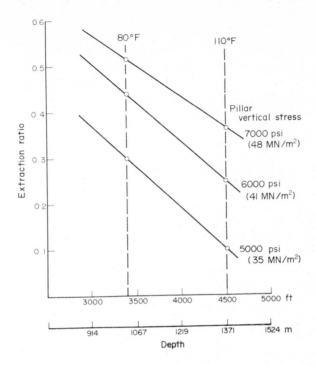

FIG. 13.16. Extraction ratio vs. depth for different vertical stresses in Saskatchewan potash (King [25]).

Pillar Collapse

Failure of the rock pillars in an underground support system may take place in one or other of two forms of collapse, other than creep. As a general rule it may occur more or less gradually, in what may be termed a stable and controlled manner. There are occasions, however, when failure takes place suddenly, taking the form of a rock-burst or a sudden collapse. Failure in such a case is violent and unstable.

Tincelin and Sinou [28] suggest a hypothesis for the sudden failure of pillars in sedimentary iron ores, based on the observation that the pillars reach their elastic limit before the stratified roof beds do so. At this point the roof beds are not yet very strongly compressed. There must therefore be a period during which the tensile and shear stresses in the roof strata balance some of the weight of the overburden. When the most resistant bed breaks, then the others must follow quickly because they are much weaker. Being thus suddenly subject to excess load the pillars collapse rapidly and violently in a kind of chain reaction which may extend over a large area.

Salamon compares the stability and instability of pillars with the stability of a laboratory compression test on a brittle rock specimen. It is only on a "stiff" testing machine that the complete stress–strain relationship can be seen, as shown in Fig. 13.17a on which OA represents the unfailed state, where an increase in strain is associated with an increase in rock resistance. The part AB represents the failing regime, during which any increase in strain is accompanied by a decrease in rock resistance.

275

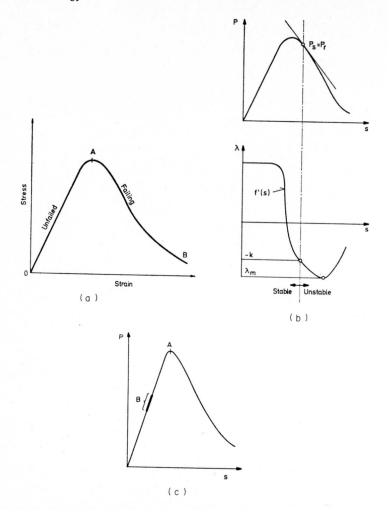

FIG. 13.17. (a) Complete stress–strain curve of brittle rock. (b) Criterion for instability (Salamon [14]). (c) Contrast between pillar strength, point *A*, and the estimated load acting upon the pillar zone *B*, if the safety factor is about 1.6 (Salamon [14]).

The testing of brittle rock on an ordinary compression machine is usually interrupted by violent failure of the specimen soon after point *A* on the stress–strain curve has been reached. But if the complete force–displacement curve is obtained, then the strength of the pillar can be observed during the failing regime, in which point *P* can represent a particular loading situation on the failing pillar. The tangent to the force–displacement characteristic at this point identifies the limiting condition for stability and instability (Fig. 13.17b). From such a point violent failure of the pillar will follow if the slope of the tangent is less than a critical value, for which Salamon [30] describes a yield criterion.

Pillar support layouts should be so designed that instability does not occur. Traditionally this aim is achieved by ensuring that the load on individual pillars is always smaller than the corresponding pillar strength, e.g., point *B* on Fig. 13.17c. The application of Salamon's criterion, however, makes it possible to design, making the utmost use of the strength of barrier pillars to take the abutment loads, such as point *A* on Fig. 13.17c. The smaller pillars inside the panels are then required merely to maintain the integrity of the roof inside the barriers.

If the layout is such that the pillars are perfectly stable, as indicated by the yield criterion, then uncontrolled pillar collapse cannot occur without prior warning. Any error in design is then likely to produce, at worst, a gradual failure, which is preceded by ample warning in the form of pillar spalling and slabbing.

The method has made it possible to estimate the probable influence of depth on the attainable percentage extraction ratio in room and pillar workings in South African coalmines. While the adoption of large barrier pillars would result in a reduction of the extraction percentage at shallow depths, considerable improvement in the extraction ratio would be achieved at greater depths. In that case it might be possible to increase the depth at which pillar support systems could be applied with an acceptably high enough percentage extraction.

References and Bibliography: Chapter 13

1. ROBERTS, A., The bord and pillar system of coal mining, MSc thesis, Univ. of Durham at Newcastle upon Tyne, Jan. 1945.
2. WALKER, A., *et al.*, Control of the strata in mining—investigations in the Durham and Northumberland coalfields, *Trans. Inst. Min. Engrs London* **113**, 83–95 (1953).
3. OTTO, H. H., Ultimate recovery from anthracite coal beds, *Trans. Am. Inst. Min. Metall. Engrs New York* **72**, 727 (1925).
4. ROBERTS, A., *Geotechnology*, pp. 35–38, Pergamon Press, Oxford, 1977.
5. EMERY, C. L., Some aspects of the use of pillars for mine support, *Proc. Symp. Rock Mech., McGill University, 1962*, Dep. Mines Tech. Surveys, Ottawa, 1963.
6. OBERT, L., and DUVALL, W. I., *Rock Mechanics and the Design of Structures in Rock*, pp. 531–3, Wiley, New York, 1967.
7. BODONYI, J. B., Some aspects of model studies for the plastic range, *Pub. Hungarian Res. Inst. Min.*, No. 8/9 (1965–6).
8. DIXON, J., Photoelastic analysis of stresses around mine openings, *King's Coll. Sch. Min. Res. Rep.* No. 2, 1954.
9. BIRON, C., Photoelastic analysis of stress distribution in pillars, *King's Coll. Sch. Min. Res. Rep.* No. 2, 1954.
10. CAUDLE, R. D., and CLARK, G. B., Stresses around mine openings, *Colliery Engng London* **33** (Jan.–Apr. 1956).
11. NAIR, O. B., and UDD, J. E., Stresses around openings due to biaxial loads—a superposition technique, *Third Canadian Symp. Rock Mech.*, Univ. of Toronto, 1965.
12. LEEMAN, E. R., and VAN HEERDEN, W. L., Stress measurements in coal pillars, *Colliery Engng London* **41** (Jan. 1964).
13. BERRY, D. S., The ground considered as a transversely isotropic material, *Int. J. Rock Mech. Min. Sci.* **1**, 159–67 (1964).
14. SALAMON, M. D. G., Two-dimensional treatment of problems arising from mining tabular deposits in isotropic or transversely isotropic ground, *Int. J. Rock Mech. Min. Sci.* **5**, 159–85 (1968).
15. CROUCH, S. L., Two-dimensional analysis of near surface single seam extraction, *Int. J. Rock Mech. Min. Sci.* **10**, 85–96 (1973).
16. DHAR, B. B., and COATES, D. F., A three dimensional method of predicting pillar stresses, *Int. J. Rock Mech. Min. Sci.* **9**, 789–802 (1972).
17. MOORE, D. R., The measurement of ground movement in bord and pillar workings, MEng thesis, Univ. of Sheffield, 1967.
18. DUVALL, W. I., *Stress Analysis Applied to Underground Mining Problems*, US Bur. Min. Rep. Invest. No. 4387, 1948.
19. EYNON, P., Rock mechanics and mine design, MSc thesis, University of Nottingham, 1966.
20. HOLLAND, C. T., The strength of coal in mine pillars, *6th Symp. Rock Mech., Univ. of Missouri, Rolla, 1964*.
21. SERATA, S., Theory and model of underground opening and support systems, *6th Symp. Rock Mech., Univ. of Missouri, Rolla, 1964*.

22. WYNN, L. E., Movement within the Jefferson Island Louisiana salt dome. *Third Canadian Symp. Rock Mech., Univ. of Toronto, 1965.*

23. POTTS, E. L. J., An investigation into the design of room and pillar workings in rock salt, *Trans. Inst. Min. Engrs London,* **124**, 27–47 (1964).

24. WRIGHT, F. D., and BUCKY, P. B., Determination of room and pillar dimensions for the oil shale mine at Rifle, Colo., *Trans. Am. Inst. Min. Metall. Engrs New York* **181**, 352–9 (1949).

25. KING, M. S., Creep in model pillars of Saskatchewan potash, *Int. J. Rock Mech. Min. Sci.* **10**, 363–71 (1973).

26. BAAR, M. S., *The Long-term Behaviour of Salt Rocks in Deep Salt and Potash Mines,* Saskatchewan Res. Coun. Rep. E72–7, May 1972.

27. BAAR, M. S., *Dangerous Application of Laboratory Parameters to the Design of Deep Potash Mines,* Saskatchewan Res. Coun. Rep. E72–6, June 1972.

28. TINCELIN, E., and SINOU, P., Collapse of areas worked by the small pillar method, *Int. Conf. on Strata Control, Paris, May 1960,* paper G6.

29. SALAMON, M. D. G., Stability, instability, and design of pillar workings, *Int. J. Rock Mech. Min. Sci.* **7**, 613–31 (1970).

30. SALAMON, M. D. G., A method of designing bord and pillar workings, *Jl. S. African Inst. Min. Metall.* **63**, 68–78 (1967).

31. WOODRUFF, S. D., *Methods of Working Coal and Metal Mines,* Pergamon Press, Oxford, 1966.

CHAPTER 14

Ground-reinforcement Techniques

THE support of excavations in soils and rocks depends upon the degree to which the structural properties of the earth materials can be marshalled to sustain the forces that are imposed upon them in the processes of nature and as a consequence of the engineering activities of man. The inherent properties of cohesion and internal friction in soils, and the strength of rocks in tension, compression, and shear, are limited. If the imposed loads generate forces that exceed those limits, then fracture and flow of the earth materials will result.

In general, earth materials are weak in tension, stronger in shear, and stronger still in compression. So far in this text we have been mainly concerned to exploit these relative weaknesses in tension and shear in order to fracture and to penetrate rocks in drilling, blasting, and breaking ground. In this chapter we are now concerned with the converse aspect of applied geotechnology—how to maintain, or possibly improve, the integrity of the soil or of the rock mass. We must then consider the support of excavations in mining and tunnelling, the improvement of foundations and abutments, and the promotion of stability in slopes, cuts, and embankments.

One basic principle is to design the support system so as to make use of the inherent strength of the earth materials, if possible diverting the imposed loads away from the perimeter of the excavation and into the interior of the surrounding walls. Another basic principle is to try to improve the engineering strength of the earth material by means of some form of reinforcement, possibly changing it from an incompetent mass into one which is competent to withstand the imposed loads without failure. An obvious example of the application of the latter principle is the process of freezing an unconsolidated earth mass with the combined objects of improving the structural properties of the material and to seal off water. An alternative is to apply one or other of the available pressure grouts that may be injected into the surrounding rock walls.

Gunite

The use of air-blown cement mixtures is now common practice in order to protect the exposed surfaces of excavations in rock where problems may exist due to atmospheric attack and weathering. Gunite is a proprietary brand name for a mixture of 1 part cement to $3\frac{1}{2}$ parts sand, mixed when dry and applied by pneumatic pressure, first into a mixing chamber where water is added, and then by high-velocity jet on to the surface being treated.

Shotcrete

Where it is required to provide a surface layer with appreciable structural strength, Gunite may be built up to a thickness of from 50 to 75 mm, but in recent years the application of an aggregate type concrete, known as shotcrete, has become more general. Added strength may be provided by incorporating wire-mesh reinforcement into the concrete for a thickness of several centimetres, thus building up a shell with considerable structural strength.

Tension Reinforcement

The relative weakness of earth materials in response to tensional forces leads logically towards the application of tension reinforcement in soil and rock structures. Such reinforcement may take the form of rock-bolts, cables, and wire mesh. Before considering the modes of application of these items in detail we should be clear as to the objects to be served. These objectives will depend upon the engineering problems concerned, which may range over a wide variety of circumstances. Our approach to all these problems will be determined by the geological structures involved, which are likely to fall into one or other of three categories: (1) discontinuous and blocky material, possibly unconsolidated, (2) bedded and stratified rocks, and (3) massive rock materials.

Rock-bolts

An example of *in situ* anchorage of sheet-steel piles in soil is illustrated in Fig. 14.1. The reinforcement here may take the form of bolted rods or sometimes cables. Reinforcement of the rock walls around underground excavations is commonly achieved by means of rock-bolts, of which many millions are in widespread use.

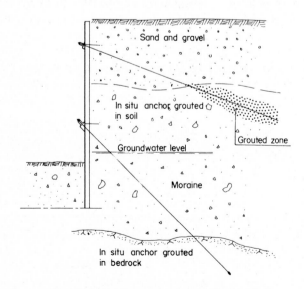

Fig. 14.1. Reinforcement of sheet piles in soil.

Generally speaking, the support of narrow excavations in rock is shared between the strata in the immediate vicinity of the excavation walls and the strata at some depth within the walls—in mining parlance, shared between the rooms and the pillars. Our present concern is with tunnels of specified width and with rooms which separate the pillars of a more extensive arrangement of workings. The support characteristics will differ considerably in laminated and bedded deposits as compared with those that exist in more massive rocks.

Types of Rock-bolt

There are various designs of rock-bolt reinforcement unit.

Slit-rod and Wedge

This is of simple design and it is used extensively in hard rock. It consists essentially of a steel rod made equal in depth to the depth of reinforcement that is desired, which may be anything from 0.7 to 6 m, but more commonly is about 2–3 m. One end of the rod is threaded over a length of about 150 mm, while at the other end is cut a slot, usually 3 mm wide by 150 mm long. In this slot is inserted a steel anchor wedge, usually 140 mm long and about 12–22 mm square at one end, tapering to 12 mm by 2 mm at the narrow end. The dimensions of the wedge are selected to suit the diameter of the bolt-hole and the hardness of the rock, the optimum choice being made in a particular installation after experimental trials.

To install one of these bolts, a hole is drilled into the rock to receive the wedge end of the bolt. The diameter of this hole is usually 32 mm in soft rocks such as shale, 38 mm in medium-strength rocks such as sandstone and limestone sediments, and 44 mm in strong igneous and crystalline rocks. The wedge is placed in the slotted end of the bolt and the unit is then inserted in the hole until the wedge makes contact with the rock at the back of the hole. A sharp tap against the rock should then be sufficient to force the wedge into the slot and so expand the slotted portion of the bolt, which then grips the side of the hole. This grip is further intensified by hammer blows on the projecting end of the bolt, on which a covering "dolly" has been placed to protect the thread. The bolt is thus driven into the wedge by pneumatic hammer, or by means of an impact wrench, until there is no further wedge penetration. The dolly is then removed from the bolt. A bearing plate, or some other form of bearing surface, such as spherical-seated washer, is next placed on the threaded end of the bolt, held in place by a 25-mm hexagonal nut, and tightened to a known value of torque by means of a heavy-duty impact wrench (Fig. 14.2a). Modifications have been made to this basic design from time to time, e.g. with the slotted end of the bolt serrated or shaped to provide a rougher surface and to offer a stronger grip. Figure 14.2b shows a modification in which the width of the slotted portion is increased and a thicker wedge introduced.

Expansion Shell

There are many varieties of bolt employing an expansion-shell anchorage. The type illustrated in Fig. 14.2c consists of a bolt which is threaded at the anchor end over a length

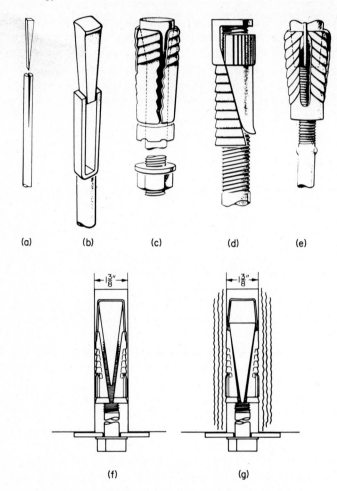

(a)   (b)   (c)   (d)   (e)

(f)   (g)

Fig. 14.2. Types of rock-bolt: (a) slit-rod and wedge; (b) slit-rod and wedge—improved type;
(c) enlarged bolt-end; (d) sliding half-shell; (e) expansion shell and plug; (f) expansion shell—
before expansion; (g) expansion shell—after expansion.

of 150 mm, the top end being wider than the initial bolt diameter (19 mm), to form an end
25 mm square. Slots are cut in the serrated shell to provide four arms, which are squared
to a thimble and retained in a closed position by a piece of thin wire.

Before the unit is inserted in the drill-hole, the shell is clipped over the bolt with the
serrated arms uppermost. The bolt is then placed inside a 25 mm diameter pipe whose
length exceeds that of the hole by 450 mm, so that the squared head of the bolt rests on the
expansion shell, which in turn rests on the pipe. Thus arranged, the pipe and bolt are
inserted into the hole. A sharp blow on the bolt-head against the top of the hole breaks the
thin wire retaining the four arms, which are expanded against the side of the hole by the
square bolt-head. The pipe is then removed and the retaining nut fastened beneath a bar
or plate.

Figure 14.2d illustrates an expansion shell rock-bolt on which both ends are threaded.
The shell consists of two wedges, the larger of which fits loosely over the bolt and is

serrated to prevent downward travel. The smaller wedge is serrated and machined with vertical serrations to prevent turning. When the bolt is screwed into the shell by impact wrench, thrust is developed against the larger wedge, which is retained at the top of the bolt, whilst the smaller wedge, being unable to rotate, travels down the inclined surface of the larger wedge and is thus expanded against the side of the hole.

Expansion Shell and Plug

As an alternative to having an enlarged head formed on the bolt, a plug may be incorporated, as shown in Fig. 14.2e. Another variation which operates in a similar fashion is shown in Fig. 14.2f and g. An advantage of this type of bolt is that the shell may be anchored at any depth in the drill-hole.

The bolt-rods are made of either 13 mm high-tensile steel or 19 mm mild steel with a breaking strength around 90 000 N and a yield strength around 45 000 N. The high tensile bolt, being more elastic, can generate a higher rock pressure per increment of yield under load.

Typical Modes of Application

While the occasional use of steel pins cemented into bore-holes in rock was not unknown in the later years of the nineteenth century, the first large-scale systematic applications of what later became known as rock-bolts was reported from Continental European coal-mines during the inter-war years 1918–40, and in North American hard-rock mines from 1920 onwards.

Bolting in Hard-rock Tunnels

Normally one might reasonably expect that tunnels in hard rock will stand with no need for support other than that which is received from the strata alone. Nevertheless, it might be that, even in a competent massive rock, skin-stress concentrations around the excavation walls may at certain points exceed the strength of the materials concerned. The conditions may be aggravated where the removal of lateral constraints permit of a degree of relaxation in the immediate rock walls. The result is likely to be seen as spalling of the rock walls in the high-stress zones, and this may call for reinforcement, as shown in Fig. 14.3.

Another application of reinforcement to the lining of excavations in hard rock comes from the need to make good the rock walls to final dimensions when the excavation has been made by blasting (Fig. 14.4). In the absence of some limit-break system, such as might be obtained by contour blasting or by pre-splitting, the rock walls will be subject to a degree of overbreak beyond the desired finished perimeter. The strength of the immediate rock walls will also be adversely affected by possible fractures and incipient breaks, generated by the explosive shock waves, to some depth beyond the surface skin. It is now general practice in such circumstances to reinforce the rock walls by a systematic bolting pattern, which often includes shotcrete and wire-mesh reinforcement (Fig. 14.5).

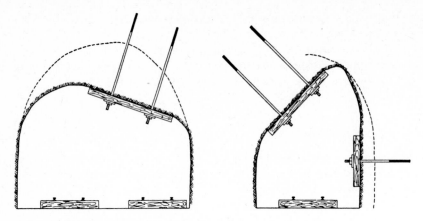

Fɪɢ. 14.3. Reinforcement of spalling rock in high-tension zones.

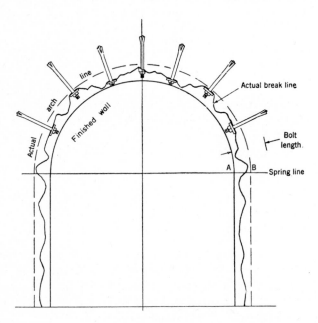

Fɪɢ. 14.4. Reinforcement of an excavation in which the walls are affected by blasting.

An example of the use of rock-bolts in the support of a hard-rock underground mining system is shown in Fig. 14.6. Here the objective is to reinforce the hanging wall against forces of shear and of compression in the wall rock and to pin a segment of ore on the hanging wall, which otherwise might break loose under tension.

Bolting in Stratified Sediments

The support requirements in excavations that are made in relatively weak sedimentary rocks are dominated by the roof conditions. The distribution of forces in the vicinity of a

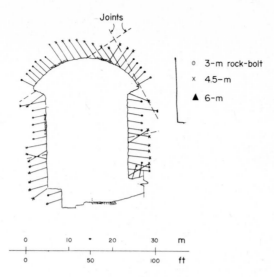

o	3-m rock-bolt
x	4.5-m
▲	6-m

FIG. 14.5. Rock-bolting pattern in an underground power station (Lang [5]).

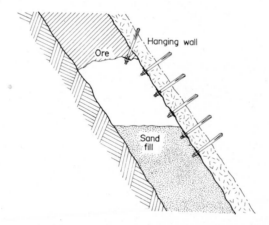

FIG. 14.6. Rock-bolt reinforcement of hanging wall in an ore mine.

narrow opening in laminated sediments is shown in Fig. 14.7. Briefly, the effects of gravitational load produce vertical bending forces and lateral compression forces in the immediate roof strata. Shear forces are generated in the roof above the sides of the excavation to an intensity and over a lateral span which depends upon the rigidity of the sides. The more rigid the sides of the opening the more intense these shearing forces will be.

Flexural bending in the roof-beds causes a loss of contact, termed "bed separation", over a progressively diminishing roof-span above the opening, so that the super-incumbent load bridges a pressure arch, with intensified vertical load being transferred to the abutments on either side.

Typical roof-reinforcement systems such as are commonly applied to such situations are shown in Fig. 14.8. Bolts with expansion shells are generally preferred here since in

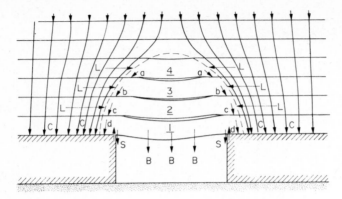

FIG. 14.7. Distribution of forces around a narrow excavation (Walker *et al.*).

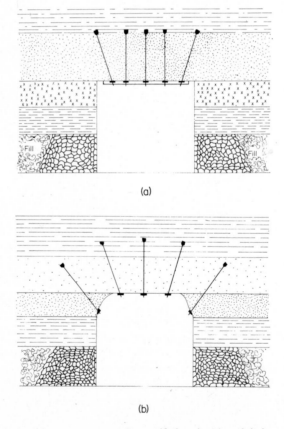

(a)

(b)

FIG. 14.8. Typical reinforcement systems in stratified rock: (a) rock-bolts attached to steel girders; (b) rock-bolts attached to patch plates.

soft rocks they will hold better than does the rod and wedge. They also have the merit that the bolt-rod and face-plate assemblies can be dismantled for re-use elsewhere if need be and where circumstances favour such an operation.

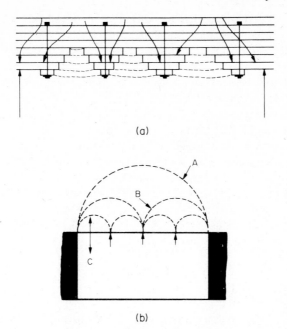

(a)

(b)

FIG. 14.9. Formation of pressure arches around bolt reinforcement: (a) reorientation of compression; (b) development of minor arching: *A*, arch above unreinforced opening; *B*, arch line with one rock-bolt at mid-span; *C*, arch lines above three bolts.

In weak ground there may be a tendency for the surface-skin strata between adjacent bolt-heads to fail by tensional spalling and so form minor arch abutments as shown in Fig. 14.9. This deterioration may be countered by using steel straps pinned by the bolts to secure the span.

Theory of Rock-bolt Action in Laminated Sediments

The results of extensive research undertaken by the US Bureau of Mines are reported by Obert and Duvall [7] and by Panek [8]. The theory applies to horizontally bedded strata in which there is friction but no bond between the beds. It is postulated that vertical bolts are installed in straight lines across the rooms, to constrain the beds in such a manner as to have equal deflections at the bolt locations. In such a system it is maintained that the reinforcement effect produced by the root-bolts consists of a friction component and a suspension component. The friction component is generated by the resistance to slip between the bedding planes which is produced by the clamping action of tensioned bolts. The suspension component is generated as a result of unequal deflections that occur in beds of different thicknesses. Such inequalities are opposed by the roof-bolts, as a result of which the suspension effect produces a transfer of load, part of the weight of the weaker, thinner, beds being taken by the stronger, thicker, sediments. The friction effect is analysed by Panek, whose analysis is also expanded to include the suspension effect. The problem is to develop mathematical expressions for the improvement of roof strength due to bolting, and this was approached by a combined theoretical and experimental stress analysis, making use of models tested in a barodynamic centrifuge. The results of the analysis are contained in mathematical formulae to give the outer fibre bending stress due to the combined friction and suspension effect, the frictional change in beam deflection, and the transferred load in the supported rock roof. These expressions may be used in idealized situations to calculate the reinforcement factor due to bolting, the resulting roof deflection, and an upper limit for the roof bolt tension that is required to prevent the roof-bolts from separating at the bolt locations.

The various parameters that will determine the friction effect are contained in a roof-bolting design chart, devised by Panek and published by the US Bureau of Mines. Obert and Duvall make the comment that, while it may be difficult to apply this treatment in a real situation, Panek's analysis brings out two important points: firstly, that a bolted roof composed of approximately equal thickness laminae will be reinforced only by the friction effect, while bolted units containing one or more thick members will be reinforced both by suspension and by friction; secondly, rock-bolts must be tensioned to develop a friction effect and the bolt tension must be high if it is to assist in closing or preventing bed separation and at the same time produce a frictional resistance to slip of the bedding planes.

Bolting in Discontinuous Ground

In a study of the stress distribution around rock-bolt anchors, Knill *et al.*[11] make the point that, since the normal stress around the limits of an excavation in jointed and blocky rocks will generally be zero, even small radial stresses applied by rock-bolts can effectively modify the overall stability of the excavation. Rock-reinforcement systems will increase the factor of safety against crack initiation, and will influence the orientation of critical existing cracks and also of any new fractures that might develop from them. They consider a jointed rock mass to be a tightly packed granular material in which the peak stress can fall to a residual value if the material is allowed to dilate and loosen during deformation. But if that dilatation is resisted, as by a system of rock-bolts, failure can only take place if new fractures can open up in the rock. In consequence, the strength of the rock mass will be reinforced.

The rock-bolt system will tend to increase the normal stress across any critical discontinuities that might exist, as described and analysed by Lang [4, 5]. Knill *et al.* maintain that as these surfaces will rarely be smooth and suitably orientated, then failure must be accompanied by dilatation and by the development of increased internal friction between the blocks. Any increase in the normal stresses across the discontinuities will tend to hold together the discrete blocks which go to make up the rock mass. Thus there will be an overall increase in the effective size of the unit blocks and they will tend to promote the development of a structural arch to span the excavation. If this is so then it implies that a rock-bolt may be effective in providing reinforcement even if the anchor is located inside the zone of fracture and not in sound rock.

As far back as 1940 Evans [14] described the concept of the voussoir beam or arch (Fig. 14.10), as applied to the broken and discontinuous rock spanning an excavation, and Corlett [15] has extended this in his concept of the "linear arch" in massive but fractured rock. In this it is apparent that rock-bolts can create pressure normal to the potential shear planes and inhibit shearing due to compressive stresses in the rock arch (Fig. 14.11). On the other hand, the observations of Nitzsch and Haas [13] on installation-induced stress for grouted rock-bolts show that, in such a bolt, tightening a nut against a bearing

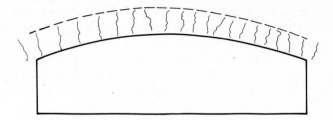

FIG. 14.10. The voussoir arch in massive fractured rock.

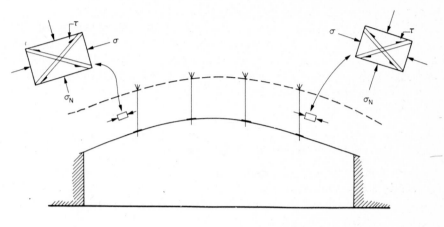

τ, shear stresses
σ, lateral compression
σ_N support compression

FIG. 14.11. Stabilizing the linear arch by rock-bolt reinforcement (Woodruff [12], after Corlett [15]).

plate will not pull separated strata together if the separation is in the interior of the rock wall. The application of a bearing plate and nut to the end of the bolt will help to support loose material near the bearing plate but will not help to stabilize weak material farther away, according to these investigators. However, rock-bolts must be tensioned if they are to provide active reinforcement, the major contribution coming from the suspension component, according to Obert and Duvall, who, when commenting on Lang's analysis of the forces acting across joint planes, point out that, whereas the friction effect in fractured and jointed rock produces only minor reinforcement, dead-weight loading can be very effective in helping to stabilize the rock. To obtain the maximum possible reinforcement from the friction component, the bolts must be tensioned and anchored to withstand the full yield strength of the bolt.

General Theory of Rock-bolt Action

Panek's [8] theoretical approach to the study of rock reinforcement by tensioned bolts was based on the assumption that the rock roof consists of a series of laminated beams. Another analysis was that put forward by McNiven and Ewoldson [16], who developed rational design procedures for choosing the geometric array of bolts for several idealized cases. They reviewed the stress distribution around a tunnel in a gravitating medium and the stress field created by a single rock-bolt. Based on this standpoint typical stress fields were computed to demonstrate the influence of various factors, such as the influence of the *in situ* state of stress without reinforcement on the stress field surrounding a tunnel in a gravitating medium, and the influence that each rock-bolt might have in altering that state of stress.

The computed stress fields were then applied to the design of rock-bolt systems with the object of removing the possibility of major failures in tension, although it was shown that superficial tension failure was likely to occur in the surface strata between the bolt-heads. They found also that high tangential compression forces were generated near the tunnel wall, and these forces were most serious at the sides of the tunnel near mid-height. The recommended design for the rock-bolt system therefore included the provision of surface-tension reinforcement in the form of wire mesh pinned to the walls by bolts, possibly treated with Gunite or with a layer of shotcrete. It was also recommended that steel straps of channel section be placed along the length of the tunnel on each side, at mid-height or near it, and held there by two rows of rock-bolts.

Empirical Approaches to the Design of Rock-bolt Systems

The importance of bolt tension on the reinforcement capability of rock-bolt systems has prompted many inquiries as to the anchorage characteristics of various bolt systems, both with the aid of laboratory models and in field studies.

Photoelastic Models

Typical photoelastic model studies of rock-bolt action are reported by Lang, by Knill *et al.*, and by Handa, the latter being illustrated in Fig. 14.12. Handa's model shows clearly

FIG. 14.12. Photoelastic model of rock-bolts under tension (Handa [17]).

the pattern of stress distribution in an elastic material in which very high stresses are generated around the bolt anchors and adjacent to the face plates. The potentially weak zones, subject to tension, near the lower surface of the roof beam, and between the rock-bolt heads, are also clearly apparent.

Field Studies

As with some other aspects of geotechnology, such as drilling and blasting, advances in rock-bolt technology have largely depended upon procedures of trial and error, accompanied by intuitive interpretations of observed phenomena, before applying any systematic instrumental measurement to assist in explaining the reasons for what has been observed. Many matters are still not fully understood, e.g., the relative suitability and mode of action of various anchor systems in a variety of rock materials are still matters of conjecture rather than fact.

At an early date in the evolution of rock-bolt technology it was obviously apparent that the devices were not functioning in accordance with what was expected of them. This applied particularly to the anchorage systems. Although when first set the bolts were given an initial torque, this rapidly diminished in a process of load shedding, so that torque needed to be re-applied, sometimes more than once, after installation. Otherwise the bolts could become loose, even to the point of falling out.

Bolt Tension Fall-off

Agapito and Parker [18] report that, in one series of tests, of 500 bolts observed, only 23 % were seen to retain torque above the recommended limit for the system of 237 N m. It was also observed that friction between the bolt-head and the face plate reduced the ratio of applied torque/bolt tension to 1/25. In order to improve this ratio, cast spherical seatings or hardened cup washers were applied at the bolt-head to reduce friction and to permit non-axial loading up to 20 degrees from the bolt axis. A torque/tension ratio of 1/70 could then be achieved.

The installed bolt tension was reduced considerably by the presence of rust at the bolt-head and in the anchor components. Rustproofing should therefore be standard procedure. The initial fall-off in setting torque was greatly increased if water was present at the anchorage. One-day torque losses averaged 8 % in dry holes but increased to 23 % in wet holes.

If it is considered that the fall-off in setting load is primarily due to strata creep around the high-stressed zones near the anchor and face plate, then the rock may be assumed to act as a standard solid viscoelastic model on which bolt tension should be re-applied a few days after initial setting and subsequently again after an interval of a few weeks. Such a procedure can only be followed if systematic observation and measurements are made. Economic considerations must necessarily limit the extent to which accurate dynamometers may be fixed permanently on the bolts, possibly instrumented for remote reading. The use of torque wrenches, applied manually, is possible if the bolts are easy of access, but there is considerable doubt as to the extent to which the torque which is applied at the bolt-head represents actual tension in the bolt.

A great deal of attention has therefore been given to the possibility of attaching some cheap and simple device to the bolt-head that would be a direct indicator of existing bolt tension. Such a device would be of considerable benefit, for not only would it help to ensure that all bolts are tensioned and taking their designed share of the total load, but the danger of inadvertently overstressing the anchorage on initial setting would be reduced. One such device on which trials have been made from time to time is the helical spring-lock washer.

Probably the most important contributory factor on anchorage performance, however, is the expansion shell design, particularly the form of the serrations that are forced into the rock wall when the shell is expanded to anchor the bolt. Sharp serrations produce intense stress concentrations in the rock that exists in immediate contact with the shell. Other high stresses are generated in the rock near the face plates. Time-dependent creep in the rock then results in a fall-off in bolt tension. The creep may be reduced if the shell is designed so as to have smooth rock-bearing surfaces, but then the anchor will have a poor response to shock loads such as are produced by blasting, and the bolt may be shaken loose.

It was as a result of these considerations that Agapito and Parker recommended a modified design of expansion shell, illustrated in Fig. 14.13. This has two small ridges on an otherwise smooth shell, arranged so that the lateral force at the anchor is sufficient to completely embed the ridges into the rock. The ridges then supplement the friction forces that are generated between the smooth shell and the rock, and serve to hold the shell secure to resist shock vibrations. The rock around the ridges is completely confined, so that there is less chance of time-dependent failure at the crest of the ridges and no cratering between them.

Grouted Bolts

The problem of anchorage creep and the associated fall-off in bolt tension has largely now been circumvented by the development of the fully grouted bolt anchor in which the anchorage stresses are more uniformly distributed along the reinforcement. Pre-stress is not applied to such a unit, but after the grout has set then an initial load may be imposed. The subsequent stress history may then be very variable. Some fall-off in tension can be

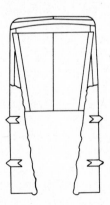

Fig. 14.13. Recommended design of expansion shell (Agapito and Parker [18]).

anticipated, but the response of the reinforced rock strata to the overall load distribution may cause the bolt tension to build up again, in the manner illustrated in Fig. 14.14.

The often unsatisfactory performance of conventional rock-bolt systems and the doubts that expansion shells and wedges can guarantee effective anchorage over a long term in soft rocks have prompted engineers to seek other forms of anchor. One modification is the arrangement shown in Fig. 14.15, in which the 25-mm diameter bolt has a welded steel air-bleeder and grout-feeder tube rolled into the bar and secured by epoxy resin. The bolt is inserted and tensioned in the normal way to seat the anchor, after which the face plate and nut are removed and a rapid hardening cement plug is inserted to seal the mouth of the hole. Cement grout, mixed with water, may then be pumped in to fill

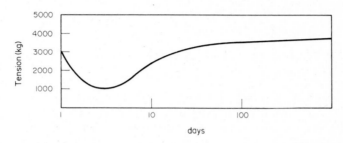

FIG. 14.14. Variation of induced tension with time on a grouted rock-bolt (after Bello and Serrano [27]).

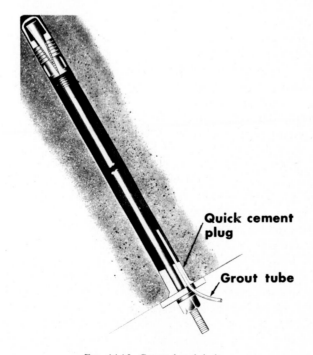

FIG. 14.15. Grouted rock-bolt.

the annulus around the bolt and to fix it along the whole of its length in the 39-mm borehole.

It is sometimes held that since the anchorage afforded by an expansion shell is necessarily suspect in weak rock it is only logical to dispense with it altogether when grouted bolts are used. If this is done, then considerable economies can be effected. In order to increase the anchorage capacity, rock-bolts have been developed with deformed shanks, such as standard concrete reinforcement bar (rebar). The use of such dowels eliminates the possibility of applying pre-stress, but tension may be applied to the bolt after the grout has set. If epoxy resin grouts are used the hardening time can be controlled, with gel times ranging from a few minutes to one hour or more, to attain full strength after a few hours. The cured strength is higher than that which may be obtained from cement grouts. Pralle and Karol [19] report tests on resin-grouted 19-mm rebars 2 m long. These bars fractured at their full tensile capacity of 88–90 tonnes before pulling out.

Theoretical Stress Distribution on Grouted Bolts

Farmer has compared the theoretical stress distribution along a loaded resin-grouted bolt, with computed distributions, as shown by tests on instrumented anchors in concrete, limestone, and chalk. He observed a shear-stress distribution on bolts set in concrete similar to the theoretical distribution, but a substantial difference in hard rock. Debonding was deduced to occur, so that the pull-out resistance might be due very largely to skin friction on the dowel. If this were so then there would be important implications in relation to resin anchorage systems in which a pull-out formula is sometimes applied:

$$P = 0.1 S_c \pi R L,$$

where P is the pull-out force required, S_c is the rock compressive strength, R is the bolt-hole radius, and L is the bolt length.

For, while in the stronger rocks the pull-out resistance will contain both residual and mobilized shear components, in the weaker rocks the resistance will depend mainly on the mobilized shear component.

Nitzsch and Haas also report on installation-induced stresses on grouted rock-bolts. They considered two types of installation: (a) a fully grouted post-tensioned bolt, and (b) a post-tensioned bolt with end-grouted anchorage. Their analysis showed that the grout annulus transferred nearly all the bolt load into the rock within a short distance from the bolt-head. Three separate events were observed: (i) the initial loading of the bolt–grout–rock structure; (ii) time-dependent and continuous loading or unloading after installation: during this period there was change in strain within the rock; (iii) discontinuous loading and unloading, with attempted strata movement along and normal to the bedding planes.

The bolt load at the collar of a fully grouted bolt was practically all transmitted into the rock within a distance of 41 cm from the collar for a 22-mm diameter bolt in a 35-mm hole. This means that the action of tightening a nut against a bearing plate will not pull separated strata together if the bed separation is far (i.e. more than about 41 cm) from the collar. Thus the application of a bearing plate and nut to the end of a fully grouted bolt anchor will help to support loose material near the bearing plate but will not help to stabilize weak rock farther away.

Optimum Length of Bolt

At first sight it might appear that the observed findings of Farmer, Nitzsch, and Karol are at variance with those of Knill *et al.*, particularly as to the implications regarding the optimum length of bolt that is required to give the required degree of reinforcement. It must not be overlooked, however, that sometimes the investigators are dealing with fully grouted bolts, with no anchorage pre-stress, in uniform and continuous rock and sometimes with a discontinuous system. In the latter case it is assumed that active friction is generated at one end of the bolt and a passive resistance at the other, while in the former it is assumed that friction stress is developed all along the bolt.

In the circumstances postulated by Knill *et al.*, in addition to the initial pre-stress, which is applied when the bolt is installed, there will be a linear increase in tension as the fractured rock mass dilates, providing that the anchor does not slip or the bolt fail. If this dilatation is constant along a radius into the rock, then the force in the bolt will be independent of its length. However, in general the dilatation and the degree of fracturing in the rock will increase in intensity near the free surface. In such a situation a short bolt will be subjected to a greater strain than would occur in a long bolt and therefore if it is desired to utilize the extra force that can be mobilized to resist dilatation of the rock mass, then short, stout, bolts with strong anchorage characteristics should be employed. The lower limit of bolt length in a discontinuous system will be determined by the dimensions of existing and potential fractures or separation surfaces, which go to make up the system, and by the length of bolt needed to build up cohesion of the mass and produce structural arching around the excavation.

Some insight as to the minimum bolt length required when using whole-length grouted bolts is provided by Bello and Serrano [27] who measured the variation of induced tension with time, and the distribution of induced tension along the bolt. The bolts were 3 m long and 19 mm in diameter, installed in holes 40 mm in diameter, and injected with a cement grout which developed a strength of 100 kg/cm^2 at 7 days. The rock was an andesite tuff with an unconfined compressive strength ranging from 39.3 to 44.1 kg/cm^2. On such a bolt 24 hours after grouting a tension force of 8 tons could be applied without causing the anchorage to fail.

Pull tests, in increments of 2 tons up to a maximum load of 20 tons, induced stresses up to and beyond the elastic limit of the bolts, but only the strain gauges within 64 cm from the free face showed any change. This confirms the findings of Nitzsch and Haas [13], that

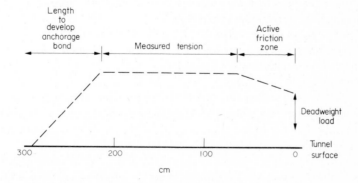

Fig. 14.16. Distribution of induced tension along a grouted rock-bolt (Bello and Serrano [27]).

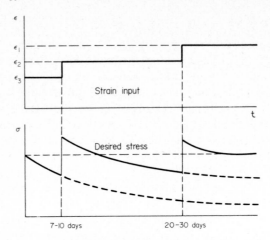

FIG. 14.17. Strain input to give desired long-term stress (McNiven and Ewoldson [16]).

the applied tension at the free end of the bolt is fully transferred to the rock within a zone very close to that end. Bello and Serrano idealize the working conditions for a grouted bolt to be as shown in Fig. 14.16, in which strata displacement near the free end generates an active friction zone on top of the dead-weight component due to fractured and discontinuous material within the pressure arch. A balancing resistant friction force is established at the far end of the bolt over the length necessary to develop the bond stresses bolt to grout and grout to rock. Between these two zones there is a uniform level of tension in the bolt. The optimum bolt length is then envisaged as one which includes the depth necessary to build up the bond stress at the far end, plus the friction stress at the near end, plus the thickness required to develop a bridging arch around the excavation.

Determination of System Dimensions

While it is possible to design a rock-bolt system on the basis of a specified bearing capacity of the reinforcement and a known or assumed imposed load, in practice it is unlikely that an extensive system would be installed without reference to a more empirical approach. Among the preliminary investigations that should be made before final decisions are taken should be, by means of optical probes, to explore visually the walls of exploratory boreholes in order to locate the extent, frequency, and orientation of important discontinuities and planes of weakness.

Indirect exploration of the excavation walls may be achieved by geophysical means to locate the depth to which wall-rock discontinuities exist, and thus help to determine the length of bolt needed. The US Bureau of Mines penetrometer can be applied to produce essential information on the rock characteristics, and this will have an important bearing on the choice of anchorage system.

Roof bolts in room and pillar workings are usually set at intervals of from 1 m by 1 m to 2 m by 2 m. In massive and hard rocks the spacing may increase to around 3 or 4 m. Experimental trials will be necessary to establish optimum spacing. Merril *et al.* [24]

describe a series of such tests in copper sediments. Based on measurements of the change in bending strain and the change in deflection of the roof-beds it was estimated that, whereas a spacing of 3 by 3 m was judged to give a factor of safety of 1, this was increased by a factor of 4 if the spacing was reduced to 1.5 by 1.5 m.

Anchorage Systems in Weak Ground

Recent developments in rock-bolting practice include systems that are specially designed for use in ground that is too weak to take pretensioned anchors. One such system, the Perfo rock-bolt, consists of two half-cylindrical sleeves which are filled with a plastic cement mortar, then tied together and inserted into the drill-hole. A steel rebar is then driven through the sleeve by means of an impact hammer, forcing the mortar through the perforations and solidly filling the hole.

Another development is the pumpable polymer bolt which makes use of a chemical polymer mix reinforced by fibreglass. The fibreglass is in the form of a rope which is threaded into the hole by an aluminium tube while at the same time the resin mix is pumped into the hole. When it has set, the completed composite bolt, 3·5 cm in diameter, has a tensile capacity around 180 kN and a shear resistance around 67 kN.

Split-set Rock-bolts

Split sets, or friction rock stabilizers, are formed by rolling a section of thin steel plate into a tube with a slit running along its entire length. Introduced by Scott [25], they vary in length up to 2 m and are usually 3.5 cm in diameter, 0.20 cm wall thickness, and 1.25 cm slit width. When they are forced into a drill-hole with a diameter slightly less than 3.5 cm, the tube is forced to close as it penetrates the hole, and a radial pressure is generated between the tube and the wall of the hole. The radial pressure distribution caused by this forced fit can be related, by the coefficient of friction between tube and wall, to a force along the length of the tube. If the split set is driven upwards through a layer of rock that is in danger of falling, and on into a stable rock region, the longitudinal friction forces that are acting upon the portion of the split set embedded in the stable rock region serve as an anchoring force. The longitudinal friction force that is present between the split set and the unstable rock region is the supporting force that helps to hold the layer of unstable rock in position.

Davis has developed a series of equations for an elastic–plastic analysis of friction rock stabilizers, and has programmed these to yield computer solutions. Typical solutions have been published in graphical form for comparison with experimental data. Plots of the effects of various parameters (rock material strength, material strain hardening, friction, thickness, radius, and angle of contact) on the holding capacity of split sets have been reproduced for use in installation design [26].

Anchorage Testing

The conventional way of testing the performance of a rock-bolt is to conduct a direct pull test in which load is applied directly on the bolt by means of a hydraulic jack. The test is repeated at loads increasing from zero up to the yield point in increments of about

4450 N. The imposition of this load causes displacement of the bolt-head, and this is measured by a dial gauge with reference to some datum (which for-convenience is usually the horizon at floor level). The total movement of the bolt-head is comprised of several components: (1) an initial component representing the taking up of slack in the system; (2) a component due to load building up in the bolt from zero to the imposed jack load; (3) a component of linear elastic load, and consequent elongation of the bolt; (4) yield of the anchorage system due to adjustment of the strata and of the anchor components to stress concentrations generated near the anchor and the face plate.

The composite result may be plotted graphically as a load–displacement characteristic in which the combined bolt elongation and anchor displacement together determine the anchorage capacity of the bolt (Fig. 14.18). The overall anchorage capacity is attained when either the bolt yields or the anchor yields, and is demonstrated by a pronounced flattening of the characteristic curve, indicating that displacement is tending to occur without a corresponding increase of load.

When assessing the results of pull tests conducted in this way it is evident that the higher the overall bolt-anchorage capacity the better the system will be, since this will determine the degree of constraint that the system can provide. Also, the steeper the slope of the linear portion of the characteristic the better, for a steep slope indicates that only small displacement of the bolt-head is necessary to generate effective build-up of support.

The maximum value of overall anchorage capacity is the yield strength of the bolt in tension. Where firm rock is available in which to anchor the bolt the overall anchorage capacity of the 25-mm bolt is much higher than that of the others by reason of its larger diameter, while the slope of its anchorage characteristic is better because of the more rigid action of the wedge anchor as compared with the more gradual build-up of anchorage resistance in an expansion shell. Also, the bolt-head displacement characteristic of an

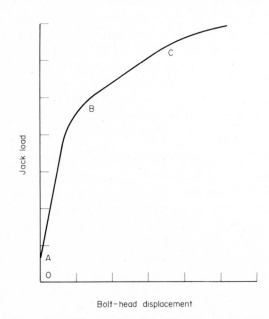

Fıg. 14.18. Rock-bolt anchoring characteristic: *OA*, slack taken up; *AB*, bolt load building up to jack load; *BC*, elastic load to failure.

expansion-type bolt tends to fall off more rapidly at the yield point, so that under a condition of increasing load the ultimate failure of the system could occur with little pre-warning. In conditions where a build-up of load on the bolt is expected to occur, care should be exercised on initial setting lest the yield strength of the bolt is approached too closely and possibly exceeded after installation. The initial setting load should not be more than 271 N m of torque on a 19-mm mild steel, or 237 N m of torque on a 13-mm high-tensile steel bolt.

References and Bibliography: Chapter 14

1. SEYMOUR, M., Rock bolting practices in Canadian metal mines, *Can. Min. J.*, pp. 62–65 (May 1953).
2. ROBESON, W. T., Rock burst incidence, *Trans. Can. Inst. Min. Metall.* **49**, 347–74 (1946).
3. THOMAS, E., Rock bolting, *Min. Engng New York* **6**, 1080–5 (1954).
4. LANG, T. A., Rock bolting in Snowy Mountains project, *Civil Engng London*, pp. 90–92 (Feb. 1958).
5. LANG, T. A., Theory and practice of rockbolting, *Trans. Am. Inst. Min. Engng* **220**, 333–48 (1961).
6. SCHMUCK, H. K., Theory and practice of rockbolting, *Colo. Sch. Mines Q.* **52**, 234–62 (1963).
7. OBERT, L., and DUVALL, W. I., *Rock Mechanics and the Design of Structures in Rock*, ch. 20, Wiley, New York, 1967.
8. PANEK, L. A., *Principles of Reinforcing Bedded Mine Roof with Bolts*, US Bur. Min., Rep. Invest. No. 6139, 1962.
9. PANEK, L. A., *The Combined Effects of Friction and Suspension in Bolting Bedded Mine Roof*, US Bur. Min., Rep. Invest. No. 5156, 1956.
10. PANEK, L. A., Anchorage characteristics of roof bolts, *Min. Congr. J.*, pp. 65–68 (Nov. 1957).
11. KNILL, J. L., FRANKLIN, J. A., and RAYBOULD, D. R., A study of the stress distribution around rock bolt anchors, *Proc. 1st Congr. Int. Soc. Rock. Mech., Lisbon 1966*, **2**, 341–5.
12. WOODRUFF, S. D., *Methods of Working Coal and Metal Mines*, ch. 7, Pergamon Press, New York, 1966.
13. NITZSCH, A., and HAAS, C. J., Installation-induced stresses for grouted rock bolts, *Int. J. Rock Mech. Min. Sci.* **13**, 17–24 (1976).
14. EVANS, W. H., The strength of undermined strata, *Trans. Inst. Min. Metall. London*, Bull. No. 438, pp. 475–500 (1940).
15. CORLETT, A. V., Rock bolts in the voussoir beam, *Trans. Can. Inst. Min. Metall.* **59**, 88–92 (1956).
16. MCNIVEN, H. D., and EWOLDSON, H. M., Rockbolting of tunnels for structural support; Part 1: Theoretical analysis; Part 2: Design of rockbolt systems, *Int. J. Rock Mech. Min. Sci.* **6**, 465–97 (1969).
17. HANDA, B. S., Model study involving photoelastic analysis of roof bolt support systems, MEng thesis, University of Sheffield, 1964.
18. AGAPITO, J., and PARKER, J., Development of a better rockbolt assembly at White Pine, *AGM Am. Inst. Min. Metall. Denver, Colo.*, preprint No. 70–AM–87, 1970.
19. PRALLE, G. E., and KAROL, R. H., Resin anchored rock reinforcing members, *Proc. 1st Congr. Int. Soc. Rock Mech., Lisbon, 1966*, **2**, 279–83.
20. COATES, D. F., and YO, H. S., *Rock Anchor Design Mechanics*, Dep. Energy, Min. Resources, Rep. R233, Ottawa, 1971.
21. FARMER, I., Stress distribution along a resin grouted rock anchor, *Int. J. Rock Mech. Min. Sci.* **12**, 347–51 (1975).
22. RABCEWICZ, L. V., and GOLSER, J., Principles of designing the support system in the new Austrian tunnelling method, *Water Power, London*, Mar. 1973.
23. STEARS, J. H., *Evaluation of Penetrometer for Estimating Rock Bolt Anchorage*, US Bur. Min., Rep. Invest. No. 6646, 1965.
24. MERRIL, R. H., MORGAN, T. A., and STEHLIK, C. J., *Determining the In-place Support of Mine Roof with Rockbolts*, US Bur. Min., Rep. Invest. No. 5746, 1961.
25. SCOTT, J. J., Friction rock stabilizers and their application to ground control problems, *Fall GM Am. Inst. Min. Metall. September 1974*.
26. DAVIS, R. L., Split set rock bolt analysis, *Int. J. Rock Mech. Min. Sci.* **16**, 1–10 (1979).
27. BELLO, A., and SERRANO, F., Measurements of the behaviour of grouted bolts used as reinforcing elements for the support of underground openings, *Proc. 3rd Congr. Int. Soc. Rock Mech., Denver, Colo., 1974*, **2**, 1189–93.
28. THOMPSON, R. R., HABBERSTAD, J. L., BATES, R. C., and WALLACE, G. R., Pumpable polymer rock bolts, *Proc. 3rd Congr. Int. Soc. Rock Mech., Denver, Colo., 1974*, **2**, 1223–8.
29. HOBST, L., and ZAJIC, J., *Anchoring in Rock*, Elsevier, Amsterdam, 1977.

CHAPTER 15

Ground Movement, Caving, and Subsidence

Surface Ground Movement and the Human Environment

While the occurrence of natural disasters such as earthquakes and landslides periodically brings public awareness of the importance of geotechnical factors in the human environment, these events are usually so short-lived and limited in their extent that their memory seldom stays in the public's mind for long, once the initial reactions have subsided. Even in or near the most earthquake-prone fault zones, or on slopes of doubtful geological integrity, towns are rebuilt and structures are re-erected, their owners and occupiers being apparently confident that the exact pattern of earlier events is not likely to repeat itself, and in any event, who can say when and where the next shock will occur, and how intense it will be?

Some surface ground movements, however, can be predicted, sometimes precisely as to their location, and often with a fair degree of certainty as to their probable extent. These ground movements are not due to natural causes, but are produced by the activities of man. They are ground movements caused by caving and subsidence of the surface overlying mining operations.

Such ground movements may take either of two general forms: (a) sudden and sometimes violent collapse, or (b) a more gradual and general movement characterized by subsidence over an observed area. Typical of the former are those due to the collapse of caverns in limestone regions from which water has been pumped or drained with the effect of lowering the water table. Similar effects are sometimes experienced over sites from which brine or sulphur have been extracted. Another possible location is one overlying old room and pillar workings when the pillars have been left too small to give permanent support, and in which the localized failure of a pillar may trigger off a more widespread collapse.

The second type of movement is characterized by fracture and flow of the strata overlying excavations of limited vertical extent, which produce a general subsidence of the affected area. But, no matter whether the movements are those of sudden local collapse, or of more general subsidence, the effects can be serious. No one familiar with the geography of mining areas can have failed to observe wide expanses of land ruined by flooding or the loss of amenity and agricultural value caused by the dislocation of drainage systems. Another consequence of subsidence in some coal-mining areas is the damage that is likely to occur due to the opening-up of air passages in the strata overlying coal-seams at shallow depth, and which have been ignited either by accident or by spontaneous combustion. Once ignited and having taken firm hold, such fires are extremely difficult, if not impossible, to quell and control.

In industrial and urban areas, ground subsidence caused by mining is a constant source of trouble, litigation, and expense not only as regards the design and construction of foundations capable of withstanding ground movement, but also in respect of repairs to the affected buildings and compensation to their owners and occupants.

Mining Systems Involving Caving

A wide variety of mining systems have been designed, or have evolved, to deal with mineral deposits of particular geology and engineering properties. In some of these systems the excavations are intended to stand, with the surrounding rock supported, and caving permanently prevented. In other cases the mining system deliberately produces caving of the orebody, and/or the overlying rock, as an essential part of the rock-breaking and excavation process. In a third group the mineral is extracted mechanically by cutting, drilling, and blasting, and the overlying rock is caused to cave, either immediately or ultimately, behind the mine workings. For a detailed description of these and of other systems of mining the reader is referred to the transactions of the professional mining institutions and to the standard references listed in the bibliography to this chapter [1–3]. Our present concern is with these systems only in so far as they affect ground movements, caving, and subsidence.

Caving in Bedded Deposits

In a mining excavation, the space from which the mineral has been extracted is known as the waste or goaf. We have seen, in Chapter 13, that the waste behind a room and pillar broken working is not supported but is allowed to cave.

Longwall Mining

An alternative arrangement in bedded deposits such as coal, which have well-defined laminar boundaries at roof and floor, is the longwall system in which the extraction faces may extend to lengths of 100 to 200 m. Access to these faces is by gateroads at either end, and possibly also at intermediate positions between the end gates. In an advancing longwall system the gateroads are supported by roadside packs of debris reinforced by roof-bolts, arches, and props. In a retreating longwall system the gateroads are taken through the solid, unworked, mineral before the face is opened up at the boundary of the panel and then brought back towards the points of entry.

In an advancing longwall layout the waste may be completely filled with debris, described then as "solid stowing" or "solid packing", or else partially filled by packs of debris running parallel with the gateroads, described as "strip packing". In retreating longwall, however, it is usual to cave all the wastes, except for strip packs that are sometimes left to protect ribside entries.

Kenny proposes a set of definitions to describe the manner of caving on longwall faces [4]. He considers the entire waste area behind a longwall face to be the caving zone in so far as the roof-beds have caved throughout this region, and he divides this in terms of an active caving zone in the region where the roof material is falling, as distinct from

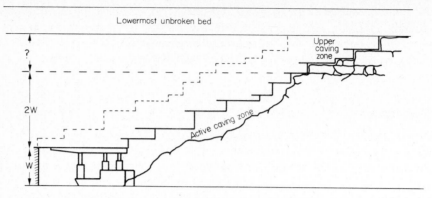

FIG. 15.1. Caving zones in bedded strata (Kenny [4]).

compacting. The active caving zone is then defined as the cavity or boundary between the hanging roof-beds and the pile of fallen debris. This zone extends to the point where the fallen debris is in contact with the hanging beds. Above this is the upper caving zone, where fracture of the roof-beds is initiated as a result of the loss of support caused by compaction of the debris below (Fig. 15.1).

The beds in the upper caving zone can only fragment at the expense of volume increase, and this can only occur if the debris below compacts. The caving limit, i.e the upper limit of the upper caving zone, extends to the lowest bed, which remains unbroken and flexes over the excavated area. How far this will be depends on the ability of the roof-bed to bend without breaking, its deflection being controlled by the support it receives from the compacting material below.

The limits of the active caving zone are set by the change in density of the fragmented roof material. Massive sandstones, for example, tend to cave in large blocks, so that the density ratio of the caved material before and after fragmentation is little greater than unity. On the other hand, more fragmented shales and mudstones occupy a caved volume about $1\frac{1}{2}$ times that of the unbroken solid. Hence if W is the thickness of mineral extracted a thickness of roof of $2W$ will be required to provide enough material to refill the space excavated by mining (Fig. 15.2).

The height to which caving will extend into the roof-beds will depend upon the mechanical characteristics of those beds, both in respect of the mode of fracture and fragmentation, and the angle of slope of the fragmented material. Jenicke and Leser showed that the boundary between the caved debris and the solid material travels upwards to a height that is limited only by the extent of maximum compaction [5]. The height of the active caving zone can extend to $4W$ if the slope of the debris pile exceeds the angle of repose, for then the debris material will roll from the top of the pile to fill the cavity completely. The active caving zone is then the boundary between the hanging roof-beds in the waste and the fallen debris.

On a vertical section normal to the working face the point where the caving roof makes contact with the material beneath it is defined as the "active caving limit", and the perpendicular distance between the waste edge and the active caving limit is the "active caving height".

Since in a practical situation it may not be possible to see the entire active caving zone

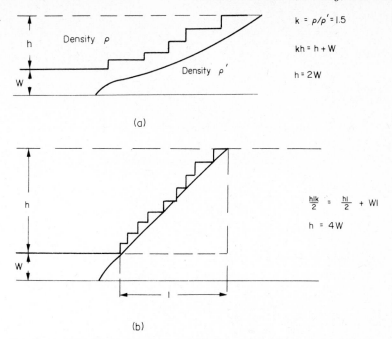

$$k = \rho/\rho' = 1.5$$

$$kh = h + W$$

$$h = 2W$$

(a)

$$\frac{hlk}{2} = \frac{hl}{2} + Wl$$

$$h = 4W$$

(b)

FIG. 15.2. Limits of the active caving zone: (a) debris piling up at a slope angle less than the angle of repose; (b) debris piling at a slope angle greater than the angle of repose, to completely fill the active zone (Kenny [4]).

by direct observation it is necessary to have corresponding definitions relating to the visibly accessible caving zone, as shown in Fig. 15.3.

Equilibrium of Supported Roof in Longwall Caving

Observation of longwall caved faces in carboniferous sediments shows a series of roof fractures which hade backwards over the waste area, as shown in Fig. 15.4. The caving angle determines the location of the centre of gravity of the roof-block, and there is a downward reaction on the roof-beds in the vicinity of the face line. These downward forces are opposed by the resultant thrust from the roof supports.

It therefore becomes possible, using Kenny's approach, to document the observed and deduced parameters of longwall caving, to yield basic information on the nature and extent of the supported roof, the support requirements, the "workability" of the mineral bed, and the mechanical characteristics of the roof strata. The method thus forms a logical corollary to the observation and measurement of subsidence effects at the ground surface as a result of which techniques of subsidence prediction and limitation may be further extended.

Top Slicing

Longwall mining is best suited to bedded deposits that are relatively thin, that is, less than about 4 m. Thicker deposits may present problems due to the weight and size of the

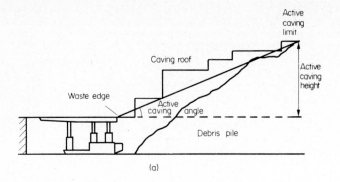

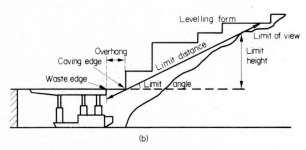

FIG. 15.3. (a) Definitions within the active caving zone; (b) the visible caving zone (Kenny [4]).

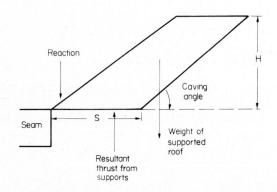

FIG. 15.4. Equilibrium of supported roof in longwall mining.

supports required, to possible difficulties of access to the roof, and to a greater danger of loose material rolling over from the face. Mineral deposits over about 4 m thick may then be removed by the method of top slicing. In this system the floor of each slice is formed of timbers, laid down to provide a protective roof for the next slice, working from the top downwards. The sequence of operations is to first remove mineral by taking drifts side by side through the ore to the limits of the block or panel to be extracted. As each drift is completed the supporting timber is blasted out, so that the overlying material caves on to the floor of the opening. A mat of timber builds up above and to the waste-side of the

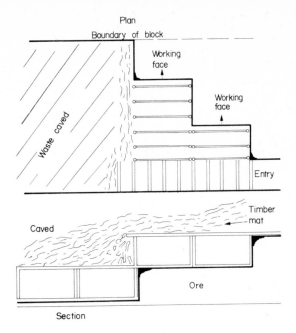

Plan

FIG. 15.5. Top slicing.

working drift, and this acts as a protective cover for the working area and also separates ore from waste on the loading horizon.

Top slicing is a high-cost mining system, owing to the quantity of timber that is required for its operation. It is best suited to wide metalliferous veins, or to thick-bedded deposits of relatively weak rock in ground that will cave easily and regularly (Fig. 15.5).

Caving in Massive Deposits

Sublevel Caving

This is an extension of top slicing. Development of the mine is by drifts as in top slicing, but in this case the drifts are driven at wider intervals apart, both in the horizontal and vertical directions, so that the height of each slice is increased to from 5 to 8 m. The method is most important in massive iron ores, notably at Kiruna, Sweden. Figure 15.6 shows the location of the sublevel drifts, and the sequence of operations which consists of (1) drifting, (2) ring drilling, (3) blasting, and (4) loading-out. The successful application of the method requires ground that will break readily and coarsely, so that it will form a capping that will arch to support itself but at the same time will cave and flow regularly to follow the ore into the draw-off points at the end of each drift (Fig. 15.7).

Within recent years considerable advances have been made in the application of theoretical approaches, and the use of models in simulated materials, to elucidate the many complex and interrelated factors that govern the fracture and flow of caved rock in underground bulk-mining processes [6, 7]. Broadly speaking, these approaches are

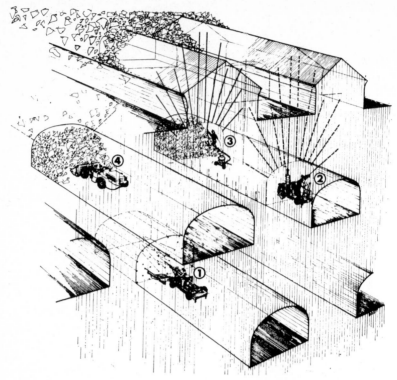

FIG. 15.6. Diagrammatic plan of sublevel caving in Kiruna: 1, drifting; 2, ring drilling (roof of slice); 3, charging of explosive; 4, ore loading (Janelid and Kvapil [6]).

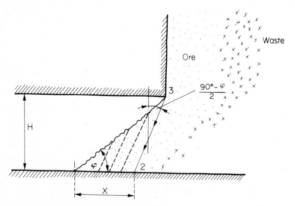

FIG. 15.7. Theoretically best digging depth of loader in ore pile at extraction drift (Janelid and Kvapil [6]).

$$x \approx H \cot \phi - H \tan \frac{90° - \phi}{2}.$$

channelled in two directions: (1) deduction of the characteristics of ground fracture and movement as evidenced by the effects of caving, observed both at the ground surface and in the underground working; (2) studies of fracture and of flow of the caved ground based on the analogy of gravity flow of the granular materials.

Approaches to sublevel caving from direction (1) form part of the general study of ground movements overlying mine workings in what is now termed subsidence engineering. Studies in group (2) are peculiar to the bulk-mining processes, which include sublevel and block caving.

The Notch Effect

The outcrop regions of wide veins and massive orebodies are likely to be worked as open pits, over which the ground is not recovered. Ground subsidence over and around the pit is then a slope stability problem, aggravated by the circumstance that the floor of the pit is not stable, but caves and subsides above the underground workings. Janelid and Kvapil describe the notch effect, which may result from slope fractures in the footwall and hanging wall of such an orebody, intersected by the horizontal plane of the extraction slices in the underground workings. Before the underground workings commence, the stress state in the ore is characterized by horizontal compressive forces that will probably exceed the vertical forces in magnitude, so that the upper strata in the orebody are subjected to lateral compression. The highest stress concentrations will then occur at the lower corners of the notch. After the first slice has been mined the stress state changes because the continuity of the orebody has been removed. The highest stress concentrations will now be transferred to the lower corners of the new working slice, and the working zone of the sublevel itself is partially stress relieved because the pressures that are generated in the new caved walls, being freed from constraint, cannot possibly attain the original compressive stress state of the virgin rock. Hence the necessary interplay of forces to produce the sublevel caving can be repeated successively for slices at progressively increasing depth (Fig. 15.8).

Gravity Flow of Caved Rock

The flow of broken rock material in sublevel caving corresponds in principle to the gravity flow of granular materials in which the particles move in accordance with an ellipsoid of motion (Fig. 15.9). At the end of each sublevel extraction face the ellipsoid of motion is cut off half-way by the vertical wall of the slice (Fig. 15.10). Janelid and Kvapil [6] describe the various factors that govern the movement of broken ore and waste,

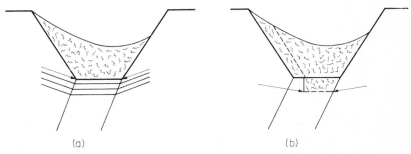

(a)   (b)

Fig. 15.8. The notch effect when starting sublevel caving: (a) stress state in the ore before starting the first slice; (b) change in the stress state when mining the first slice (Janelid and Kvapil [6]).

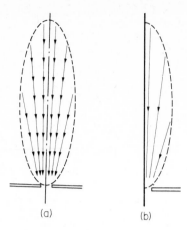

FIG. 15.9. Gravity flow of granular materials: (a) trajectory of particles in gravity flow; (b) the ellipsoid of motion when the granular materials runoff through an opening to one side (Janelid and Kvapil [6]).

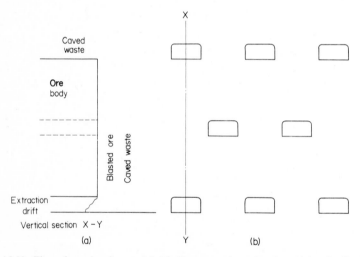

FIG. 15.10. Flow of caved rock material: (a) diagrammatic section through longitudinal axis of extraction drift; (b) diagrammatic view of slice wall and extraction drifts (after Janelid and Kvapil [6]).

defined as the burden, height of slice, digging depth of the loader, runoff width, width of the extraction drift, and gradient of the slice. Just *et al.* [7] also used full-scale tests and models in order to prepare design graphs for different geometrical layouts in sublevel caving. Mathematical models were developed to determine the ore-recovery and dilution relationships for various designs.

Block Caving

Massive orebodies of medium-strong to weak rock are sometimes mined by block caving. In this system large blocks of ore, sometimes from 60 to 180 m high by 30 to 60 m

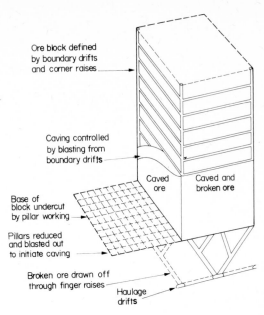

Ore block defined
by boundary drifts
and corner raises

Caving controlled
by blasting from
boundary drifts

Caved
ore

Caved and
broken ore

Base of
block undercut
by pillar working

Pillars reduced
and blasted out
to initiate caving

Broken ore drawn off
through finger raises

Haulage
drifts

FIG. 15.11. Block caving layout.

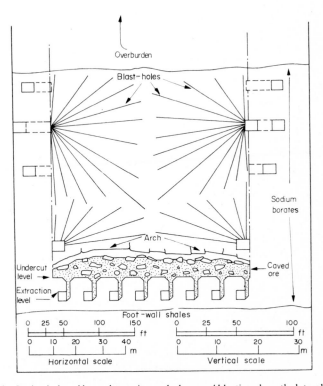

Overburden

Blast-holes

Sodium
borates

Arch

Undercut
level

Caved
ore

Extraction
level

Foot-wall shales

0	25	50	100	150		0	25	50	100	
Horizontal scale						Vertical scale				
0	10	20	30	40	m	0	10	20	30	m

FIG. 15.12. Caving induced by undercutting at the base and blasting along the lateral boundaries
(Obert and Long [9]).

309

square, are cut out by means of excavations along the lateral boundaries and by a system of horizontal drives and cross-cuts to cut off the base of each block. Below the undercutting level is constructed a series of raises, the purpose of which is to collect broken ore from the base of the block and feed it through a system of transfer tunnels to the main haulage and access level below. After the support given by pillars at the undercut level has been removed by blasting, caving of the overlying ore is induced, and is controlled by drilling and blasting at the lateral boundaries (Figs. 15.11 and 15.12) [8, 9].

Rock Mechanics Aspects of Block Caving

As with sublevel caving the rock mechanics aspects of block caving are concerned with (a) the promotion of fracture and collapse of the solid rock in a controlled manner, and (b) the flow of broken ore from the collection points and through the transfer raises to the haulage system. Coates compares the conditions at the caving underface of each block with the idealized concept of a horizontal plate, uniformly loaded and simply supported on four sides. On this basis he produces curves relating the variation of roof stresses with the span of the opening and with the height of caving and of arching (Fig. 15.13) [10].

Woodruff points out that when the ore to be caved is strong in tension, then the cave line may not produce a vaulted arch but the ore may subside as an integral block, with fracture occurring in the lower portions, under compressive stress. As the block settles after being undercut, the concentration of compressive stress at random points of contact within the subsiding mass causes the rock to break up by shearing and crushing. The broken material must then be drawn off as fast as it is produced, otherwise it will be compacted again by the weight of the overlying block [11].

The average pressure at the base of a subsiding block is estimated by Woodruff to be equivalent to about 22 % of the rock depth, the remainder of the weight of the block being transferred by arching and friction to the rock around the boundaries of the block (Fig. 15.14).

Ground Movement Theories

The Arch Theory

The pressure-arch theory, as expounded by Spruth, was adapted by Alder *et al.* [16]. to explain the observed characteristics of strata control in coal-mines worked by the room and pillar system. If this theory is accepted, certain questions then arise, on which the experts do not all agree. Such questions as to the precise location of the abutment zones and their lateral and vertical dimensions, the relationship between the strata pressure in the abutment zones and the original undisturbed strata pressure, the effects of such factors as depth, extraction thickness, and the geological and engineering properties of the strata concerned, all remain to be answered.

In theory the integral resultant pressure in the overlying strata, before and after mining, must be the same. It is the pressure distribution which alters (Fig. 15.15). Spruth's [13] original pressure distribution diagram showed a front abutment load of 65 % and a rear abutment load of 45 % while van Iterson asserts that the front abutment pressure cannot

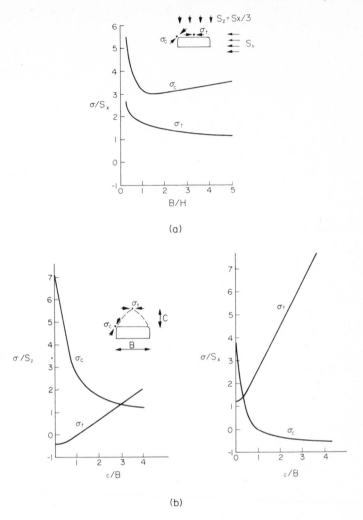

(a)

(b)

FIG. 15.13. (a) Variation of roof stresses with span of opening; (b) variation of roof stresses
with height of caving or arching (Coates [10]).

exceed 1.7 times the vertical pressure. However, perhaps we should not take these figures
too literally. Indeed, both Labasse [15] and Seldenrath [14] reject completely the idea that
a pressure arch exists at all over longwall workings.

Spruth suggests that the pressure arch increases in size when caving is substituted for
solid stowing. In such a case he asserts that the abutment zones are deflected farther away
from the working areas, the arch core is larger, and the support loads heavier. However, he
does not support the view, put forward by Alder *et al.* [16], that the width of the pressure
arch increases with increase in depth. This view was based on the observation, frequently
made in coal-mines, that a zone of intensified strata pressure appears to exist in advancing
longwall gateroads at a certain distance behind the face. In this zone the rate of roof and
floor convergence, and of consequent damage, increases after a period of apparent regular
convergence, to give what is called a "second weight". They interpreted this region of

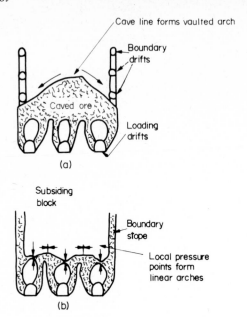

FIG. 15.14. Fracturing produced by block caving: (a) in weak ground; (b) in strong rock
(Woodruff [11]).

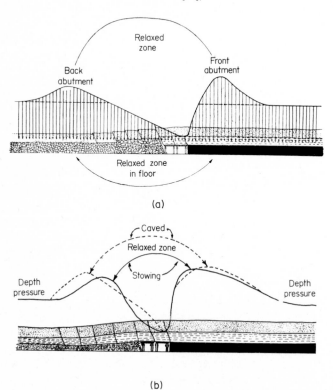

FIG. 15.15. (a) Front and rear abutment pressure zones around a longwall face; (b) relative
extent of pressure zones with caving and solid stowing (Spruth [13]).

damage as the rear abutment zone. Based on this interpretation of observed phenomena underground, Alder *et al.* estimated that the approximate width of the pressure arch increases with depth from about 40 m wide at 122 m depth to about 120 m at 610 m depth. In subsequent work, Alder, Potts, and Walker quote a span of at least 100 m at a depth of 500 m. Nevertheless, several authorities in Continental Europe refute this and claim that no effects that can be directly attributed to depth are likely to be apparent beyond about 400 m of cover [13–16].

Dome Theories

In the general case the pressure arch that is imagined to exist above an excavation will in reality be a three-dimensional dome. Denkhaus [17] analyses the conditions around such a dome, a cross-section of which is shown in Fig. 15.16, for two conditions, which he describes as "sufficiently cohesive" and "insufficiently cohesive" rock material.

If the strata are sufficiently cohesive the material within the dome does not separate at the boundary, which must then take all the weight of the material within the dome. With increasing span of the dome the weight of the core strata will ultimately exceed their cohesive resistance, and a sudden collapse of the core material will occur.

On the other hand, if the cohesion of the rock material is less, defined as "insufficient", then sections of the core rock will fracture and break off more easily at the boundary contact. It then becomes possible to regulate the process of caving within the dome by ground-control technique. With increasing span the core material may support itself by forming smaller domes within the main dome. Resistance will increase as the yielding rock periodically builds up to fracture point, to be relieved at fracture. The support system will adjust and readjust to take the loads imposed by the subsiding mass [17].

In practice it is probable that the real situation will be one which lies between the extremes of "sufficient" and "insufficient" cohesion. For the case of insufficient cohesion the relationship between span and height of the dome is given by

$$S = \sqrt{\frac{8\sigma_c d}{w}\left(1 - \frac{h}{d}\right)\log\left(1 - \frac{h}{d}\right)},$$

where S is the span, h is the height of the dome, d is the depth below surface, w is the specific weight, and σ_c the uniaxial compressive strength of the rock.

The largest possible height for the dome to remain stable, and the corresponding span, are then

$$h_{max} = 0.63d,$$

$$S_{max} = \sqrt{\frac{2.96\sigma_c d}{w}}.$$

For a dome with "sufficient cohesion" the corresponding relationships are:

$$S = \sqrt{\frac{8\sigma_c h}{w}\left(1 - \frac{h}{d}\right)},$$

$$h_{max} = 0.5d,$$

$$S_{max} = \sqrt{\frac{2\sigma_c d}{w}}.$$

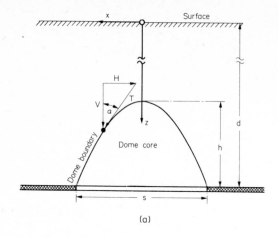

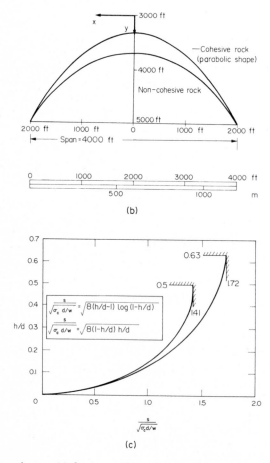

FIG. 15.16. Dome theory: (a) forces at dome boundary; (b) shapes of rigid domes; (c) relationship between height and span for rigid domes (Denkhaus [17]).

Beam Theories

In the pressure-dome concepts the virgin ground is imagined to be homogeneous and isotropic, which is difficult to justify when the ground is obviously laminated or stratified. This has caused some investigators to consider alternative theories based on beam and plate analogies.

The roof-beam is usually considered to be loaded by its own weight and built in to rigid abutments at both ends. Maximum compressive stress acts at the abutments, and maximum tensile stress at mid-span. If the abutments are not rigid but elastic, the points of maximum compressive stress and the points of inflexion of the deflection curve all move towards mid-span.

Having treated the first roof-beam to such an analysis the next overlying beam may be considered. Due to the bending of the lower beds the span of the second beam will be less than that of the first, and so on into the higher strata. The deflection of each beam depends upon its thickness and its elastic modulus. If, as is likely, the lower beam deflects more than the next overlying beam, a gap, termed the Weber cavity, or bed-separation cavity, occurs between the two beams. By joining the abutments of successive beams a dome-shaped curve is produced in the roof strata. Lateral pressure then intensifies the roof-beam bending and, in weak strata, causes uplift bulging of the floor layers. Hence the dome can extend from the roof and into the floor area (Fig. 15.17).

The concept of flexing roof-beams is greatly favoured by Continental European engineers, and is propounded as an explanation of the mechanism by which the observed underground fractures of the roof overlying longwall faces in coal-mines, and seen to be hading backwards over the caved wastes, can be correlated with the forward-hading fractures which produce the typical subsidence profile at the surface. It is difficult to see how, in rock of uniform character, planes of weakness or of sliding can suddenly change their direction. Accordingly, Seldenrath suggests the hypothesis that above an advancing face, somewhere between 100 and 200 m above the seam, the laminated strata bend to produce a flexure that can extend to a distance of around 100 m on either side of the face line.

In a review of geotechnical aspects of longwall face caving Sovinc describes subsidence phenomena overlying a seam of lignite up to 80 m thick, which is mined by successive top slicing. It is shown that here it is not a question of the tearing of layers under tensile stress, as might happen when extracting thin layers, but one of plastic deformation or plastic displacement in the boundary zone [18, 19].

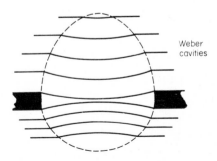

FIG. 15.17. Formation of Weber cavities and the resultant pressure domes in roof and floor.

Plastic Flow Theory

The combination of beam flexure and plastic flow of the ground mass was pictured by Grond (Fig. 15.18) and extended by Labasse and by Seldenrath. Considering the stress around an underground excavation at various depths, Seldenrath imagines the conditions to be as depicted in Fig. 15.19 on which three pressure zones are represented: (1) a zone of relaxed strata immediately adjacent to the boundary of the excavation; (2) a zone of high pressure which reaches a maximum along the enveloping surface *SE*, and (3) a zone of decreasing pressure which extends to the limits of influence.

In the zone *SE* there is a stress situation in which the rock demonstrates its strength limit, where the Mohr circle touches the intrinsic curve for the material. Labasse assumes a similar effect in the roof strata overlying a longwall excavation, as depicted in Fig. 15.20. He postulates a flow of material towards the open space. Fracture planes are formed continuously in advance of and parallel to the face, both in the roof and in the floor strata.

Trough Theories

The concept of the strata acting as a granular material in which plastic flow extends to the limits of a zone of influence is applied by Seldenrath to describe the mechanism of

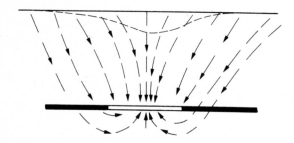

FIG. 15.18. Ground movement around an excavation (Grond [19]).

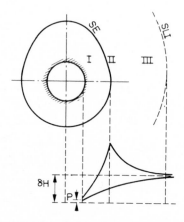

FIG. 15.19. Pressure zones around an underground excavation (Seldenrath [14]).

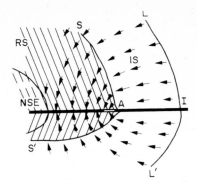

F<small>IG</small>. 15.20. Pressure zones and plastic flow of strata around an excavation (Labasse [15]).

subsidence spreading from an excavation to the ground surface in the form of a trough of movement. The lateral boundaries of such a trough extend to limit planes of influence which enclose zones of fracture and flow, within which vertical and lateral movements occur (Fig. 15.21).

On a lateral cross-section of the ground overlying a mine extraction face the boundaries of the limit plane of influence at each end of the trough appear as straight lines inclined at an angle β to the horizontal. β is termed the *limit angle of influence* and its complement y is termed the *angle of draw*. Thus the angle of draw may be defined as the angle, measured from the vertical at the edge of the underground extraction to the point on the surface at which the limit of subsidence movement exists.

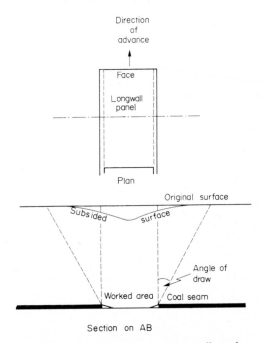

F<small>IG</small>. 15.21. Subsidence over a longwall panel.

A longitudinal section would show a subsidence profile in advance of an extraction panel similar to that over one of the sides, but over the extraction area or goaf we would expect the ground towards the centre of the panel ultimately to attain a new level, at a lower elevation than its original value. The surface would thus experience a flexural wave of movement, starting some distance ahead of the face and advancing at the same rate as the face advance [20–24].

The Subsidence Profile

The trough of ground movement and fracture takes the form depicted in Fig. 15.22 in which, within the zone of influence *ABBA*, various subzones exist. On the surface, from *B* to *C* vertical and lateral ground movements occur and gradually increase in magnitude, with the lateral movement in tension. At *C* the vertical movement increases sharply, and the lateral tensile strain reaches a maximum. This is considered to be due to movement of the ground along the *plane of fracture CA*. The dip of this plane is α, the *angle of slide*, measured from the horizontal.

In the zone of movement from *C* to *D* the subsidence profile is still concave downwards, but with a more pronounced curvature. In this zone also the lateral tensile strain diminishes, to reach zero at point *D*, vertically above *A*. *D* is thus the point of inflection on the subsidence profile.

From *D* to *E* the vertical displacement increases further, to reach a maximum at *E* and within the surface zone *EE*. In this area the subsidence profile becomes a straight line, subsidence has attained its maximum value, and the lateral strain is zero.

Within the zone *DE* the lateral strain on the surface is compressional, and beyond this zone the surface extent over which maximum subsidence is attained is determined by second limit planes of influence *AE*, inclined at the same angle as the first but on the opposite side of the vertical.

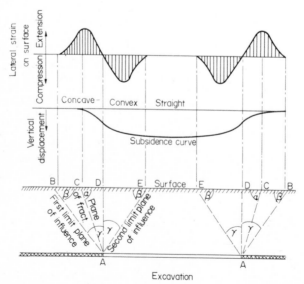

FIG. 15.22. Trough of ground movement and fracture, with the associated subsidence profile and curves of surface strain.

Effects of Width of Extraction and Depth of Workings

The effects of width of extraction and depth of working are interrelated. Referring to Fig. 15.23a, if, from the point E on the surface, lines EA are drawn at the limit angle of influence to meet the seam at points AA, then these points denote the width of a panel that, when extracted, will produce maximum subsidence at E. The base area, in the seam, of the cone generated by rotation of the angle of draw on the vertical through E is termed the *critical area* or the *area of influence* of the seam on the surface point E.

The extraction of a narrower panel than that depicted in Fig. 15.23a, or of a *subcritical area*, will produce less than the maximum subsidence at E (Fig. 15.23b). On the other hand, the extraction of a panel wider than the critical value, or of a *supercritical area*, will not produce greater subsidence in amplitude but it will widen the lateral extent over which the maximum subsidence appears at the ground surface (Fig. 15.23c).

Angle of Draw

A commonly quoted value for the angle of draw in Britain is 35°. This is probably an absolute limit or ultimate limit. For all practical purposes subsidence is likely to tail out over active workings at an angle of about 25° draw.

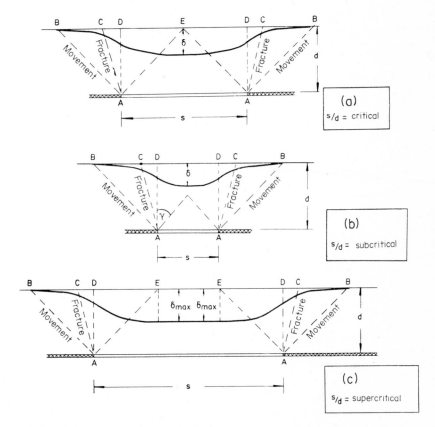

FIG. 15.23. (a) Critical, (b) subcritical, and (c) supercritical subsidence profiles.

There are indications that the angle of draw varies with depth and inclination, and also with the nature of the strata. Generally speaking, the harder and more brittle the strata, the less will be the angle of draw. Each individual stratum will thus have its own angle of draw, the final total angle being the cumulative effect. In Britain the overburden in coal-mining areas consists generally of carboniferous sandstones, mudstones, and shales, on top of which, in the concealed coalfields, are the Permian and Triassic marls, limestones, and sandstones. It is usually assumed that the Permo-Triassic overburden has little influence on the angle of draw, but conclusive evidence has not yet been quoted. In Belgium and Holland, substantial thicknesses of running sand in the overburden increase the angle of draw to about 45°.

TABLE 15.1. *Angles of draw for European coal-mines*

Area	Limit angle β	Draw angle γ	Investigator
Netherlands	45–55°	35–45°	Drent
Britain	50–58°	32–40°	Wardell, Orchard
Germany	51–61°	29–39°	Flaschentrager
Poland	56–71°	19–34°	Knothe

Angle of Slide

The angle of slide may be predicted on the basis of a failure hypothesis for granular materials and is calculated by Denkhaus as

$$\alpha = 45° + \frac{\rho}{2},$$

where ρ is the angle of internal friction of the material.

Values of ρ and α for various types of rock are listed by Denkhaus in Table 15.2.

TABLE 15.2. *Angles of internal friction and angles of slide for various types of rock (after Seldenrath and Labasse [14, 15])*

Rock	Angle of internal friction, ρ	Angle of side, α
Clay	15–20°	52.5–55°
Sand	35–45°	62.5–67.5°
Moderate shale	37°	63.5°
Hard shale	45°	67.5°
Sandstone	50–70°	70–80°
Coal average	45°	67.5°

Effects of Faulting on the Angle of Draw

A fault may either increase or else decrease the angle of draw, depending upon its position and hade (Fig. 15.24). The fault often acts as a plane of weakness along which much of the subsidence movement takes place. Severe dislocations can be expected at the surface along the outcrop of a fault.

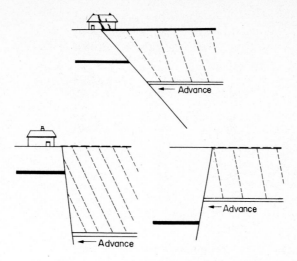

FIG. 15.24. Effects of faulting on the angle of draw.

Effects of Inclination on the Angle of Draw

The angle of draw is generally decreased to the rise and increased to the dip. However, the value of the limit angle does not vary to dip and rise in direct proportion to the seam inclination.

Effect of Thickness of Seam Extracted and of Goaf Support

In Britain, observations from numerous field studies have enabled empirical relationships to be formulated, relating subsidence to width, depth, and thickness extracted. It is customary to refer to the amplitude of subsidence in terms of thickness extracted, since the two parameters are directly proportional to one another. The subsidence factor is the ratio subsidence/thickness extracted or S/m ratio. This ratio is often quoted as a percentage, which depends on the depth of working and the method of goaf support.

Typical figures for maximum subsidence over caved, strip-packed, and solid-stowed panels (assuming that a critical area has been extracted) are:

Caved	90%
Strip-packed	85–90%
Solid-stowed	45–40%

These figures apply only to seams at a depth of about 210 m. Figures for other depths will be very different. For example, over a strip-packed face subsidence may vary from 55% over seams at 274 depth to 85% over seams at 61 m depth.

Even with solid stowing, a considerable amount of roof and floor convergence may occur before the stowing material is packed, and the time interval between extraction and stowing should therefore be as small as possible. Better conditions are likely to ensue over a face that is fast-moving with a regular turn-over.

Figures, quoted for strip packing at various depths of working, are as Table 15.3 shows.

TABLE 15.3. *Observed subsi-dence factors at various depths with strip-packing (National Coal Board)*

Depth		Subsidence
(yd)	(m)	factor (S/m)
45	41	0.69
90	82	0.60
110	100	0.71
148	134	0.58
170	155	0.63
230	210	0.66
573	510	0.73
870	795	0.78

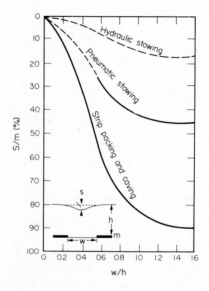

Fig. 15.25. Relationship between width/depth ratio and subsidence for various goaf treatments.

Under present-day conditions of working in Britain the general weight of evidence tends to show little difference in the subsidence factor as between caving and strip packing except in very shallow workings (Fig. 15.25).

Prediction of the Complete Subsidence Profile

For many years the complete subsidence profile was estimated by sketching in a smooth curve between the limit points and the maximum, using the half-subsidence points as the points of inflection. In recent years the results of research have been applied to more precise methods, an empirical approach being favoured in Britain, while on the continent of Europe and in Japan analytical techniques are applied, based on mathematical, photoelastic, and equivalent material models. The NCB method is now to plot the profile in terms of dimensionless coordinates. These coordinates are simply ratios of the distance of various points on the profile (measured from the transition or inflection points and expressed in terms of depth) and subsidence (expressed in terms of maximum subsidence). Details of the method, with examples of its application, are contained in the NCB Information Bulletins on the subject, and in the *NCB Subsidence Engineer's Handbook*. Using the NCB method, prediction of mining subsidence can be made to within an accuracy of 10 % for the majority of cases.

Ground Strain Profiles

The ground strain which accompanies surface subsidence is a function of the differential movements associated with shortening and lengthening of the ground. The transverse tensile strain profiles are similar for different ratios of width/depth but the compressive strain profile differs. For ratios of width/depth (w/h) less than 0.45 the compressive strain profile gradually increases to reach its peak value at the centre of the trough. When the ratio w/h is greater than 0.45, two peak values of compressive strain occur. No compressive strain occurs at the centre of the trough when $w/h =$ is greater than 1.4. The peak values of tensile and of compressive strain are directly related to S/h, where S is the maximum subsidence.

Ground tilt due to mining subsidence is related to the strain profile. The maximum tilt occurs at the transition point, which is the position where the strain profile passes from tensile to compressive and where half the maximum subsidence occurs. The value of maximum tilt for a critical width of extraction is $2.75S/h$, the value of the numerical constant being dependent upon the w/h ratio [24–32].

The Time Factor in Subsidence Ground Movements

It is important to know how long subsidence takes to develop and how long it continues to occur in relation to the extraction of the mineral, since it is necessary for us to know when structures are likely to be first affected, when the likelihood of damage is most acute, and, finally, when the risk of damage is past, so that, with the ground once more stable, repairs may be carried out.

Wardell summarizes the position very clearly. Very simply it is that so long as mineral is being removed from the critical area below a given surface point, that point will continue to subside.

If we consider the effect of an extraction face advancing below a surface point, it is clear that once the face has approached close enough for the point to be within the angle of draw for the workings, that point begins to subside. It will continue to subside while the face moves below the point and beyond it. By the time the face reaches a position beyond the point equal to the distance of draw, the limit of influence is reached. The surface point is then beyond the range of influence from the workings. Subsidence due to those workings will then cease except for that due to consolidation of the strata above the extracted area [33–35].

Wardell's observations on the relationship between subsidence and the movement of an advancing working below are reproduced in graphical form in Fig. 15.26. The observed subsidence as the face advances is plotted vertically above the face positions plotted

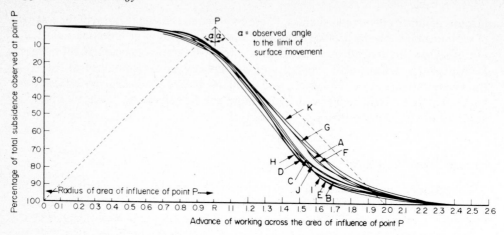

FIG. 15.26. Advance of a working face across the area of influence of a surface point *P*, plotted against the percentage of observed subsidence at that point (Wardell [33]).

horizontally, so that the curve shows the development of subsidence in relation to the position of the face. It can be seen that when the face is vertically below the surface point *P*, that point has subsided approximately 15 % of the amount that it is ultimately going to subside. Also, by the time that the face has reached the limit of the critical area the point *P* has subsided by approximately 95 % of its ultimate vertical movement. The 5 % residual subsidence occurs fairly soon after this, and only the residual subsidence is affected by such factors as depth and strata characteristics.

It therefore becomes apparent that the time taken for surface subsidence to occur depends entirely on the time taken for the face to be worked through the critical area underground, and this in turn depends on the depth and the angle of draw, and also on the rate of advance. In seams at shallow depth the critical area is also small, so that for a given rate of face advance the subsidence is completed much sooner than it is over deeper workings. It is also necessary to bear in mind the layout of the underground workings in relation to the critical area. One face may only remove a part of the area, and one or more succeeding faces may be required to complete the critical extraction. Each face on passing through the critical area will cause more subsidence.

Exactly how long the ground movement takes for it to be transmitted from the underground face to the surface, and how long subsidence will continue after the workings have ceased, are matters upon which there is still much scope for research.

Effects of the Travelling Wave on the Surface

The development of subsidence on the surface overlying an advancing face takes the form of a travelling wave in which a tensional phase precedes a compressional phase. The strains over the ribside areas will eventually stabilize the residual tension and compression zones, whereas over the middle of the panel we can expect a final zero equilibrium to be attained when all subsidence is complete.

It can be expected that a building on the surface will first show signs of damage when the workings enter its area of influence. The damage will probably be at its worst when the tensile strain reaches its maximum, that is, when the extraction face is below the point concerned. Thereafter, tensile strain will decrease to zero and compressional strains will ensue. However, the tensile strains will probably have initiated damage which subsequent compressive strain cannot remedy. Repairs may reasonably be undertaken when the face has gone beyond the area of influence.

Permanent repairs should not normally be carried out if a damaged structure is due to be undermined again by another extraction face, either in another seam or in the area of influence in the same seam.

If a site is required for development an estimate has to be made of the earliest date at which building can safely commence. This can be done by plotting a subsidence development (time–subsidence) curve. The possible 5 % or so residual subsidence may often be ignored since most of this will probably be covered by the site preparation period.

Restricted Workings

The expansion of urban areas and the establishment of new towns in what have hitherto been rural districts around our main industrial centres means that, inevitably, a greater proportion of future mineral resources must lie in regions where subsidence may cause damage to surface works and property. In these circumstances the methods of mining that are adopted must be such as to cause a minimum of strata movement above the extraction area. While in some cases solid stowing of the mined cavities may permit complete extraction of the mineral, in others only a certain percentage of the mineral may be taken out, the remainder being left to support the surface.

Should partial extraction only be possible, a practice often followed is to adopt the room and pillar system of working, taking the whole working only. The pillars must be of adequate size to take the loads imposed upon them, with a considerable safety margin if subsequent failure is to be avoided. The relationship between the permissible percentage extraction and depth, in such a situation, has already been described. Broadly speaking it may be said that as the depth of working increases to about 460 m the required pillar dimensions increase from about 15–45 m centres, with rooms 3.5–4.5 m wide. Beyond a depth of 460 m or so, pillar dimensions increase, but at a slower rate, to from 50 to 60 m centres at 610 m depth, the width of the rooms increasing, in some cases, from 6 to 8 m wide.

In circumstances where difficulty may be experienced in mining and maintaining narrow rooms with large pillars at depth, an alternative layout is the block-pillar or strip system in which partial extraction of the mineral deposit is achieved by means of faces somewhat wider than the narrow work of the room and pillar system, leaving the required percentage of mineral necessary for support either in large block pillars or in strips. Examples of such layouts in British coal-mines at depths of up to 900 m give an extraction of 40–50 % from semi-longwall panels 30–80 m wide and strip pillars 35–100 m wide (Fig. 15.27) (Table 15.4).

Support Afforded by Block Pillar Workings

Some early investigators and theorists on subsidence adhered to the "harmless depth" theory. This theory, based on the fact that broken strata after subsidence will fill a greater

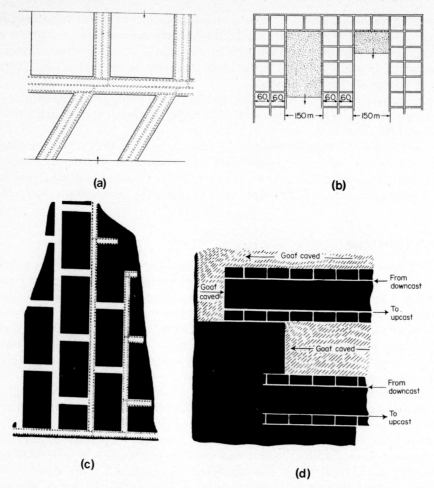

FIG. 15.27. (a) Semi-longwall or block-pillar layout; (b) strip and pillar workings; (c) retreat longwall partial extraction; (d) advancing–retreating strip caving.

TABLE 15.4. *Strip pillars and semi-longwall panels—percentage extraction and maximum subsidence (National Coal Board)*

Maximum subsidence (%)	Depth (m)	Panel width (m)	Pillar width (m)	% extraction
3.21	120	30	$\left\{\begin{array}{c}35\\40\\50\end{array}\right\}$	43
5.1	192	48	48	50
6.48	240	50	50	50
9	600	80	80	50
14	750	80	120	40
15.7	900	70	100	40

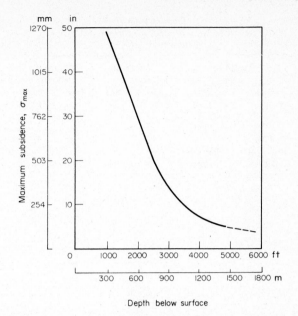

Fɪɢ. 15.28. Maximum subsidence as a function of the depth of an excavation (Seldenrath [14]).

volume than they originally occupied when whole, held that after a certain depth had been exceeded the amplitude of surface subsidence would be nil. However, experience shows that subsidence takes place over the deepest workings when the excavations are wide, and although the amplitude of surface subsidence decreases with depth the lateral extent of the surface affected is increased (Fig. 15.28).

For narrow workings it may be true to say that, at a given locality, for a given width of excavation there will be a depth below which the extraction will produce an amplitude of subsidence so slight as to be negligible. Conversely, at any given depth there will be a maximum width of excavation if subsidence is not to be apparent at the surface. This maximum width will be dependent mainly upon the depth of the seam and the nature of the overlying strata, the essential feature being that the zone of broken and yielding rock within the pressure domes above each strip should not extend to the surface.

The occurrence of strong rocks in the overlying strata, if they are able to flex and bridge the excavations, will restrict the vertical extent of relaxed and fractured strata. In weaker rocks, fracture and flow may take place above the excavations until the pressure domes have developed to their full extent. In either case, if the dome limits do not extend to the surface, subsidence of the surface can only occur by virtue of the yield of the blocks and strips. If these are adequate in size then surface movement should be small.

The proportions of height to width of the pressure dome will determine the minimum depth at which an excavation of given width can safely be made. Theorists who have approached this question show wide divergence of opinion. Alder *et al.* [16] stress the importance, in national economic interest, that the width of excavation should be the maximum possible in the circumstances of each particular case. They suggest that this width can be estimated, in a particular seam, from observations of the distance between the front and back abutments in longwall faces in the same seam, quoting the figures listed in Table 15.5.

TABLE 15.5. *Ratio of width of excavation
to width of coal strip at various depths*
(*Alder* et al. [16])

Depth to seam		Width of excavation		Width of coal strip	
(ft)	(m)	(yd)	(m)	(yd)	(m)
300	92	25	23	30	27
750	230	30	27	40	37
1000	305	40	37	50	46
1600–1800	490–550	40	37	80	74
2500	760	70	64	100	92

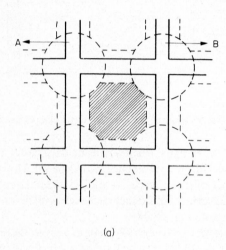

(a)

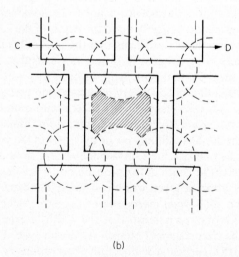

(b)

FIG. 15.29. (a) Pressure domes above strip junctions; (b) pressure domes above staggered strips.

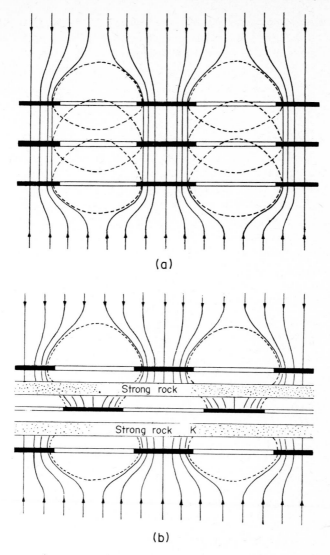

FIG. 15.30. (a) Seams in close proximity in weak ground: pressure domes fully developed;
(b) seams in close proximity: pressure domes spanned by strong beds.

Effects of Alternative Layout

Two alternative arrangements of the strip junctions are shown in Fig. 15.29. In Fig. 15.29b the blocks are situated with their sides staggered, while in Fig. 15.29a the branches are set off in both directions at one and the same place. In the latter case, if the pillar blocks are of adequate size, the abutments of the pressure domes above the strip junctions do not overlap, and a sufficient width of ore will remain between them in the interior of the pillars to carry the abutment loads. On the other hand, if the strips are staggered the pressure domes above the strip junctions, although possibly a little smaller in size, will occur at intervals which will be about one-half of those in Fig. 15.29b. The conditions may now be such that the domes overlap and as a result of this the effective size of the pillar is reduced. Also an extent of caved and yielding ground is now produced above each strip for practically the full width of the main pressure domes, although no more mineral has been extracted.

Workings in Adjacent Seams

Subsidence over an area from which more than one seam is extracted is cumulative. If the lateral boundary of the extraction faces coincides, then the curvature at the edges of the subsidence profile becomes sharper, and fissuring and fracturing become more severe. Nevertheless, such a juxtaposition of workings might be necessary in weak ground in order to derive maximum support from the pillars (Fig. 15.30a). On the other hand, if the strata include strong rocks it is possible that more rigid and stable support to the surface might result from an arrangement such as that depicted in Fig. 15.30b.

References and Bibliography: Chapter 15

1. FRITZSCHE, C. H., and POTTS, E. L. J., *Horizon Mining*, Allen and Unwin, 1954.
2. MASON, E. (ed.), *Practical Coal Mining*, Virtue, London, 1959.
3. WOODRUFF, S. D., *Methods of Working Coal and Metal Mines*, Pergamon, New York, 1966.
4. KENNY, P., The caving of the waste on longwall faces, *Int. J. Rock Mech. Min. Sci.* 6, 541–55 (1969).
5. JENICKE, A. W., and LESER, T., Caving and underground subsidence, *Trans. Am. Inst. Min. Metall. Engrs* 223, 67–73 (1962).
6. JANELID, I., and KVAPIL, R., Sublevel caving, *Int. J. Rock Mech. Min. Sci.* 3, 129–53 (1966).
7. JUST, G. D., FREE, G. D., and BISHOP, G. A., Optimization of ring burdens in sublevel caving, *Int.J. Rock Mech. Min. Sci.* 10, 1–13 (1973).
8. FLETCHER, J. B., Ground movement and subsidence from block caving at Miami Mine, *Trans. Inst. Min. Mech. Engng, New York* 217, 413–21 (1960).
9. OBERT, L., and LONG, A. E., *Underground Borate Mining*, US Bur. Min., Rep. Invest. No. 6110, 1962.
10. COATES, D. F., *Rock Mechanics Principles*, ch. 5, Dep. Energy, Mines, Resources, Mines Branch Monograph 874, Ottawa, 1967.
11. WOODRUFF, S. D., Rock mechanics of block caving, *Proc. Int. Symp. on Mining Res., Rolla, Missouri, 1962* 2, 509–20, Pergamon Press, New York, 1964.
12. VANDERWILT, J. W., Ground movement adjacent to block caving at Climax, *Trans. Am. Inst. Min. Metall. Engrs* 181, 350–70 (1949).
13. SPRUTH, F., Distribution of pressure up, on, and near the coal face, *Proc. Int. Conf. on Rock Pressure, Liege, 1951*.
14. SELDENRATH, T. R., Theory and practice of rock pressure on an advancing face, *J. Leeds Univ. Min. Soc.*, pp. 57–65 (1956).
15. LABASSE, A. B., Le pression de terrain dans les mines de houille, *Revue Univlle Mines* 92–95, Pt. 7 (1949–52).
16. ALDER, H., WALKER, A., and WALKER, L., Subsidence and its bearing on mining methods, *Trans. Inst. Min. Engrs London* 102, 302–15 (1943).
17. DENKHAUS, H. G., Critical review of strata movement theories and their application to practical problems, *J. S. African Inst. Min. Metall.* 64, 310–32 (1964).
18. SOVINC, I., Geotechnical aspects of longwall face caving, *Proc. 3rd Congr. Int. Soc. Rock Mech., Denver, Colo., 1974* 2b, 1096–1101.
19. GROND, G. J. A., Ground movements due to mining, *Colliery Engng, London* 34, 157–97 (1957).
20. FORRESTER, D. T., and WHITTAKER, B. N., Effects of mining subsidence on colliery spoil heaps, *Int. J. Rock Mech. Min. Sci.* 13, 113–20 (1976).
21. RANKILOR, P. R., The construction of a photoelastic model simulating mining subsidence, *Int. J. Rock Mech. Min. Sci.* 8, 423–44 (1971).
22. ANON., *Investigation of Mining Subsidence Phenomena*, NCB Information Bull. No. 52/78.
23. ANON., *Principles of Subsidence Engineering*, NCB Information Bull. No. 63/240.

24. ANON., *Subsidence Engineers' Handbook*, NCB Production Dep., London, 1966.
25. ZENC, M., Comparison of Bals and Knothes methods of calculating surface movement due to underground mining, *Int. J. Rock Mech. Min. Sci.* **6**, 159–90 (1969).
26. HIRAMATSU, Y., and OKA, Y., Precalculation of ground movements caused by mining, *Int. J. Rock Mech. Min. Sci.* **5**, 399–414 (1968).
27. PARK, D. W., SUMMERS, D. A., and AUGHENBAUGH, N. B., Model studies of subsidence and ground movement using laser holographic interferometry, *Int. J. Rock Mech. Min. Sci.* **14**, 235–45 (1977).
28. VAN OPSTAL, G. H. E., The effect of base rock rigidity on subsidence due to reservoir compaction, *Proc. 3rd Congr. Int. Soc. Rock Mech., Denver, 1974* **2b**, 1102–11.
29. HESLOP, T. G., Failure by overturning in ground adjacent to cave mining at Havelock Mine, *Proc. 3rd Congr. Int. Soc. Rock Mech., Denver, Colo., 1974* **2b**, 1085–9.
30. ROPSKI, S. T., and RAMA, R. D., Subsidence in the near vicinity of a longwall face, *Int. J. Rock Mech. Min. Sci.* **10**, 105–18 (1973).
31. CORDEN, C. H. H., and KING, H. J., A field study of the development of surface subsidence, *Int. J. Rock Mech. Min. Sci.* **2**, 43–55 (1965).
32. WHITTAKER, B. N., and PYE, J. H., Ground movements associated with the construction of a mine drift, *Int. J. Rock Mech. Min. Sci.* **14**, 67–75 (1977).
33. WARDELL, K., Some observations on the relationship between time and mining subsidence, *Trans. Inst. Min. Engrs London* **113**, 471–86 (1953).
34. ORCHARD, R. J., The control of ground movements in undersea mining, *Trans. Inst. Min. Engrs London* **128**, 259–71 (1968).
35. WHETTON, J. T., and KING, H. J., The time factor on mining subsidence, *Proc. Int. Symp. on Mining Res., Rolla, Missouri, 1962* **2**, 521–39, Pergamon Press, 1964.
36. KUMAR, R., and SINGH, B., Mine subsidence over a longwall working: prediction of subsidence parameters for Indian mines, *Int. J. Rock Mech. Min. Sci.* **16**, 1–18 (1973).
37. BAUMGARTNER, P., and STIMPSON, B., A tiltable base frame for kinematic studies of caving, *Int. J. Rock Mech. Min. Sci.* **16**, 265–7 (1979).

Epilogue

New Horizons in Geotechnology

The Future of Mining Technology

We depend on the mining industry for industrial raw materials and for our basic energy needs. Future developments in mining are likely to include moves towards the replacement of manual labour by automation and control techniques wherever possible. With the progressive exhaustion of richer mineral deposits must come trends towards the recycling of old and disused waste materials, and the wider adoption of bulk mining and earth-moving techniques, so that the extraction of low-grade deposits will become economically attractive.

The need to replace the world's limited resources of oil and natural gas will promote increasing activity in two directions: (i) the revival of research and development in processes to produce oil from coal, shale, and tar sands, and (ii) the development of reserves located in regions that are progressively more difficult of access and in operation, such as the polar regions. We are also likely to see, in the very near future, a re-examination of techniques to produce oil and gas from coal and shale *in situ* by underground gasification. A concurrent development of *in situ* mineral extraction is that of solution mining or leaching, particularly in low-grade ores of copper and uranium. The use of bacteria, in bacterial leaching, is another associated development.

The Utilization of Underground Space

The use of underground space is especially advantageous when providing suitable and secure housing for libraries, art galleries, and valuables generally. Besides offering greater security, underground housing facilitates the provision of a controlled atmospheric environment where this is necessary for the stored materials. Many industrial processes, too, require a closely controlled atmospheric environment, which can be more easily provided underground, and at the same time leave the surface free for other purposes, especially in urban areas where land values are high, or where urban populations must live in close proximity to industrial installations. Underground construction may also be of advantage in mountainous regions, where the surface contour and topography may render development above ground difficult and expensive, while at the same time facilitating access by underground tunnels.

Civil engineers have long been accustomed to tunnelling underground in connection with transport systems, sewerage, and water supply, and when constructing pressure tunnels and machinery halls in hydroelectric developments. To these must be added underground pumped storage schemes, and the use of underground chambers in which to store compressed air and hot water. The compressed air chambers are, in effect, large air receivers which are charged by the electric power system during periods of off-peak demand, to be drawn upon for power generation during subsequent peak periods. The use of rock tunnels storing hot water in district heating systems is another way of saving waste heat from power stations for subsequent urban distribution.

Underground excavations offer potential sites for the secure containment of nuclear reactors, which are becoming increasingly unpopular on the surface, and another development of great potential is that of providing housing for superconductive energy magnets for power generation at zero resistance operating under cryogenic conditions.

Underground construction offers protection in circumstances where adverse climates exist, such as in the tropics and in the polar regions. One proposal for developing the polar oilfields, as yet untouched, is to excavate under the ocean bed from access stations situated on offshore islands. Large chambers would be excavated at intervals along the ocean bed in which to site the production drills and store the equipment and house the personnel. The recovered product would be piped back through the access tunnels to the on-shore base.

Author Index

Subject Index